T0324181

MECHANICAL MEASUREMENTS

(2nd Edition)

MECHANICAL MEASUREMENTS
(2nd Edition)

S.P. Venkateshan
Professor Emeritus
Department of Mechanical Engineering
Indian Institute of Technology Madras
Chennai, INDIA

WILEY

John Wiley & Sons Ltd.

Athena
ACADEMIC

Mechanical Measurements (2nd Edition)

© 2015 S.P. Venkateshan

First Edition: 2008
Reprint: 2009, 2010, 2013
Second Edition: 2015

This Edition Published by

John Wiley & Sons Ltd
The Atrium, Southern Gate
Chichester, West Sussex
PO19 8SQ United Kingdom
Tel : +44 (0)1243 779777
Fax : +44 (0)1243 775878
e-mail : customer@wiley.com
Web : www.wiley.com

For distribution in rest of the world other than the Indian sub-continent

Under licence from:

Athena Academic Ltd.
Suit LP24700, Lower Ground Floor
145-157 St. John Street, London,
ECIV 4PW. United Kingdom
Email: athenaacademic@gmail.com
Web: athenaacademic.com

ISBN : 978-11-1911-556-4

Library Congress Cataloging-in-Publication Data

A catalogue record for this book is available from the British Library.

Dedicated
to the
Shakkottai Family

Preface

Preface to the second edition

The second edition of the book has been thoroughly revised and all errors that have come to my notice have been corrected. Additions have been made at various places in the book. Notable additions are in the statistical analysis of measured data in Module I. Important questions regarding normality of deviations and identification of outliers have been discussed in great detail. These should interest the advanced reader who is looking for an understanding of these issues. Thermistors have been described in greater detail in Chpater 4. Also, the line reversal technique of measuring gas temperature has been described in greater detail. Theory of the integrating sphere has been discussed in detail in Chapter 12. Module V has been augmented with more examples from laboratory practice. Exercises are now positioned at the end of each module. Many new exercise problems have been added in this edition.

The modules have been rearranged with the number of chapters going up by one to a total of sixteen chapters in this edition. Many references are indicated as footnotes in the text apart from the bibliography and the list of references given in the Appendix. All illustrations have been redrawn for this edition using 'tikz' - a program environment compatible with 'latex'. All graphs have been replotted for this edition using QtiPlot. In general these were done to improve the quality of the illustrations as also to bring uniformity in the format.

It is hoped that the second edition will be received with the same enthusiasm as the original edition by the student community.

S.P. Venkateshan

Preface to the first edition

In recent times there have been rapid changes in the way we perceive measurements because new technologies have become accessible to any one who cares to use them. Many of the instruments that one takes for granted now were actually not there when I started my engineering studies in the 1960's. Training we received in those days, in Mechanical Engineering did not include a study of "Mechanical Measurements". Whatever was learnt was purely by doing experiments in various laboratory classes! Electrical Engineers were better off because they studied "Electrical Measurements" for a year. The semester system was to be introduced far in the future. Even when "Mechanical Measurements" was introduced as a subject of study the principles of measurements were never discussed fully, the emphasis being the descriptive study of *instruments*! In those days an average mechanical engineer did not have any background in measurement errors and their analysis. Certainly he did not know much about regression, design of experiments and related concepts.

At that time the integrated chip was to appear in the future and the digital computer was in its infancy. We have seen revolutionary changes in both these areas. These developments have changed the way we look at experiments and the art and science of measurements. The study of measurements became divorced from the study of instruments and the attention shifted to the study of the measurement process. The emphasis is more on knowing how to make a measurement rather than with what. One chooses the best option available with reasonable expense and concentrates on doing the measurement well.

I have been teaching a course that was known as "Measurements in Thermal Science" for almost 20 years. Then the title changed to "Measurements in Thermal Engineering"! The emphasis of the course, however, has not changed. The course is one semester long and the student learns about the *measurement process* for almost third of this duration. After he understands the principles he is ready to learn about measurement of quantities that are of interest to a mechanical engineer. The course stresses the problem solving aspect rather than the mundane descriptive aspects. The student is asked to use library and web resources to learn about instruments on her/his own.

In the mean while I have produced a video series (40 lectures each of 55 minutes duration on "Measurements in Thermal Science") that has been widely circulated. Thanks to the NPTEL project (National Program for Technology Enhanced Learning) I had an opportunity to bring out another video lecture series (50 lectures of 55 minutes duration each, this time called "Mechanical Measurements"). This is being broadcast over the 'Technology

Channel'. Also I have prepared a five module web course with the same title. Interested reader can access the web course through the IIT Madras web site. This effort has encouraged me to write a more detailed book version of "Mechanical Measurements" that is now in your hands.

I have arranged this book in five parts, each part being referred to as a module. Details of what is contained in each module is given in an abstract form at the beginning of each module. It has taken me close to three years to produce this book. Over this period I have improved the readability of the text and weeded out unnecessary material and have tried to give to the reader what I believe is important. I have tried to give a balanced treatment of the subject, trying hard to keep my *bias* for thermal measurements!

The text contains many worked examples that will help the reader understand the basic principles involved. I have provided a large number of problems, at the end of the book, arranged module wise. These problems have appeared in the examination papers that I have set for students in my classes over the years. The problems highlight the kind of numerics that are involved in practical situations. Even though the text is intended to be an undergraduate text book it should interest practicing engineers or any one who may need to perform measurements as a part of his professional activity! I place the book in the hands of the interested reader in the hope that he will find it interesting and worth his while. The reader should not be content with a study of the book that contains a large number of *line drawings* that represent instruments. He should spend time in the laboratory and learn how to make measurements in the real world full of hard ware!

S.P. Venkateshan

Acknowledgements

The writing of the book has involved support from several people. My research scholars have extended cooperation during the recording of the video lectures. The feedback received from them - Dr. Rameche Candane, Mr. M. Deiveegan, Mr. T.V.V. Sudhakar and Mr. G. Venugopal - helped in correcting many errors. Their feed back also helped me in improving the material while transforming it to the book form. The photographs used in the book have been taken by Mr. M. Deiveegan with assistance from G. Venugopal in the Heat Transfer and Thermal Power, Internal Combustion Engines and the Thermal Turbo machines Laboratories, Department of Mechanical Engineering, IIT Madras. I am grateful to Prof. B.V.S.S.S. Prasad for permitting me to take pictures of the heat flux gages. Mr. T.V.V. Sudhakar and Mr. G. Venugopal have also helped me by sitting through the classes in "Measurements in Thermal Engineering" and also by helping with the smooth running of the course. The atmosphere in the Heat Transfer and Thermal Power Laboratory has been highly conducive for the book writing activity. The interest shown by my colleagues has been highly encouraging.

Many corrections were brought to my notice by Mr. Renju Kurian and Mr. O.S. Durgam who went through the first edition very carefully. I thank both of them for this help. I thank Dr. Eng. Mostafa Abdel-Mohimen, Benha university, Egypt for pointing out mistakes in the figure and correspondingly the description of diaphragm type pressure gauge. Corrections have been made in the revised second edition. I acknowledge input to this book of Dr. Prasanna Swaminathan who designed a class file called 'bookspv.cls' which has made it possible to improve the aesthetic qulaity of the book.

I particularly thank Athena Academic Ltd and Wiley, UK for bringing out this text in an expeditious manner. I thank my wife for all support during my book writing activity.

S.P. Venkateshan

Nomenclature

Note: (1) **Symbols having more than one meaning are context specific**
(2) **Sparingly used symbols are not included in the Nomenclature**

Latin alphabetical symbols

a Acceleration, m/s^2
or Speed of sound, m/s
or Any parameter, appropriate unit

A Area, m^2

c Callendar correction, $°C$
or Linear damping coefficient, $N \cdot s/m$
or Gas concentration, m^{-3}
or Speed of light, $3 \times 10^8 m/s$

C Specific heat, $J/kg°C$
or Capacitance of a liquid system, m^2
or Capacitance of a gas system, $m \cdot s^2$
or Electrical capacitance, F

C_d Coefficient of discharge, no unit

C_D Drag coefficient, no unit

C_p Specific heat of a gas at constant pressure, $J/kg°C$

C_V Specific heat of a gas at constant volume, $J/kg°C$

D Diameter, m

d Diameter, m
or Degrees of freedom
or Piezoelectric constant, $Coul/N$

E Electromotive force (emf), V
or Emissive power, W/m^2
or Young's modulus, Pa

E_b Total emissive power of a black body, W/m^2

$E_{b\lambda}$	Spectral emissive power of a black body, $W/m^2\mu m$
E_s	Shear modulus, Pa
$\dot{E}$	Enthalpy flux, W/m^2
f	Frequency, s^{-1} or Hz
	or Friction factor, no unit
f_D	Doppler shift, Hz
F	Force, N
FA	Fuel air ratio, $kg(fuel)/kg(air)$
g	Acceleration due to gravity, standard value $9.804m/s^2$
G	Gain, Numerical factor or in dB
	or Gauge constant, appropriate units
	or Bulk modulus, Pa
Gr	Grashof number, no unit
h	Heat transfer coefficient, $W/m^{2\circ}C$
	or Head, m
	or Enthalpy, J/kg
$\bar{h}$	Overall heat transfer coefficient, $W/m^{2\circ}C$
HV	Heating value, J/kg
HHV	Higher Heating Value, J/kg
LHV	Lower Heating Value, J/kg
I	Electrical current, A
	or Influence coefficient, appropriate unit
	or Moment of inertia, m^4
I_λ	Spectral radiation intensity, $W/m^2 \cdot \mu m \cdot ste$
J	Polar moment of inertia, m^4
k	Boltzmann constant, $1.39 \times 10^{-23}, J/K$
	Number of factors in an experiment, no unit
	or Thermal conductivity, $W/m^\circ C$
$\bar{kA}$	Thermal conductivity area product, $W \cdot m/^\circ C$
K	Flow coefficient, no unit
	or Spring constant, N/m
L	Length, m
m	Fin parameter, m^{-1}
	or Mass, kg
	or Mean of a set of data, appropriate unit
$\dot{m}$	Mass flow rate, kg/s
M	Mach number, no unit
	or Molecular weight, g/mol
	or Moment, $N \cdot m$
	or Velocity of approach factor, no unit
n	Index in a polytropic process, no unit
	or Number of data in a sample, no unit
n_i	Number of levels for the ith factor, no unit
N	Number of data in the population, no unit
	or Number count in analog to digital conversion, no unit
N_{St}	Strouhal number, no unit

Nu	Nusselt number, no unit
p	Pressure, Pa
	or Probability, no unit
ppm_V	Gas concentration based on volume, m^{-3}
P	Pressure, Pa
	Perimeter, m
	Power, W
P_D	Dissipation constant, W/m
p_0	Stagnation pressure, Pa
Pe	Peclet number $= Re \cdot Pr$, no unit
Pr	Prandtl number, v/α, no unit
q	Electrical charge (Coulomb), $Coul$
	or Heat flux, W/m^2
Q	Any derived quantity, appropriate unit
	or Heat transfer rate, W
	or Volume flow rate, m^3/s etc.
$\dot{Q}_P$	Peltier heat (power), W
$\dot{Q}_T$	Thomson heat (power), W
R	Electrical resistance, Ω
	or Fluid friction resistance, $1/m \cdot s$
	or radius, m
	or Thermal resistance, $m^{2\circ}C/W$
R_g	Gas constant, $J/kg \cdot K$
$\Re$	Universal gas constant, $J/mol \cdot K$
Re	Reynolds number
s	Entropy, J/K or Entropy rate, W/K
	or Spacing, m
S	Surface area, m^2
Stk	Stoke number, no unit
S_e	Electrical sensitivity, appropriate unit
S_t	Thermal sensitivity, appropriate unit
t	Time, s
	or Temperature, $^\circ C$ or K
	or t - distribution
	or Thickness, m
t_{Pt}	Platinum resistance temperature, $^\circ C$
t_{90}	Temperature according to ITS90, $^\circ C$
T	Period of a wave, s
T	or Temperature, K
	or Torque, $N \cdot m$
T_B	Brightness temperature, K
T_c	Color temperature, K
T_{st}	Steam point temperature, K
T_t	Total or Stagnation temperature, K or $^\circ C$
T_{tp}	Triple point temperature, K
T_{90}	Temperature according to ITS90, K

u	Uncertainty in a measured quantity, Appropriate units or ratio or percentage
V	Potential difference(Volts) or Volume, m^3 or Velocity, m/s
V_P	Peltier voltage, μV
V_S	Seebeck voltage, μV
V_T	Thomson voltage, μV
W	Mass specific heat product, $J/^\circ C$ or Weight of an object, N
x	Displacement, m
$\bar{X}$	Indicated mean or average value of any quantity X
X_C	Capacitive reactance, Ω
X_L	Inductive reactance, Ω
Y	Expansion factor, no unit
Z	Electrical impedance, Ω

Acronyms

ac	Alternating current
dc	Direct currebt
ADC	Analog to Digital Converter
APD	Avalnche Photo Diode
BSN	Bosch Smoke Number
DAC	Digital to Analog Converter
DAQ	Data Acquisition
DAS	Data Acquisition System
DIAL	Differential Absorption LIDAR
DOE	Design Of Experiment
DPM	Digital panel meter
FID	Flame Ionization Detector
GC	Gas Chromatography
GC IR	GC with Infrared spectrometer
GC MS	GC with Mass spectrometer
HC	Hydro Carbon
ISA	Instrument Society of America
IR	Infra Red
LASER	Light Amplification by Stimulated Emission of Radiation
LDV	Laser Doppler Anemometer
LIDAR	Light Detection and Ranging
LVDT	Linear Voltage Differential Transformer
MS	Mass Spectrometer
NDIR	Non Dispersive Infrared Analyzer
NO_x	Mixture of oxides of nitrogen
Op Amp	Operational Amplifier
PC	Personal Computer

PRT or PT	Platinum Resistance Thermometer
RTD	Resistance Temperature Detector
SRM	Standard Reference Material
USB	Universal Serial Bus

Greek symbols

α	Area (fractional) of the tail of the χ^2 distribution
	or Coefficient of linear expansion, $/^\circ C$
	or Pitch angle in a multi-hole probe, rad or $^\circ$
	or Seebeck coefficient, $\mu V/^\circ C$
	or Shock angle in wedge flow, rad or $^\circ$
	or Temperature coefficient of resistance of RTD, $^\circ C^{-1}$
β	Constant in the temperature response of a thermistor, K
	or Diameter ratio in a variable area meter, no unit
	or Extinction coefficient, m^{-1}
	or Isobaric coefficient of cubical expansion, $1/K$
	or Yaw angle in a multi-hole probe, rad or $^\circ$
γ	Ratio of specific heats of a gas, C_p/C_V
δ	Thickness, mm or μm
	or Displacement, m
Δ	Change or difference or error in the quantity that follows Δ
ϵ	Strain, m/m or more usually $\mu m/m$
ε	Emissivity, no unit
$\varepsilon_{\lambda h}$	Spectral Hemispherical emissivity, no unit
ε_h	Total Hemispherical emissivity, no unit
η	Similarity variable in one dimensional transient conduction
ϕ	Non-dimensional temperature
	or Phase angle, rad or $^\circ$
κ	Dielectric constant, F/m
λ	Wavelength, μm
μ	Dynamic viscosity, $kg/m \cdot s$
	or Mean of data
	or Micro (10^{-6})
ν	Kinematic viscosity, m^2/s
	or Poisson ratio, no unit
π	Mathematical constant, 3.14159...
	or Peltier emf, μV
ρ	Density, kg/m^3
	or Correlation coefficient (linear fit)
	or the index of correlation (non-linear fit)
	or Reflectivity, no unit
σ	Stress, Pa (more commonly Mpa or Gpa)
	or Stefan Boltzmann constant, $5.67 \times 10^{-8} W/m^2 K^4$
	or Thomson coefficient, $\mu V/^\circ C$
	or Standard deviation, appropriate unit

σ_e	Estimated standard distribution, appropriate unit
σ_a	Absorption cross section, m^2
σ_s	Scattering cross section, m^2
σ_t	Total cross section, m^2
σ_x	Standard deviation of the x's
σ_y	Standard deviation of the y's
σ_{xy}	Covariance
θ	Temperature difference, $^\circ C$
τ	Shear stress, Pa
	or Time constant, s
	or Transmittance, no unit
ω	Circular frequency, rad/s
ω_n	Natural circular frequency, rad/s
Ω	Electrical resistance (Ohms)
χ^2	Chi squared distribution, appropriate unit
ζ	Damping ratio for a second order system, no unit

Contents

III   Measurement of Pressure, Fluid velocity, Volume flow rate, Stagnation and Bulk mean temperatures

Module **I**

Measurements, Error Analysis and Design of Experiments

Module I is a comprehensive introduction to any measurement including Mechanical Measurements - that are of interest to a mechanical engineer. This module consists of three chapters. Chapter 1 deals with basics of measurement along with the description of errors that almost always accompany any measurement. Chapter 2 considers a very important topic, viz. regression analysis. Regression helps in summarizing the measurements by evolving relationships between measured quantities that represent outcome of experiments. Chapter 3 discusses about design of experiments (DOE). Goal of DOE is to reduce the number of experiments that need to be performed and yet get the most amount of information about the outcome of the experiments.

Chapter *1*

Measurements and Errors in measurement

This chapter introduces measurement errors and methods of describing them so that measured data is interpreted properly. Statistical principles involved in error analysis are discussed in sufficient detail. Concepts such as precision and accuracy are clearly explained. Different statistics useful in experimental studies are discussed. Tests of normality of error distribution and procedure for rejection of data are discussed. Results from sampling theory are discussed because of their use in the interpretation of sparse experimental data that is almost the rule.

1.1   Introduction

We recognize three reasons for making measurements as indicated in Figure 1.1. From the point of view of the present book measurements for commerce is outside its scope. Engineers design physical systems in the form of machines to serve some specified functions. The behavior of the parts of the machine during the operation of the machine needs to be examined or analyzed or designed such that it functions reliably. Such an activity needs data regarding the machine parts in terms of material properties. These are obtained by performing measurements in the laboratory.

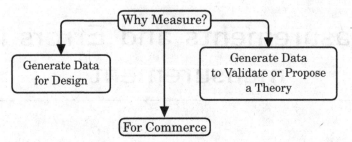

Figure 1.1: *Why make measurements?*

The scientific method consists in the study of nature to understand the way it works. Science proposes hypotheses or theories based on observations and these need to be validated with carefully performed experiments that use many measurements. When once a theory has been established it may be used to make predictions which may themselves be confirmed by further experiments.

1.1.1   Measurement categories

Broadly measurements may be categorized as given below.

- **Primary quantity**: It is possible that a single quantity that is directly measurable is of interest. An example is the measurement of the diameter of a cylindrical specimen. It is directly measured using an instrument such as vernier calipers. We shall refer to such a quantity as a primary quantity.
- **Derived quantity**: There are occasions when a quantity of interest is not directly measurable by a single measurement process. The quantity of interest needs to be estimated by using an appropriate relation involving *several* measured primary quantities. The measured quantity is thus a derived quantity. An example of a derived quantity is the determination of acceleration due to gravity (g) by finding the period (T) of a simple pendulum of length (L). T and L are the measured primary quantities while g is the derived quantity.
- **Probe or intrusive method**: It is common to place a *probe* inside a system to measure a physical quantity that is characteristic of the system. Since a probe invariably affects the measured quantity the measurement process is referred to as an intrusive type of measurement.

- **Non-intrusive method**: When the measurement process does not involve insertion of a probe into the system the method is referred to as being non-intrusive. Methods that use some naturally occurring process, like radiation emitted by a body to measure a desired quantity relating to the system, may be considered as non-intrusive. The measurement process may be assumed to be non-intrusive when the probe has negligible interaction with the system. A typical example for such a process is the use of Laser Doppler Velocimeter (LDV) to measure the velocity of a flowing fluid.

1.1.2 General measurement scheme

Figure 1.2 shows the schematic of a general measurement process. Not all the elements shown in Figure 1.2 may be present in a particular case.

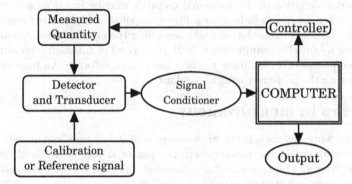

Figure 1.2: *Schematic of a general measurement system with controller*

The measurement process requires invariably a detector that responds to the measured quantity by producing a measurable change in some property of the detector. The change in the property of the detector is converted to a measurable output that may be either mechanical movement of a pointer over a scale or an electrical output that may be measured using an appropriate electrical circuit. This action of converting the measured quantity to a different form of output is done by a transducer. The output may be manipulated by a signal conditioner before it is recorded or stored in a computer. If the measurement process is part of a control application the computer can be used to drive the controller. The relationship that exists between the measured quantity and the output of the transducer may be obtained by calibration or by comparison with a reference value. The measurement system requires external power for its operation.

1.1.3 Some issues

- Errors - *Systematic or Random*
- Repeatability
- Calibration and Standards
- Linearity or Linearization

Any measurement, however carefully conducted, is subject to measurement errors. These errors make it difficult to ascertain the true value of the measured quantity. The nature of the error may be ascertained by repeating the measurement a number of times and looking at the spread of the values. If the spread in the data is small the measurement is repeatable and may be termed as being good. If we compare the measured quantity obtained by the use of any instrument and compare it with that obtained by a standardized instrument the two may show different performances as far as the repeatability is concerned. If we add or subtract a certain correction to make the two instruments give data with similar spread the correction is said to constitute a systematic error. Then the spread of data from each of the instruments will constitute random error.

The process of determining the systematic error is calibration. The response of a detector to the variation in the measured quantity may be linear or non-linear. In the past the tendency was to look for a linear response as the desired response. Even when the response of the detector was non-linear the practice was to make the response linear by suitable manipulation. With the advent of automatic recording of data using computers this practice is not necessary since software can take care of this aspect during the post-processing of the data.

1.2   Errors in measurement

Errors accompany any measurement, however well it has been conducted. The error may be inherent in the measurement process or it may be induced due to variations in the way the experiment is conducted. The errors may be classified as systematic errors and random errors.

1.2.1   Systematic errors *(Bias)*

Systematic error or bias is due to faulty or improperly calibrated instruments. These may be reduced or eliminated by careful choice and calibration of instruments. Sometimes bias may be linked to a specific cause and estimated by analysis. In such a case a correction may be applied to eliminate or reduce bias.

Bias is an indication of the accuracy of the measurement. Smaller the bias more accurate the data.

1.2.2   Random errors

Random errors are due to non-specific causes like natural disturbances that may occur during the measurement process. These cannot be eliminated. The magnitude of the spread in the data due to the presence of random errors is a measure of the precision of the data. Smaller the random error more precise is the data. Random errors are statistical in nature. These may be characterized by statistical analysis.

We shall explain these through the familiar example shown in Figure 1.3. Three different individuals with different skill levels are allowed to complete a round of target[1] practice. The outcome of the event is shown in the figure.

---

[1]Target shown in Figure 1.3 is non-standard and for purposes of illustration only. Standard targets are marked with 10 evenly spaced concentric rings.

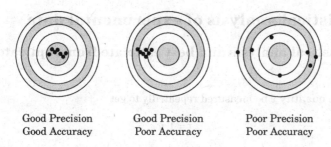

Good Precision   Good Precision   Poor Precision
Good Accuracy   Poor Accuracy   Poor Accuracy

Figure 1.3: *Precision and accuracy explained through a familiar example*

It is evident that the target at the left belongs to a highly skilled shooter. This is characterized by all the shots in the inner most circle or the 'bull's eye'. The result indicates good accuracy as well as good precision. A measurement made well must be like this! The individual in the middle is precise but not accurate. Maybe it is due to a faulty bore of the gun. The individual at the right is an unskilled person who is behind on both counts. Most beginners fall into this category. The analogy is quite realistic since most students performing a measurement in the laboratory may be put into one of the three categories. A good experimentalist has to work hard to excel at it! The results shown in Figure 1.4 compare the response of an

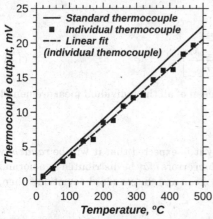

Figure 1.4: *Illustration of presence of systematic and random errors in data*

individual thermocouple (that measures temperature) and a standard thermocouple. The measurements are reported between room temperature (close to) $20°C$ and an upper limit of $500°C$. That there is a systematic variation between the two is clear from the figure that shows the trend of the measured temperatures indicated by the individual thermocouple. The systematic error appears to vary with the temperature. The data points indicated by the full symbols appear also to hug the trend line (we look at in detail at trend lines while discussing regression analysis of data in Section 2.2), which is a linear fit to the data. However the data points do not *lie* on it. This is due to random errors that are always present in any measurement. Actually the standard thermocouple would also have random errors that are *not* indicated in the figure. We have deliberately shown only the trend line for the standard thermocouple to avoid cluttering up the graph.

1.3   Statistical analysis of experimental data

1.3.1   Statistical analysis and best estimate from replicate data

Let a certain quantity x be measured repeatedly to get

$$x_i, \quad i = 1, n \tag{1.1}$$

Because of random errors these *will* all be different. How do we find the best estimate x_b for the true value of x? It is reasonable to assume that the best value be such that the measurements are *as precise* as they can be! In other words, the experimenter is confident that he has conducted the measurements with the best care and he is like the skilled shooter in the target practice example presented earlier! Thus, we minimize the variance with respect to the best estimate x_b of x. Thus we minimize:

$$S = \sum_{i=1}^{n} [x_i - x_b]^2 \tag{1.2}$$

This requires that

$$\frac{\partial S}{\partial x_b} = 2 \sum_{i=1}^{n} [x_i - x_b](-1) = 0 \tag{1.3}$$

or

$$x_b = \frac{\sum_{i=1}^{n} x_i}{n} \tag{1.4}$$

The best estimate is thus nothing but the mean of all the individual measurements!

1.3.2   Error distribution

When a quantity is measured repeatedly it is expected that it will be *randomly* distributed around the best value. The random errors *may* be distributed as a normal distribution. If μ and σ are respectively the mean and the standard deviation,[2] then, the normal probability density is given by

$$f(x) = \frac{1}{\sigma\sqrt{2\pi}} e^{-\frac{1}{2}\left\{\frac{x-\mu}{\sigma}\right\}^2} \quad \text{or} \quad f(z) = \frac{1}{\sigma\sqrt{2\pi}} e^{-\frac{z^2}{2}} \tag{1.5}$$

The distribution is also represented some times as $N(\mu, \sigma)$. Normal distribution is also referred to as Gaussian distribution [3] or "bell shaped curve". The probability that the error around the mean is $(x - \mu)$ is the area under the probability density

---

[2]The term standard deviation was first used by Karl Pearson, 1857-1936, an English mathematician, who has made outstanding contributions to the discipline of mathematical statistics

[3]Named after Johann Carl Friedrich Gauss, 1777-1855, a German mathematician and physical scientist

function between $(x - \mu) - \dfrac{dx}{2}$ and $(x - \mu) + \dfrac{dx}{2}$ represented by the product of the probability density and dx. The probability that the error is anywhere between $-\infty$ and x is thus given by the following integral:

$$F(x) = \frac{1}{\sigma\sqrt{2\pi}} \int_{-\infty}^{x} e^{-\frac{1}{2}\{\frac{x-\mu}{\sigma}\}^2} dx \text{ or } F(z) = \frac{1}{\sqrt{2\pi}} \int_{-\infty}^{z} e^{-\frac{z^2}{2}} dz \quad (1.6)$$

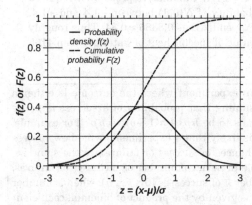

Figure 1.5: *Normal probability density function and the cumulative probability*

This is referred to as the cumulative probability. It is noted that if $x \to \infty$ the integral tends to 1. Thus the probability that the error is of all possible magnitudes (between $-\infty$ and $+\infty$) is unity! The integral is symmetric with respect to $z = 0$ where $z = \dfrac{x - \mu}{\sigma}$, as is easily verified. The above integral is in fact the error integral that is a tabulated function. A plot showing $f(z)$ and $F(z)$ with respect to z is given in Figure 1.5.

Many times we are interested in finding out the chances of error lying between two values in the form $\pm p\sigma$ around the mean or $\pm z$. This is referred to as the "confidence interval" and the corresponding cumulative probability specifies the chances of the error occurring within the confidence interval. Table 1.1 gives the confidence intervals that are useful in practice. A more complete table of confidence intervals is given in Appendix B as Table B.4.

Table 1.1: *Confidence intervals according to normal distribution*

p	0	± 1	± 2	± 1.96	3	± 2.58	± 3.30
CP	0	0.6827	0.9545	**0.95**	0.9973	**0.99**	**0.999**

CP = Cumulative Probability

Example 1.1

A certain measurement gave the value of C, the specific heat of water, as $4200\ J/kg^{\circ}C$. The precision of measurement is specified by the standard deviation given by $25\ J/kg^{\circ}C$. If the measurement is repeated what is the probability that the value is within $4200 \pm 35\ J/kg^{\circ}C$? You may assume that the error is normally distributed.

Solution :

The maximum and minimum values that are expected for the specific heat of water are given by

$$C_{min} = 4200 - 35 = 4165, \quad \text{and} \quad C_{max} = 4200 + 35 = 4235 \ J/kg^{o}C$$

Scaling the values in terms of σ the spread of the readings should be $\pm\dfrac{35}{25}\sigma = \pm1.4\sigma$ with respect to the mean or $z = \pm1.4$. Cumulative probability required is nothing but the cumulative probability of $N(0,1)$ between -1.4 and $+1.4$. This is obtained from Table B.4 of Appendix B as $0.8385 \approx 0.84$. Thus roughly 84% of the time we should get the value of specific heat of water between 4165 and 4235 $J/kg^{o}C$.

Gaussian distribution: Consider an experiment where the outcome is either a success $s = 1$ or a failure $f = 0$. Assuming that probability of a success is p, we expect the value of successes in n trials to be $nps + n(1-p)f = np$. For example, if the number of experiments is 8, it is likely that 4 experiments will show success as the outcome, assuming that the chances of success or failure are the same i.e. $p = 0.5$. Since the outcome is countable in terms of number of successes or failures, the probability $B(k)$ of a certain number k of successes ($0 \leq k \leq n$), when n number of experiments have been performed, is given by the product of binomial coefficient $\dfrac{n!}{k!(n-k)!}$ (this represents the number of combinations that yield k successes) and $p^{k}(1-p)^{n-k}$. For example, if $n = 8$ and $p = 0.5$ we have $0.5^{k}(1-0.5)^{8-k} = 0.5^{8} = \dfrac{1}{256}$ where the denominator 256 represents total number of possible ways of getting all possible combinations of outcomes when 8 experiments are performed. Thus we have the binomial distribution function

$$B(k;n,p) = \frac{n!}{k!(n-k)!}p^{k}(1-p)^{n-k}$$

where n and p are indicated as parameters, by placing them after a semicolon following k. The expression given above is one of the terms in the binomial $[p+(1-p)]^{n}$. In the special case $p = \dfrac{1}{2}$ of interest to us (we are expecting equal likelihood of positive and negative errors in measurements) this becomes

$$B(k;n,0.5) = \frac{n!}{2^{n}k!(n-k)!}$$

Binomial distribution is a function of a discrete variable k and satisfies the requirement that $\sum\limits_{k=0}^{k=n} B(k;n,0.5) = 1$. Mean value of k can be obtained as

$$\bar{k} = \sum_{k=0}^{k=n} kB(k;n,0.5) = 0.5n$$

The variance may be shown to be given by

$$\sigma_{k}^{2} = \sum_{k=0}^{k=n} (k-\bar{k})^{2}B(k;n,0.5) = 0.25n = \frac{\bar{k}}{2}$$

For sufficiently large n the Binomial distribution is closely approximated by the Normal distribution or the Gaussian curve. Thus we have

$$B(k;n,0.5) \approx N(\mu,\sigma) = N\left(\bar{k}, \sqrt{\frac{\bar{k}}{2}}\right)$$

For example, with $n = 30$ we have $\mu = 15$, $\sigma = 2.739$, the Binomial distribution very closely resembles the Normal distribution as shown in Figure 1.6. Note that the Normal distribution function is a function of continuous variable $\dfrac{k - \bar{k}}{\sigma_k}$ when $n \gg 1$.

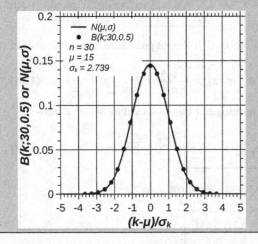

Figure 1.6: *Comparison of Binomial and Normal distributions for $n = 30$*

1.3.3 Principle of Least Squares

Earlier we have dealt with the method of obtaining the best estimate from replicate data based on minimization of variance. No mathematical proof was given as a basis for this. We shall now look at the above afresh, in the light of the fact that the errors are distributed normally, as has been made out above.

Consider a set of replicate data x_i. Let the best estimate for the measured quantity be x_b. The probability for a certain value x_i within the interval $x_i - \dfrac{dx_i}{2}$, $x_i + \dfrac{dx_i}{2}$ to occur in the measured data is given by the relation

$$p(x_i) = \frac{1}{\sigma\sqrt{2\pi}} e^{-\frac{(x_i - x_b)^2}{2\sigma^2}} dx_i \tag{1.7}$$

The probability that the particular values of measured data are obtained in replicate measurements must be the compound probability given by the product of the individual probabilities. Thus

$$p = \left(\frac{1}{\sigma\sqrt{2\pi}}\right)^n \prod_{i=1}^{n} e^{-\frac{(x_i - x_b)^2}{2\sigma^2}} dx_i = \left(\frac{1}{\sigma\sqrt{2\pi}}\right)^n e^{-\sum_{i=1}^{n}\frac{(x_i - x_b)^2}{2\sigma^2}} \prod_{i=1}^{n} dx_i \tag{1.8}$$

The reason the set of data was obtained as replicate data is that it was the most probable! Since the intervals dx_i are arbitrary, the above will have to be maximized by the proper choice of x_b and σ such that the exponential factor is a maximum. Thus we have to choose x_b and σ such that

$$p' = \left(\frac{1}{\sigma}\right)^n e^{-\sum_{i=1}^{n} \frac{(x_i - x_b)^2}{2\sigma^2}} \tag{1.9}$$

has the largest possible value. As usual we set the derivatives $\dfrac{\partial p'}{\partial x_b} = \dfrac{\partial p'}{\partial \sigma} = 0$ to get the values of the two parameters x_b and σ . We have:

$$\frac{\partial p'}{\partial x_b} = \frac{1}{2}\left(\frac{1}{\sigma}\right)^{n+2} e^{-\sum_{i=1}^{n} \frac{(x_i - x_b)^2}{2\sigma^2}} \cdot \underline{\sum_{i=1}^{n} 2(x_i - x_b)\cdot(-1)} = 0 \tag{1.10}$$

The term shown with underline in Equation 1.10 should go to zero. Hence we should have

$$\sum_{i=1}^{n}(x_i - x_b) = 0 \quad \text{or} \quad x_b = \frac{\sum_{i=1}^{n} x_i}{n} = \bar{x} \tag{1.11}$$

It is thus clear that the best value is nothing but the mean value! We also have:

$$\frac{\partial p'}{\partial \sigma} = \underline{\left[-\frac{n}{\sigma^{n+1}} + \frac{1}{\sigma^{n+3}}\sum_{i=1}^{n}(x_i - x_b)^2\right]} e^{-\sum_{i=1}^{n} \frac{(x_i - x_b)^2}{2\sigma^2}} = 0 \tag{1.12}$$

or, again setting the underlined term to zero, we have

$$\boxed{\sigma^2 = \frac{\sum(x_i - x_b)^2}{n}} \tag{1.13}$$

Equation1.13 indicates that the parameter σ^2 is nothing but the variance of the data with respect to the mean! Thus the best value of the measured quantity and its spread are based on the minimization of the squares of errors with respect to the mean. This embodies what is referred to as the "Principle of Least Squares". We shall be making use of this principle while considering regression.

1.3.4   Error estimation - single sample

In practice measurements are expensive in terms of cost and time. Hence it is seldom possible to record a large number of replicate data. In spite of this one would like to make generalizations regarding the parameters that describe the population from the parameters like mean and variance that characterize the sample. Population represents the totality of experiments that could have been conducted had we the resources to do it. For example, if we would like to study the characteristics of student performance under a schooling system, all the students in the schooling system would make up the population. In order to reduce the cost a statistician would draw a random sample of students, characterize this sample, and extrapolate by statistical theory to get parameters that describe the entire population. In engineering we may want to estimate a physical quantity based on a small number of experiments. The problem is not trivial and hence is discussed in some detail here.

Variance of the means

The problem is posed as given below:

- Replicate data is collected with n measurements in a set or a sample
- Several (possibly) such sets of data are collected
- Sample mean is m_s and sample variance is σ_s^2
- What is the mean and variance of the mean of all samples?

Population mean:

Let N be the total number of data in the entire population, if indeed, a large number of samples have been collected. Without loss of generality we assume that the population mean is zero. Hence we have

$$m = \sum_{i=1}^{N} \frac{x_i}{N} = 0 \tag{1.14}$$

Consider sample s whose members are identified as $x_{k,s}$ with $1 \le k \le n$. Mean of the sample then is

$$m_s = \sum_{k=1}^{N} \frac{x_{k,s}}{n} \tag{1.15}$$

The number of samples n_s each comprising n data, drawn out of the population N is given by $n_s = {}^N C_n.$[4] Mean of all sample means - $\bar{m}_s$ - is then given by

$$\bar{m}_s = \sum_{s=1}^{{}^N C_n} \frac{m_s}{{}^N C_n} \tag{1.16}$$

A particular data x_i will occur in ${}^{N-1}C_{n-1}$ samples as may be easily seen. Hence the summation in the above equation may be written as

$$\bar{m}_s = \sum_{i=1}^{N} \frac{{}^{N-1}C_{n-1}x_i}{n\,{}^N C_n} = \sum_{i=1}^{N} \frac{x_i}{N} = m = 0 \tag{1.17}$$

Thus the mean of all samples is also the population mean.

Population variance:

The population variance is given by

$$\sigma^2 = \sum_{i=1}^{N} \frac{(x_i - m)^2}{N} = \sum_{i=1}^{N} \frac{x_i^2}{N} \tag{1.18}$$

since the population mean is zero.

---

[4]This result is analogous to filling n bins with one object each, drawing objects from a container with N objects, without replacing them. Number of samples is just the number of ways the bins may be filled.

Variance of the means:

Let the variance of the means be σ_m^2. By definition we then have

$$^{N}C_n \sigma_m^2 = \sum_{s=1}^{^{N}C_n} (m_s - \bar{m})^2 = \sum_{s=1}^{^{N}C_n} m_s^2 \tag{1.19}$$

Using Equation 1.15 we have, for the sth sample

$$m_s^2 = \left[\sum_{k=1}^{n} \frac{x_{h,s}}{n} \right]^2 = \frac{(x_1 + x_2 + ... + x_n)_s^2}{n^2}$$

On expanding squares, the right hand will contain terms such as x_l^2 and $2x_l x_m$ with $l \neq m$. Both l and m are bounded between 1 and n. Hence the summation over s indicated in Equation 1.19 will have x_l^2 appearing $^{N-1}C_{n-1}$ times and $2x_l x_m$ appearing $^{N-2}C_{n-2}$ times. Hence we have

$$^{N}C_n \sigma_m^2 = \frac{1}{n^2} \left[\sum_{i=1}^{N} {}^{N-1}C_{n-1} x_i^2 + 2 \underbrace{\sum_{i=1}^{N} \sum_{j=1}^{N} {}^{N-2}C_{n-2} x_i x_j}_{i \neq j} \right] = 0 \tag{1.20}$$

However, since the population mean is zero, we have

$$\left(\sum_{i=1}^{N} x_i \right)^2 = \sum_{i=1}^{N} x_i^2 + 2 \underbrace{\sum_{i=1}^{N} \sum_{j=1}^{N} x_i x_j}_{i \neq j} = 0$$

Hence we have $2 \sum_{i=1}^{N} \sum_{j=1}^{N} x_i x_j = - \sum_{i=1}^{N} x_i^2$. Substitute this in Equation 1.20 and simplify to get

$$\sigma_m^2 = \frac{1}{n^2} \underbrace{\left(\frac{^{N-1}C_{n-1}}{^{N}C_n} \right)}_{=\frac{n}{N}} \underbrace{\left[1 - \left(\frac{^{N-2}C_{n-2}}{^{N-1}C_{n-1}} \right) \right]}_{=1-\frac{n-1}{N-1}} \sum_{i=1}^{N} x_i^2 = \frac{(N-n)}{n(N-1)} \sigma^2 \tag{1.21}$$

If $n << N$ and N is large, the above relation may be approximated as

$$\sigma_m^2 = \frac{1 - \frac{n}{N}}{n\left(1 - \frac{1}{N}\right)} \sigma^2 \approx \frac{\sigma^2}{n} \tag{1.22}$$

Estimate of variance:

- Sample variance - how is it related to the population variance?
- Let the sample variance from its own mean m_s be σ_s^2
- i.e. $\sigma_s^2 = \frac{1}{n} \sum_{i=1}^{n} x_{i,s}^2 - m_s^2$

The mean of all the sample variances may be calculated by summing all the sample variances and dividing it by the number of samples. Since the total number of samples is $^{N}C_n$

$$\bar{\sigma}_s^2 = \frac{1}{{}^{N}C_n}\left[\frac{{}^{N-1}C_{n-1}}{n}\sum_{i=1}^{N}x_i^2 - \sum_{j=1}^{{}^{N}C_n}m_{j,s}^2\right] = \sigma^2 - \sigma_m^2 \tag{1.23}$$

Combine this with Equation 1.21 and simplify to get

$$\sigma_s^2 = \frac{N(n-1)}{n(N-1)}\sigma^2 \tag{1.24}$$

If $n << N$ the above relation will be approximated as

$$\boxed{\sigma_s^2 \approx \sigma^2\left(1 - \frac{1}{n}\right)} \tag{1.25}$$

Error estimator σ_e:

The last expression may be written down in the more explicit form

$$\boxed{\sigma_e^2 = \sum_{1}^{n}\frac{(x_i - m_s)^2}{n-1}} \tag{1.26}$$

Essentially the experimenter has only one sample and the above formula tells him how the variance of the single sample is related to the variance of the population!

Physical interpretation

Equation 1.26 may be interpreted using physical arguments. Since the mean (the best value) is obtained by one use of all the available data, the degrees of freedom available (units of information available) is one less than before. Hence the error estimator *should* use the factor $(n-1)$ rather than n in the denominator! The estimator thus obtained is referred to as unbiased variance.

Example 1.2

Resistance of a certain resistor is measured repeatedly to obtain the following data.

Expt. No.	1	2	3	4	5	6	7	8	9
$R, k\Omega$	1.22	1.23	1.26	1.21	1.22	1.22	1.22	1.24	1.19

What is the best estimate for the resistance? What is the error with 95% confidence?

Solution :
Best estimate is the mean of the data.

$$\bar{R} = \frac{4 \times 1.22 + 1.23 + 1.26 + 1.21 + 1.24 + 1.19}{9} = 1.223\,k\Omega \approx 1.22\,k\Omega$$

Standard deviation of the error σ_e:

Unbiased Variance $\sigma_e^2 = \dfrac{\left[\begin{array}{c} 4 \times 1.22^2 + 1.23^2 + 1.26^2 \\ +1.21^2 + 1.24^2 + 1.19^2 \end{array}\right]}{8} - 1.223^2 = 3.75 \times 10^{-4} \, k\Omega^2$

Hence

$$\sigma_e = \sqrt{3.75 \times 10^{-5}} = 0.0194 \, k\Omega$$

The corresponding error estimate based on 95% confidence interval is

$$\text{Error} = 1.96 \times \sigma_e = 1.96 \times 0.0194 = 0.038 \, k\Omega$$

Example 1.3

Thickness of a metal sheet (in mm) is measured repeatedly to obtain the following replicate data. What is the best estimate for the sheet thickness? What is the variance of the distribution of errors with respect to the best value? Specify an error estimate to the mean value based on *99%* confidence.

No.	1	2	3	4	5	6
t, mm	0.202	0.198	0.197	0.215	0.199	0.194
No.	7	8	9	10	11	12
t, mm	0.204	0.198	0.194	0.195	0.201	0.202

Solution :

The best estimate for the metal sheet thickness is $\bar{t}$, the mean of the 12 measured values. This is given by

$$\bar{t} = \dfrac{\left[\begin{array}{c} 2 \times 0.202 + 2 \times 0.198 + 0.197^2 + 0.215 \\ +0.199 + 2 \times 0.194 + 0.204 + 0.195 + 0.201 \end{array}\right]}{12} = 0.200 \, mm$$

The variance with respect to the mean or the best value is then given by

$$\sigma_e^2 = \dfrac{\left[\begin{array}{c} 2 \times 0.202^2 + 2 \times 0.198^2 + 0.197^2 + 0.215^2 \\ +0.199^2 + 2 \times 0.194^2 + 0.204^2 + 0.195^2 + 0.201^2 \end{array}\right]}{11} - 0.2^2$$

$$= 3.3174 \times 10^{-5} \, mm^2$$

The corresponding standard deviation is given by

$$\sigma_e = \sqrt{3.3174 \times 10^{-5}} = 0.0058 \, mm$$

The corresponding error estimate based on 95% confidence is

$$\text{Error} = 2.58 \times \sigma_e = 2.58 \times 0.0058 = 0.0149 \, mm$$

1.3.5 Student t distribution

We have seen above that, an estimate for the variance of the population, based on a single sample of n values is given by $\sigma_e^2 = \dfrac{n(N-1)}{N(n-1)}\sigma_s^2$. We also know that the variance of the sample means is given by the expression $\sigma_m^2 = \dfrac{\sigma^2}{n}$. In practice only one sample of n values may have been obtained experimentally. The population variance σ^2 is, in fact, not known and hence we use σ_e^2 for σ^2 and hence we have $\sigma_m^2 = \dfrac{\sigma_e^2}{n}$. Thus the standard deviation of the means is given by $\sigma_m = \dfrac{\sigma_e}{\sqrt{n}}$. The advantage of doing this is that the standard deviation of the population may be calculated based on this expression even though the population variance is not known. Now we consider the following function given by

$$T_n = \frac{m_s - m}{\frac{\sigma_e}{\sqrt{n}}} \tag{1.27}$$

Note that σ_e is a random variable and hence the function T given by Equation 1.27 is not standard normal. The distribution is referred to as the "Student t distribution" and is defined as $T_n = t_{n-1}$. This distribution depends on n and is given by the following expression

$$t(y,d) = \frac{\Gamma\left(\frac{d+1}{2}\right)}{\sqrt{\pi d}\,\Gamma\left(\frac{d}{2}\right)}\left[1 + \frac{y^2}{d}\right]^{-\frac{(d+1)}{2}} \tag{1.28}$$

Here d is the degrees of freedom given by $d = n - 1$ and Γ is the Gamma function

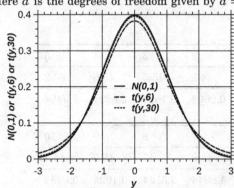

Figure 1.7: Comparison of t - distributions with the Normal distribution

also referred to as the Generalized Factorial function. The argument $y = m_s - m$ represents the difference between the sample mean and the population mean. This distribution was discovered by a British mathematician who published his work under the pseudonym "Student".[5] For large n (or d) the t - distribution approaches the normal distribution with 0 mean and unit variance given by $N(0,1)$. For small n the distribution is wider than the Normal distribution with larger areas in the tails of the distribution. Plots in Figure 1.7 show these trends. We see that for $d = 30$,

---

[5] real name William Sealy Gosset, 1876-1937, well known statistician

the t -distribution is quite close to the normal distribution. If the number of samples available is more than about 30 one may simply use the N distribution.

In statistical analysis of data what are more important are the confidence intervals that are appropriate with the t - distributions. These are, in deed, larger than the corresponding values for the Normal distribution. A short table useful for analysis is given as Table B.5. Notice that the 95% confidence interval tends asymptotically to ±1.96σ, characteristic also of the Normal distribution.

Example 1.4

The temperature of a controlled space was measured at random intervals and the spot values are given by the following *10* values:

Trial	1	2	3	4	5
Temperature °C	45.3	44.2	45.5	43.5	46.2
Trial	6	7	8	9	10
Temperature °C	46.4	43.8	47	45.5	44.4

The control was expected to maintain the temperature at 44.5°C. How would you describe the above observations?

Solution :

The number of data in the sample is $n = 10$.
The number of degrees of freedom is $d = n - 1 = 9$.
Tabulation of data helps in pursuing the statistical analysis of data. Sample mean is the arithmetic mean of all the spot values of temperature while the estimated variance is based on d.

Trial No.	1	2	3	4	5
Temperature,°C	45.3	44.2	45.5	43.5	46.2
Square of error with respect to mean	0.0144	0.9604	0.1024	2.8224	1.0404
Trial No.	6	7	8	9	10
Temperature,°C	46.4	43.8	47	45.5	44.4
Square of error with respect to mean	1.4884	1.9044	3.3124	0.1024	0.6084

Sample mean m_s= 45.2°C
Estimated variance σ_e^2=1.373
Estimated standard deviation σ_e=1.172°C
The population mean is specified to be $m = 44.5°C$.
Hence the t - value based on the data may now be calculated as

$$t = \frac{m - m_s}{\sigma_e} \sqrt{n} = \frac{45.2 - 44.5}{1.172} \sqrt{10} = 1.835$$

The 95% Confidence Interval for the t - distribution with $d = 9$ is read from Table B.5 in Appendix B as 2.262. Since the t - value is less than the 95% Confidence Interval we conclude that the sample indicates satisfactory functioning of the controller.

Example 1.5

A sample of 6 resistors is picked up from a lot during a manufacturing process. The resistances are measured in the laboratory and the values are found to be 1020, 1040, 995, 1066, 970 and 992 Ω. The manufacturer will label all the resistors as being equal to a mean value of 1000 Ω. Is this justified? Also specify a tolerance for the resistors from this lot.

Solution :

It is convenient to make a spreadsheet as in Table 1.2.

The table shows the sample mean, the population mean, estimated standard error and finally the value of t calculated for the sample of 6 measured resistance values with degrees of freedom of $d = 6 - 1 = 5$. The 95% confidence interval for t with $d = 5$ is 2.571 (Table B.5). Since the t calculated from the sample is less than this the manufacturer is justified in labeling the resistors as having a mean value of 1000 Ω. We may now use the 95% confidence interval to specify the tolerance. We thus have

$$\text{Tolerance} = \pm 2.571 \times 35.2 = \pm 90.5 \, \Omega$$

The resistors from this lot may be labeled as 1000 Ω nominal with 10% tolerance.

Table 1.2: *Spreadsheet for Example 1.5*

Resistor Number	Resistance Value Ω	Square of Error
1	1020	38.44
2	1040	686.44
3	995	353.44
4	1066	2724.84
5	970	1918.44
6	992	475.24
$m_s =$	1013.8	
$\sigma_e^2 =$	1239.368	
$\sigma_e =$	35.2	
$t =$	0.963	

1.3.6 Test for normality

Distribution of random errors in measurements are obtained by repeated measurements of a physical quantity. These are done by keeping the conditions under which the experiments are conducted invariant. For example, in most measurements, pressure and temperature may affect the outcome and hence these need to be kept fixed during the experiment. Once replicate data is collected we should like to ascertain the error distribution so that we may draw conclusions on the quality of the measurement based on the distribution of errors. Specifically we would like to ascertain whether random errors are distributed normally.

Box and whisker plot

Many methods are available for test of normality of a sample distribution. Shape of the histogram may indicate whether the distribution is close to being normal. Alternately, we look for symmetry, lower and upper quartile values and the minimum and maximum values to make a "box and whisker plot" to check for normality as shown in Figure 1.8. The box and whisker plot shown here is for a sample of data that follows closely a normal distribution. If the sample size is large the values on

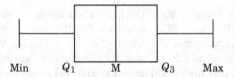

Min Q_1 M Q_3 Max

Figure 1.8: *Box and whisker plot: M- Median, Q_1 - Lower quartile, Q_3 - Upper quartile, Min - Lower extreme, Max - Upper extreme*

the box and whisker plot are like those indicated here: $Min = -2.33$, $Q1 = -0.67$, $Median = 0$, $Q3 = 0.67$, $Max = 2.33$. The values shown are the z values for $N(0,1)$.

Box and whiskers plot is a construct introduced by Tukey[6] and gives a summary of the distribution. Obvious asymmetry, outliers and sharpness of the distribution may be gleaned by looking at the plot.

Example 1.6

A sample data consists of 15 values shown in the table. i is the serial number of data and v_i is the corresponding value. Make a box and whisker plot and comment on the nature of the distribution.

i	1	2	3	4	5	6	7	8
v_i	-1.829	-1.259	-1.187	-0.884	-0.854	-0.745	-0.343	-0.130

i	9	10	11	12	13	14	15	
v_i	0.097	0.224	0.844	1.098	1.112	1.806	2.050	

Solution :

[6]J. W. Tukey, "Box-and-Whisker Plots", Section 2C in Exploratory Data Analysis, Reading, Mass, Addison-Wesley, pp. 39-43, 1977.

We note that the data is already arranged in the ascending order. Hence the minimum and maximum values are the first and last entries in the table. Thus $Min = -1.829$ and $Max = 2.050$.

Since there are an odd number of data values, the median is the value corresponding to the eighth data point i.e. $Median = -0.130$. Lower quartile Q_1 may be calculated as the median of first eight data points and hence we have $Q_1 = \dfrac{-0.884 - 0.854}{2} = -0.869$. Similarly the upper quartile Q_3 is obtained as the median of data points 8 through 15. Thus $Q_3 = \dfrac{0.844 + 1.098}{2} = 0.971$. The standard deviation may easily be calculated as $\sigma_e = 1.171$. It may be verified that the mean of all data is 0. With these values we make a box and whisker plot as shown in Figure 1.9. It is seen that the mean is slightly

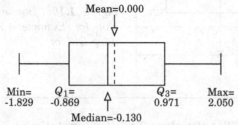

Figure 1.9: Box and whisker plot for Example 1.6

larger than the median and hence the distribution is heavier to the right and marginally skewed.

QtiPlot software has an option to make a menu driven Box and Whisker plot. To demonstrate its use we solve a typical problem below.

Example 1.7

Repeated measurement of a certain quantity gives replicate data. The deviate from mean is calculated to yield deviate data which has zero mean. Scaling the deviates with respect to the standard deviation gives the data shown rank ordered in the table. Use QtiPlot to make a Box and Whisker plot and comment on the quality of the data.

Rank Order	$\dfrac{d_i}{\sigma}$	Rank Order	$\dfrac{d_i}{\sigma}$	Rank Order	$\dfrac{d_i}{\sigma}$	Rank Order	$\dfrac{d_i}{\sigma}$
1	-1.946	6	-0.442	11	0.275	16	0.520
2	-1.057	7	-0.193	12	0.305	17	0.585
3	-0.829	8	-0.164	13	0.316	18	1.102
4	-0.675	9	-0.091	14	0.398	19	1.919
5	-0.660	10	-0.028	15	0.489	20	2.520

Solution :

Key in the given data in two columns of QtiPlot table. Invoke the option "Box plot" under "Statistical graphs" menu to obtain the Box and Whisker plot shown below (Figure 1.10).

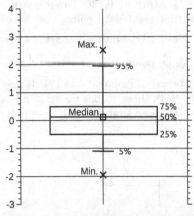

Figure 1.10: *Box and Whisker plot for Example 1.7 obtained using QtiPlot*

QtiPlot also gives the following statistics:

28/04/14 9:12 PM	Statistics for Table 1:
Min = −1.95	$D1$ (1st decile) = −0.85
$Q1$ (1st quartile) = −0.50	Median = 0.12
$Q3$ (3rd quartile) = 0.50	$D9$ (9th decile) = 1.18
Max = 2.52	Size = 20

All the calculations required for the plot are automatically done by QtiPlot and the box plot is the output. In this plot the minimum and maximum are shown by '×' and the median by □. Other percentiles are as explained earlier.

In the present example there is a slight amount of skewness in the distribution. However there is no indication that the distribution is not normal.

Q-Q plot

Another useful graphical method, that helps in identifying match or mismatch with a normal distribution, is a Quantile-Quantile plot or a Q-Q plot. This plot is made with z values for a normal distribution against the z values for the sample. The sample data is arranged in ascending order and the deviates with respect to the sample mean or divided by the unbiased estimate of the sample standard deviation. The rank order of the samples are used for calculating the probability as $p(i) = \dfrac{i - 0.5}{n}$ where i is the rank order and n is the number of data in the sample. The corresponding z values for a normal distribution are calculated consulting a table.

If the sample is close to being normal the data points will lie close to a 45° line (called the parity line) assuming that the same scale is used along the two axes. Departures from linearity will be clear indication of non normal behavior of the sample. We consider the sample in Example 1.6 and demonstrate how a Q-Q plot is made and what conclusions we may draw from it.

Example 1.8

Consider the sample of data of Example 1.6 and make a Q-Q plot. Draw conclusions regarding the distribution underlying the sample.

Solution :

Since the data is already presented in ascending order rank order is the same as the number in column 1 (see table below). With the mean being zero and estimated standard deviation being $\sigma_e = 1.171$ the data is recalculated as $z(i) = \dfrac{v_i - \bar{v}}{\sigma_e}$ and is in the third column of the table. The probabilities are calculated based on the rank order and are in column 4. $z_N(i)$ is calculated using built in function NORMSINV($p(i)$) in the spreadsheet program.[7]

Q-Q plot is obtained by plotting z_N along the abscissa and z along the ordinate. The parity line is obtained by plotting a 45° line passing through the origin. In the present case the Q-Q plot is as shown in Figure 1.11.

The data points in the Q-Q plot lie close to the parity line. Also the data points are distributed evenly around the parity line and do not present any systematic variation. Hence it is safe to conclude that the sample of data is from a normal distribution.

Median value of $v(i)$ is calculated as -0.130. With the mean being zero, skewnwss in the sample distribution is represented by skewness = $\dfrac{3(mean - median)}{\sigma_e} = \dfrac{3(0 - 0.130)}{1.171} = 0.333$ (Note that skewness is bounded between -3 and +3). This is less than critical value of 0.863 and hence is considered to be not significant.[8] Skewness in the sample is purely due to chance and not due to non normality of the distribution.

i	$v(i)$	$z(i)$	$p(i)$	$z_N(i)$
1	-1.829	-1.561	0.033	-1.834
2	-1.259	-1.074	0.100	-1.282
3	-1.187	-1.013	0.167	-0.967
4	-0.884	-0.755	0.233	-0.728
5	-0.854	-0.729	0.300	-0.524

··· Continued on next page

---

[7]Most spreadsheet programs have built in functions useful for statistical analysis. The reader may familiarize herself/himself with these.

[8]D.P. Doane and L.E. Seward, Measuring Skewness: A Forgotten Statistic?, Journal of Statistics Education, Vol. 19, No.2, 2011. Critical values are taken from this reference.

		continued from previous page$\cdots$		
i	$v(i)$	$z(i)$	$p(i)$	$z_N(i)$
6	-0.745	-0.636	0.367	-0.341
7	-0.343	-0.293	0.433	-0.168
8	-0.130	-0.111	0.500	0.000
9	0.097	0.083	0.567	0.168
10	0.224	0.192	0.633	0.341
11	0.844	0.721	0.700	0.524
12	1.098	0.937	0.767	0.728
13	1.112	0.949	0.833	0.067
14	1.806	1.541	0.900	1.282
15	2.050	1.750	0.967	1.834
$\bar{v} =$	0.000			
$\sigma_e =$	1.171			

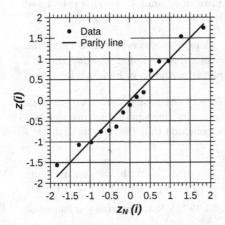

Figure 1.11: *Q-Q plot for sample data in Example 1.8*

Jarque-Bera test for normality

It is well known that the normal distribution is symmetric with respect to the mean. In other words the distribution is not skewed and hence the third moment defined as

$$g_1 = \sum_{i=1}^{n} \left(\frac{x_i - \bar{x}}{\sigma} \right)^3 \quad \text{or} \quad g_1 = \frac{n}{(n-1)(n-2)} \sum_{i=1}^{n} \left(\frac{x_i - \mu}{\sigma_e} \right)^3 \qquad (1.29)$$

Unbiased estimator, small sample

is zero. The fourth moment, known as the Kurtosis, is defined by the relation

$$g_2 = \sum_{i=1}^{n} \left(\frac{x_i - \bar{x}}{\sigma}\right)^4 - 3 \text{ or } g_2 = \underbrace{\frac{n(n+1)}{(n-1)(n-2)(n-3)} \sum_{i=1}^{n} \left(\frac{x_i - \mu}{\sigma_e}\right)^4 - \frac{3(n-1)^2}{(n-2)(n-3)}}_{\text{Unbiased estimator, small sample}}$$

(1.30)

also has a value of zero. Any departure of the sample data from normality would mean that these two quantities are non-zero. Note that g_1 and g_2 are available as functions in spreadsheet programs invoked by SKEW(Sample Data) and KURT(Sample Data) where the sample data is a column of numbers invoked as with the other functions. In case of Jarque-Bera test the statistic JB is defined as

$$JB = \frac{n}{6} \left[g_1^2 + \frac{g_2^2}{4} \right]$$

(1.31)

The critical values for JB are obtained by simulations and useful critical value tables are available from references.[9]

Example 1.9

Fifteen deviates arranged in ascending order forms a sample (second column of the spreadsheet). The sample is expected to follow a normal distribution. Test it using JB test.

Solution :

Extract of spreadsheet used is shown below as a table. The first two columns show the deviates in ascending order. The appropriate statistical parameters have been calculated using the functions available in the spreadsheet program and presented in the last column. The JB value that characterizes the sample and the critical value taken from the cited reference is also shown in the last column.

Data		Data		Parameters	
No.	Deviate	No.	Deviate	$\mu =$	0.000
1	-1.829	9	0.097	$\sigma_e =$	1.171
2	-1.259	10	0.224	$g_1 =$	0.321
3	-1.187	11	0.844	$g_2 =$	-0.916
4	-0.884	12	1.098	$JB =$	0.781
5	-0.854	13	1.112	$\alpha =$	0.05
6	-0.745	14	1.806	$JB_{crit} =$	3.768
7	-0.343	15	2.050		
8	-0.130				

Since $JB < JB_{crit}$ the hypothesis that the deviates are normally distributed is valid.

---

[9]Thadewald T., and Büning H, Working Paper - "Jarque-Bera test and its competitors for testing normality: A power comparison", accessed at www.econstor.eu/bitstream/10419/49919/1/668828234.pdf

χ^2 test for normality

Another useful test for normality is the χ^2 test that is recommended to be used for large samples with $n \geq 50$. We work with the frequencies of occurrence rather than the magnitude of the data values. Essentially, the test compares the observed frequencies with the expected frequencies according to normal distribution, to test for normality. We work with "binned" data such that each bin contains at least five values.

Let the data consist of n values v_i arranged in ascending order. We create bins by defining lower and upper bound values for the data to group the data. Each group fills a bin of certain width. The number of data values which are within a particular bin is the observed frequency f_O for that particular bin. Let there be k bins. The sum of frequencies in all the bins is thus equal to n i.e. $\sum_{i=1}^{k} f_{O,i} = n$.

Arrange the data now by calculating the z values defined, as usual as, $z_i = \dfrac{v_i - \bar{v}}{\sigma_e}$. Compute the z values that bound a particular bin and calculate the cumulative probability that the chosen values are within the particular bin. Multiply this probability with the number of data n to get the expected frequency f_E. We see that $\sum_{i=1}^{k} f_{E,i} = n$.

The statistic χ^2 (refer Chapter 2 for a more detailed discussion about the χ^2 distribution) defined as

$$\chi^2 = \sum_{i=1}^{k} \frac{(f_{O,i} - f_{E,i})^2}{f_{E,i}} \qquad (1.32)$$

follows the χ^2 distribution which is one sided and has a range of $0 \leq \chi^2 \leq \infty$. The critical χ^2 value is calculated based on the chosen α for a particular number of degrees of freedom ν. In the case of normality test of a sample there are c columns and n rows of data. Hence $\nu = (n-1) \times (c-1)$. Tables of χ^2 is available in the Appendix as Table B.6 in Appendix B.

Example 1.10
A set of 50 deviates arranged in ascending order has been obtained and shown in Table 1.4. Perform χ^2 test of normality on this sample.

Solution :

Step 1 We calculate the mean and variance of the sample and hence order the data according the z values. The bounds are rephrased in terms of z values. We create bins using the lower and upper bounds as shown in Table 1.5. The corresponding observed frequencies are also shown.

Step 2 The cumulative probabilities are obtained by subtracting the cumulative probabilities between $z = -\infty$ to $z =$ left bound from the cumulative probability between $z = -\infty$ to $z =$ right bound for each bin. In the spreadsheet

Table 1.4: Deviates data for Example 1.10

i	d_i	i	d_i	i	d_i	i	d_i
1	-2.803	14	-0.972	27	-0.084	40	0.660
2	-1.781	15	-0.928	28	-0.020	41	0.677
3	-1.571	16	-0.815	29	0.070	42	0.749
4	-1.555	17	-0.701	30	0.132	43	0.753
5	-1.512	18	-0.605	31	0.151	44	0.868
6	-1.422	19	-0.576	32	0.156	45	0.967
7	-1.388	20	-0.562	33	0.236	46	1.070
8	-1.213	21	-0.556	34	0.375	47	1.255
9	-1.204	22	-0.393	35	0.398	48	1.335
10	-1.128	23	-0.358	36	0.405	49	1.552
11	-1.049	24	-0.346	37	0.543	50	1.841
12	-0.987	25	-0.277	38	0.547		
13	-0.982	26	-0.249	39	0.654		

we use the function NORMSDIST(z) for this purpose. The expected frequencies are then obtained by multiplying the probability of occurrence within each bin and the total number of data n. The expected frequencies are indicated in the 5th column of the table.

Step 3 We calculate the value of χ^2 as $\chi^2 = 5.877$ from the observed and expected frequencies from the binned data as shown by the sum of entries in the last column of the table. Calculations have been performed using a spreadsheet program. The critical χ^2 for $v = (9-1) \times (2-1) = 8$ and $\alpha = 0.1$ is calculated using the function CHISQINV(0.9,8)[10] as 13.362. Since the calculated χ^2 is less than the critical value there is no reason to doubt the normality of the sample data.

Table 1.5: Expected frequencies and evaluation of χ^2 for data in Example 1.10

Bin No. i	Bin bounds	$f_{O,i}$	$f_{E,i}$	$\dfrac{(f_{O,i} - f_{E,i})^2}{f_{E,i}}$
1	$-\infty$ to -1.2	5	5.753	0.099
2	-1.2 to -1	4	2.179	1.521
3	-1 to -0.7	5	4.165	0.167
4	-0.7 to -0.2	7	8.939	0.421
5	-0.2 to 0.2	6	7.926	0.468
6	0.2 to 0.4	7	3.808	2.675
7	0.4 to 0.7	4	5.131	0.249
8	0.7 to 1	5	4.165	0.167
9	1 to ∞	7	7.933	0.110
Total number of data =		50	$\chi^2 =$	5.877

[10]Argument 0.9 is the probability which is nothing but $1 - \alpha$ with $\alpha = 0.1$.

Example 1.11

5 machines are used to produce identical parts in a factory setting. Number of parts made by each machine and the number rejected during inspection are shown in the table. Test the hypothesis that all machines are of equal quality.

Machine No. i	1	2	3	4	5
No. of parts made n_i	300	400	250	150	200
No. of parts rejected, O_i	24	13	10	16	8

Solution :

The hypothesis that all the machines are of similar quality requires that the number of parts rejected be the same proportion of the number of parts made by each machine. Thus the expected number of parts rejected are given by

$$E_i = \frac{n_i \times \sum_{i=1}^{5} O_i}{\sum_{i=1}^{5} n_i}.$$

We round the E_i to whole numbers and have the following:

i	O_i	E_i	$\frac{(O_i - E_i)^2}{E_i}$
1	24	16	4.0000
2	13	22	3.6818
3	10	14	1.1429
4	16	8	8.0000
5	8	11	0.8182
$\chi^2 =$			17.643
$v =$			4
$\chi^2(\alpha = 0.05, v) =$			9.4877

Since $\chi^2 > \chi^2_{critical}$ the hypothesis is not sustained. The machines are not all of the same quality.

1.3.7 Nonparametric tests

In experimental studies, we often need to compare samples of data and decide whether they follow the same distribution. What the distribution itself is, as long as it is continuous, may be secondary and hence the parameters that characterize the distributions are not important for the proposed test. We essentially devise a nonparametric test for comparing two samples of data. For example, if two samples of data have been collected at different times, we would like to know whether there are significant changes between them.

A commonly used nonparametric test is the Kolmogorov Smirnov[11] or KS two sample test.

---

[11]Andrey Nikolaevich Kolmogorov 1903 - 1987, Russian mathematician; Nikolai Vasilyevich Smirnov 1900 - 1966, Russian mathematician

Kolmogorov Smirnov two sample test

Let the first sample consist of n_1 data and the second sample n_2 data. Order both the data in ascending order. We assume that each data is independent and identically distributed. Assume that an empirical distribution function (EDF) is assumed such that the probability is defined by a uniform distribution. Hence the cumulative empirical probability distribution (CEDF) jumps by $\dfrac{1}{n_1}$ at each data point for the first sample and by $\dfrac{1}{n_2}$ for the second sample. We make a plot of CEDF vs data value for the two samples on the same graph. The supremum of the difference between the two cumulative probabilities D_{max} is the statistic of interest to us. Critical value of $D_{crit}(n_1, n_2, \alpha)$ depends on the number of data and the significance level α.[12] The null hypothesis H_0 is that the two samples follow the same distribution. If $D_{max} < D_{crit}$ the hypothesis is accepted. Otherwise it is concluded that the two samples do not follow the same distribution.

An example is presented below to demonstrate the KS two sample test.

Example 1.12

Two samples of data were collected by two batches of students taking part in a laboratory class. Each batch had 20 students (i.e. $n_1 = n_2 = 20$) and the task consisted of measuring a physical quantity d using the same method. The data has been arranged in ascending order and presented in the table below. Perform a KS test to determine whether the two samples follow the same distribution.

Data No.i	Batch 1 d_i	Batch 2 d_i	Data No.i	Batch 1 d_i	Batch 2 d_i
1	4.541	4.543	11	5.040	4.930
2	4.611	4.586	12	5.053	4.940
3	4.708	4.781	13	5.099	4.965
4	4.737	4.808	14	5.122	5.031
5	4.739	4.817	15	5.164	5.033
6	4.777	4.835	16	5.171	5.096
7	4.912	4.836	17	5.215	5.181
8	4.939	4.865	18	5.249	5.345
9	4.939	4.882	19	5.266	5.431
10	5.031	4.905	20	5.311	5.489

Solution :

Step 1 Since both samples have the same number of data the CEDF is the same for the two samples. As we pass each data point the CEDF increases by $\dfrac{1}{20} = 0.05$. We make a plot of CEDF versus the data value as shown in Figure 1.12.

---

[12]Table of critical values may be downloaded from "www.soest.hawaii.edu/wessel/courses/gg313/Critical_KS.pdf"

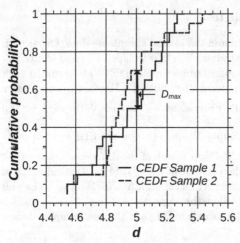

Figure 1.12: *Cumulative probability plot for two samples in Example 1.12*

Step 2 The maximum value of D is equal to 0.2 as shown in Figure 1.12.

Step 3 Critical value (based on reference cited) is obtained for $n_1 = n_2 = 20$

and $\alpha = 0.05$ as $D_{crit} = 1.36\sqrt{\dfrac{20 + 20}{20 \times 20}} = 0.430$.

Step 4 Since $D_{max} < D_{crit}$ the hypothesis that the two samples are from the same distribution holds. This means that the observed differences between the two samples are solely due to chance.

Kolmogorov-Smirnov test for normality

The KS two sample test has to be modified to use it as a test for normality of a single sample. In that case the test is also known as Kolmogorov-Smirnov *goodness of fit* test. We assume that the mean of the sample μ is known and calculate the standard deviation with respect to the mean σ_e, the usual way, based on an unbiased estimator. We compare the ECDF with the normal distribution function based on $z = \dfrac{x - \mu}{\sigma_e}$. The cumulative normal probability is then calculated, after transforming the measured values to corresponding z values. The maximum difference between the two cumulative probabilities is the statistic that is used to test H_0.

Since the standard deviation has been calculated by using the sample data the critical values need a correction as given by Lilliefors[13].

Example 1.13

Consider the first sample data in Example 1.12. Test whether it is distributed normally using KS test for normality.

Solution :

---

[13]Lilliefors, H., "On the Kolmogorov-Smirnov test for normality with mean and variance unknown", Journal of the American Statistical Association, Vol. 62. pp. 399-402, 1967

Step 1 Calculations are best performed using a spreadsheet as shown below. The mean μ and standard deviation σ_e of the sample are calculated using the data d_i for $1 \leq i \leq 20$. Based on these the z values are obtained. The corresponding cumulative normal probabilities P_i are then calculated. The D_i and D'_i are obtained as indicated. The maximum of the last two columns is the D_{max} required to perform the KS normality test (shown as bold entry).

i	d_i	$z_i = \dfrac{d_i - \mu}{\sigma_e}$	P_i	$p_i = \dfrac{i-1}{n}$	$p'_i = \dfrac{i}{n}$	$D_i = |P_i - p_i|$	$D'_i = |P_i - p'_i|$
1	4.541	-1.914	0.028	0	0.05	0.028	0.022
2	4.611	-1.611	0.054	0.05	0.1	0.004	0.046
3	4.708	-1.188	0.117	0.1	0.15	0.017	0.033
4	4.737	-1.063	0.144	0.15	0.2	0.006	0.056
5	4.739	-1.055	0.146	0.2	0.25	0.054	0.104
6	4.777	-0.890	0.187	0.25	0.3	0.063	0.113
7	4.912	-0.301	0.382	0.3	0.35	0.082	0.032
8	4.939	-0.184	0.427	0.35	0.4	0.077	0.027
9	4.940	-0.180	0.429	0.4	0.45	0.029	0.021
10	5.031	0.218	0.586	0.45	0.5	**0.136**	0.086
11	5.040	0.257	0.602	0.5	0.55	0.102	0.052
12	5.053	0.313	0.623	0.55	0.6	0.073	0.023
13	5.099	0.510	0.695	0.6	0.65	0.095	0.045
14	5.122	0.613	0.730	0.65	0.7	0.080	0.030
15	5.164	0.795	0.787	0.7	0.75	0.087	0.037
16	5.171	0.825	0.795	0.75	0.8	0.045	0.005
17	5.215	1.016	0.845	0.8	0.85	0.045	0.005
18	5.249	1.166	0.878	0.85	0.9	0.028	0.022
19	5.266	1.239	0.892	0.9	0.95	0.008	0.058
20	5.311	1.432	0.924	0.95	1	0.026	0.076
$\mu =$	4.981						
$\sigma_e =$	0.230						

Step 2 A plot is made, with sample values along the abscissa and the cumulative probabilities along the ordinate, as shown in Figure 1.13. The maximum D_{max} is identified on the figure.

Step 3 We note that the maximum deviate is 0.136. The critical value for comparison is obtained from a recent paper[14] as 0.192 for $\alpha = 0.05$.

Step 4 Since $D_{max} < D_{crit}$ the hypothesis that the sample data is from a normal distribution is valid.

---

[14] Hervé Abdi ansd Paul Molin, Lilliefors/Van Soest's test of normality, In: Neil Salkind (Ed.), Encyclopedia of Measurement and Statistics, Thousand Oaks (CA): Sage, 2007

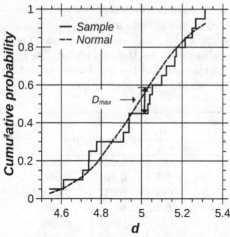

Figure 1.13: *Cumulative probability plot for sample data in Example 1.13*

1.3.8 Outliers and their rejection

We have seen that most data should lie within the bracket $\pm 3\sigma$ (if a large number of data has been collected) around the mean granting that the error distribution is normal. Those data that lie outside this range are called outliers. Even though there may be special cases where an outlier may be physically meaningful it is unusual to get such errors in normal practice. Since the outlier will change the mean and standard deviation it is best to reject such outliers unless there is a reason to believe them to be important. There are a large number of statistical tests that may be used to determine or reject outliers. From the point of view of the present book we shall discuss a few of the more useful ones.

Chauvenet's criterion for discarding outliers

Chauvenet's [15] criterion states that outliers (one or more than one) may be discarded as spurious or suspicious data if the data is outside a range on either side of the mean with probability less than $\dfrac{1}{2n}$ where n is the number of data. For example, if the number of data is 20, the probability we are looking for is $\dfrac{1}{2 \times 20} = 0.025$ and this corresponds to the region outside the cumulative probability interval 0.025 to 0.975 of the standard normal distribution $N(0,1)$. The corresponding critical value of the confidence interval is ± 2.24. Similarly we may evaluate critical values for different number of data as shown in Table 1.7. This table is useful in applying the Chauvenet criterion for data sets with different number of data points. Chauvenet's test is recommended to be used only once. After rejection of data the statistical parameters calculated using the rest of the sample is accepted.

An example is given below to show the effect of outliers and also the improvement in the deduced results when outliers are discarded.

Example 1.14

A certain experiment has been conducted by 12 students using the same

---

[15] after William Chauvenet, 1820-1870, American mathematician

Table 1.7: Critical value for outlier according to Chauvenet's criterion

n	$\left\|\dfrac{d_{max}}{\sigma}\right\|$	n	$\left\|\dfrac{d_{max}}{\sigma}\right\|$	n	$\left\|\dfrac{d_{max}}{\sigma}\right\|$
3	1.38	14	2.10	80	2.74
4	1.54	16	2.15	100	2.81
5	1.65	18	2.20	150	2.93
6	1.73	20	2.24	200	3.02
7	1.81	25	2.33	300	3.14
8	1.86	30	2.39	400	3.23
9	1.91	40	2.49	500	3.29
10	1.96	50	2.57	1000	3.48
12	2.04	60	2.64		

Note: d_{max} = maximum deviation in the data set

experimental set up, to determine the acceleration due to gravity g in m/s^2. The value of g estimated by various students is given in the following table. What is the best estimate for the value of the measured quantity? Specify a suitable error bar. Would you like to discard any data? If so which ones and why? What are the mean and error bar when you discard spurious data?

Student No.	1	2	3	4	5	6
Value g	9.628	9.813	9.729	9.81	9.836	9.718
Student No.	7	8	9	10	11	12
Value g	9.666	9.725	9.615	9.999	9.701	8.245

Solution :

Step 1 Calculate first with all data included Calculation has been made, using a spreadsheet program which calculated the mean and the variance using in-built functions, AVERAGE(number 1,number 2,...) or AVERAGE(B1:B12) to calculate the mean, VAR(number 1,number2,...) or VAR(B1:B12) to calculate the variance or STDEV(number 1,number2,...) or STDEV(B1:B12) to calculate the standard deviation [16] assuming that the data is entered in column B and occupies rows 1 to 12. Error with respect to the mean may be calculated for each entry as shown in column C. Absolute value of the error divided by the standard deviation is shown in column D. It is seen from Table 1.7 that this ratio is more than the critical value 2.04 for the data collected by student number 12. It represents an outlier which is suspect data. We may delete this data and recalculate the statistical parameters.

Step 2 Calculate next after discarding data number 12

The calculations after dropping the data point 12 are shown in the columns E - H. The entries are self explanatory. We see that the mean has changed

---

[16]Spreadsheet uses the unbiased estimate for the variance and the standard deviation

significantly and is more representative of the data. The standard deviation or the spread with respect to the mean has also changed significantly.

The revised calculations paint a much better picture of the data collected by the students in the class.

Step 3 It is interesting to make Q-Q plots for the two samples - (a) before rejecting outlier and (b) after rejecting outlier.

The single outlier has the effect of introducing a pattern to the departure from the parity line. The single outlier indicated by the arrow in Figure 1.14, of course, is far away from the parity line. However when the outlier is removed all the data points move close to the parity line and do not show any specific pattern of variation. Hence one may conclude that errors in the experimental data are normally distributed.

Column identifier							
A	B	C	D	E	F	G	H
	Value	Error			Value	Error	
No.	g	$g - \bar{g}$	$\left\| \dfrac{g - \bar{g}}{\sigma} \right\|$	No.	g	$g - \bar{g}$	$\left\| \dfrac{g - \bar{g}}{\sigma} \right\|$
1	9.628	0.004	0.010	1	9.628	-0.121	1.098
2	9.813	0.189	0.424	2	9.813	0.064	0.580
3	9.729	0.105	0.236	3	9.729	-0.020	0.182
4	9.81	0.186	0.417	4	9.81	0.061	0.552
5	9.836	0.212	0.475	5	9.836	0.087	0.788
6	9.718	0.094	0.211	6	9.718	-0.031	0.282
7	9.666	0.042	0.095	7	9.666	-0.083	0.753
8	9.725	0.101	0.227	8	9.725	-0.024	0.218
9	9.615	-0.009	0.020	9	9.615	-0.134	1.216
10	9.999	0.375	0.840	10	9.999	0.250	2.266
11	9.701	0.077	0.173	11	9.701	-0.048	0.436
12	8.245	-1.379	3.086				
$\bar{g}$	9.624			$\bar{g}$	9.749		
σ_e	0.447			σ_e	0.110		

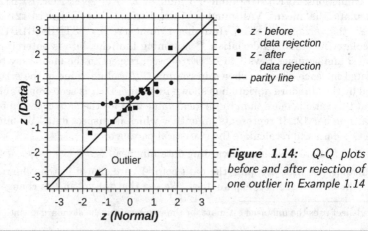

Figure 1.14: Q-Q plots before and after rejection of one outlier in Example 1.14

Pierce's criterion for discarding outliers

Pierce's[17] criterion is useful if we have to discard more than one outlier. Rejection of data is based on the principle - to quote the author -

> "that the proposed observations should be rejected when the probability of the system of errors obtained by retaining them is less than that of the system of errors obtained by their rejection multiplied by the probability of making so many, and no more, abnormal observations"

In a recent publication Ross[18] has discussed in detail the use of Pierce's criterion for rejection of abnormal data. He has also provided a table of critical values that correspond to the probabilities of obtaining "so many, and no more, abnormal observations". The detailed method given in this paper (the reader should read this paper) is summarized below:

1. Number of data = n. Calculate m_s and σ_s
2. Assume one data is suspect
3. Read off R from table in paper by Ross
4. If any deviate (absolute value) $> R\sigma_s$ discard the corresponding data.
5. Assume a second data may be suspect.
6. Read off R from table in paper by Ross. No change in n for now.
7. If any deviate (absolute value) $> R\sigma_s$ discard the corresponding data.
8. Continue till no more data needs to be rejected.
9. Use the reduced number of data after rejection of all suspect data. Let number of data equal to n_1.
10. Calculate m_s and σ_s with n_1.
11. Repeat 2 - 9 with new parameters.
12. Break the loop whenever there is no scope for rejecting any more data.

An example is worked out now to demonstrate the above method.

Example 1.15

In a laboratory class students were asked to measure a certain physical quantity and came up with the readings given in the table. Use Pierce's test to discard abnormal data. How would you summarize the data after rejecting the abnormal data points?

Is it reasonable to assume that the deviates are normally distributed? Base your decision making a box and whisker plot.

---

[17]after Benjamin Peirce, 1809 - 1880, an American mathematician who authored the paper "Criterion for rejection of doubtful observations", The Astronomical Journal, Vol. II, No.21, pp. 161-163, 1852.

[18]Stephen M. Ross, Peirce's criterion for the elimination of suspect experimental data, Journal of Engineering Technology, Fall 2003, pp. 38-41

i	v_i	i	v_i	i	v
1	4.590	8	4.952	15	5.062
2	4.764	9	4.78	16	5.129
3	4.854	10	5.106	17	4.452
4	5.254	11	5.039	18	4.959
5	4.894	12	5.143	19	4.903
6	4.998	13	5.442		
7	5.114	14	4.813		

Solution :

Step 1 Consider all the data in the sample i.e. $n = 19$. We calculate the mean and standard deviation of the sample as $m_s = 4.960$ and $\sigma_e = 0.230$.

Step 2 We would like to check if a single data is abnormal. From table critical value is $R = 2.185$ for $n = 19$ and one abnormal data. We calculate the maximum possible deviate as $R \times \sigma_e = 2.185 \times 0.230 = 0.502$. We calculate absolute values of all the 19 deviates and pick the maximum value. The maximum value is found to correspond to student No. 17 and is 0.508. Since this value is greater than the critical value of 0.502 we discard this data.

Step 3 Now assume that there may be a second data that is abnormal. From table critical value is $R = 1.890$ for $n = 19$ and two abnormal data. We calculate the maximum possible deviate as $R \times \sigma_e = 1.890 \times 0.230 = 0.434$. The second largest value is found to correspond to student No. 13 and is 0.482. Since this value is greater than the critical value of 0.434 we discard this data also.

Step 4 Now assume that there may be a third data that is abnormal. From table critical value is $R = 1.707$ for $n = 19$ and three abnormal data. We calculate the maximum possible deviate as $R \times \sigma_s = 1.707 \times 0.230 = 0.392$. The third largest value is found to correspond to student No. 1 and is 0.370. Since this value is less than the critical value of 0.392 we conclude that there are no more abnormal data.

Step 5 We discard the two abnormal data found above and end up with a sample containing 17 data points. The mean and standard deviation of this data set are given by $m_s = 4.962$ and $\sigma_e = 0.170$.

Step 6 We would like to check if a single data is abnormal in the reduced set obtained by discarding the outliers found previously. From table critical value is $R = 2.134$ for $n = 17$ and one abnormal data. We calculate the maximum possible deviate as $R \times \sigma_e = 2.134 \times 0.170 = 0.362$. We calculate absolute values of all the 17 deviates and pick the maximum value. The maximum value is found to correspond to student No. 1 and is 0.372. Since this value is greater than the critical value of 0.362 we discard this data.

Step 7 We would like to check if a second data is abnormal in the reduced set obtained by discarding the outliers found previously. From table critical value is $R = 1.836$ for $n = 17$ and one abnormal data. We calculate the maximum possible deviate as $R \times \sigma_s = 1.836 \times 0.170 = 0.311$. The second largest deviate is found to correspond to student No. 4 and is 0.292. Since this value is less than the critical value of 0.311 we conclude that there are no more abnormal data.

Step 8 We break the loop and accept three data as abnormal and hence have a sample containing 16 values. A repeat of the Pierce's test for these data shows that no more abnormal data are present in the sample. The pruned sample is shown in the following table.

i	v_i	i	v_i	i	v
2	4.764	8	4.952	15	5.062
3	4.854	9	4.78	16	5.129
4	5.254	10	5.106	18	4.959
5	4.894	11	5.039	19	4.903
6	4.998	12	5.143		
7	5.114	14	4.813		

The mean and standard deviation values characterizing the pruned sample are $m_s = 4.985\,cm$ and $\sigma_e = 0.144\,cm$.

Step 9 We now calculate the parameters required for nmaking a Box and Whisker plot.

Mean =	4.985	Median =	4.979
Sigma =	0.144	Quartile 3 =	5.108
Minimum =	4.764	Maximum =	5.254
Quartile 1 =	4.884		

We make a Box and Whisker plot as Figure 1.15. We have also included the three discarded data points in this plot. The plot shows good symmetry and spread close to a normal distribution. Hence it is reasonable to assume that the data is distributed normally. Also, it seems the three outliers that were discarded, had some calculation errors!

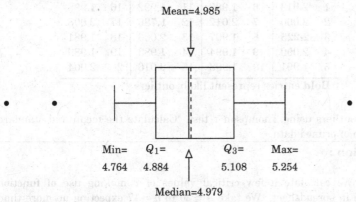

Figure 1.15: *Box and whisker plot for pruned data of Example 1.15. Discarded outliers are shown by • in the plot.*

Thompson τ for discarding outliers

Another test useful for discarding abnormal data from a sample is the Thompson τ test. Consider a data of n samples in which we would like to discard abnormal data. Calculate the mean and the standard deviation of the sample, the usual way. Calculate the absolute value of the difference between data and the mean i.e. calculate $d_i = |v_i - \bar{v}|$. Determine the largest, $d_{i,max}$ among these. The Thomson statistic τ is given by

$$\tau(\alpha, n) = \frac{t_{\alpha/2}(n-1)}{\sqrt{n}\sqrt{n-2+t^2_{\alpha/2}}} \tag{1.33}$$

The Student t value is calculated based on a chosen α and the number of data n. For example, if $\alpha = 0.1$, we calculate t as TINV(0.05,n), a function available in spreadsheet programs. Corresponding critical τ value may be calculated based on the definition given above. If $d_{i,max} > \sigma_e \tau$ discard the data.

If a data has been discarded as abnormal, redo the above steps with $n - 1$ data to discard a second outlier, if it exists.

Continue the process till no more outliers are found.

Example 1.16

The period of a simple pendulum was repeatedly measured and the replicate data is tabulated below.

i	T_i	i	T_i	i	T_i	i	T_i
1	2.013	6	1.987	11	1.992	16	1.998
2	2.000	7	2.012	12	**1.786**	17	2.008
3	**2.225**	8	1.997	13	2.003	18	1.981
4	2.000	9	1.994	14	1.983	19	1.989
5	1.991	10	1.998	15	2.016	20	2.004

Bold entries represent likely outliers

Discard outliers using Thompson τ test. Calculate the mean and standard deviation of pruned data.

Solution :

Step 1 We calculate a few critical values of τ making use of function available in spreadsheet. We take $n = 20$ to $n = 17$ expecting no more than 4 abnormal data. The critical values are tabulated below.

n	17	18	19	20
$\tau(0.1, n)$	1.871	1.876	1.881	1.885

We note in passing that as $n \to \infty$ the critical value of τ tends to 1.96. Note also that all calculations have been rounded to 3 digits after decimals.

Step 2 The mean and standard deviation of the given sample consisting of 20 replicate data is calculated as $\bar{T} = 1.999$ and $\sigma_e = 0.072$. The biggest absolute deviate is identified as data No. 3 of $T_3 = 2.225$ and $d_3 = |2.225 - 1.999| = 0.226$. With $n = 20$, we have, $\tau \times \sigma_e = 1.885 \times 0.072 = 0.136$. Since $d_3 > 0.136$ we discard this data as an outlier.

Step 3 After discarding T_3 perform the above calculations with the rest of the 19 data points. The mean and standard deviation are given by $\bar{T} = 1.987$ and $\sigma_e = 0.050$. Identify the maximum deviation (absolute value) as that corresponding to T_{12} in the original sample (or T_{11} in the pruned sample). The maximum deviate of 0.201 is compared with critical value of $\tau(0.1, 19) \times \sigma_e = 0.093$. Discard T_{12} in the original sample.

Step 4 Repeat the calculations with 18 data and show that no more data is to be rejected.

Step 5 The statistical parameters that represent the pruned data are calculated with $n = 18$ and are given by $\bar{T} = 1.998$ and $\sigma_e = 0.010$.

Dixon's Q test

Ordered sample data is characterized by the difference between the extreme values, also known as the range. If abnormal data is present, it may be at either end of the table. We may identify such an outlier by comparing the difference between the suspected outlier and its nearest neighbor to the range and decide whether to retain or reject data. In the Dixon's Q (Rejection Quotient) test we calculate the ratio of the difference alluded to above with the range and compare this ratio with a critical Dixon Q value to make a decision. Table of critical Q values are presented in a paper by Rorabacher.[19]

Several Q values are defined as given in the cited reference. Correspondingly different Q's are used to make a decision regarding retention or rejection of data. We shall demonstrate the method by taking a simple example.

Example 1.17

A certain experiment has been conducted by 12 students using the same experimental set up. The value of a measured quantity estimated by various students is given in the following table. Would you like to discard any data? If so which ones and why? Base your decision on Dixon's test. What are the mean and error estimator for the pruned data.

i	v_i	i	v_i
1	7.962	7	8.000
2	8.147	8	8.059
3	8.063	9	7.949
4	8.144	10	8.664
5	8.170	11	8.035
6	8.052	12	6.579

[19] D. B. Rorabacher, Statistical Treatment for Rejection of Deviant Values: Critical Values of Dixon's "Q" Parameter and Related Subrange Ratios at the 95% Confidence Level, Analytical Chemistry, Vol. 63, NO. 2, pp. 139-146, 1991

Solution :

Step 1 Arrnage the data in ascending order as shown below.

i	v_i	i	v_i
1	**6.579**	7	8.059
2	7.949	8	8.063
3	7.962	9	8.144
4	8.000	10	8.147
5	8.035	11	8.170
6	8.052	12	**8.664**

Supected outliers are shown **bold**. The difference between adjacent readings in ordered data is the basis for this suspicion. Thus we suspect v_1 or v_{12}.

Step 2 We use Dixon's ratio defined as $r_{10} = \dfrac{v_2 - v_1}{v_{12} - v_1}$ or $r_{10} = \dfrac{v_{12} - v_{11}}{v_{12} - v_1}$, following the cited reference.

$$r_{10} = \frac{7.949 - 6.579}{8.664 - 6.579} = 0.657 \quad \text{or} \quad r_{10} = \frac{8.664 - 8.170}{8.664 - 6.579} = 0.237$$

Step 3 Critical value for r_{10} with $n = 12$ is read from the table as $r_{10,\text{critical}} = 0.426$. At once we see that v_1 is to be discarded.

Step 4 The pruned data at this stage will consist of 11 values from $i = 2$ to $i = 12$ from the original set. The data may be renumbered so that i spans from 1 to 11. The last data point is the next suspect data.

Step 5 The reader may redo the test with the 11 data and show that the last data point is also to be discarded.

Step 6 Finally the following data set with 10 data points is obtained.

i	v_i	i	v_i
1	7.949	6	8.059
2	7.962	7	8.063
3	8.000	8	8.144
4	8.035	9	8.147
5	8.052	10	8.170

Step 7 We report the mean and standard deviation for the pruned data as $\bar{v} = 8.058$ and $\sigma_e = 0.077$.

1.4   Propagation of errors

Replicate data collected by measuring a single quantity, enables us to calculate the best value and characterize the spread by the variance with respect to the best value, using the principle of least squares. Now we look at the case of a derived quantity that is estimated from the measurement of several primary quantities. The question that needs to be answered is the following:

"A derived quantity D is estimated using a formula that involves the primary quantities $a_1, a_2...a_m$. Each one of these is available in terms of the respective best values $\bar{a}_1, \bar{a}_2...\bar{a}_m$ and the respective variances $\sigma_1^2, \sigma_2^2...\sigma_m^2$. What is the best estimate for D and what is the corresponding variance σ_D?"

We have, by definition

$$D = D(a_1, a_2...a_m) \tag{1.34}$$

It is *obvious* that the best value of D should correspond to that obtained by using the best values for the a's. Thus, the best estimate for D is given by $\bar{D}$ as

$$\bar{D} = D(\bar{a}_1, \bar{a}_2...\bar{a}_m) \tag{1.35}$$

Again, by definition, we should have:

$$\sigma_D^2 = \sum_{i=1}^{n} \frac{[D_i - \bar{D}]^2}{n-1} \tag{1.36}$$

In the above expression n represents the number of measurements that have been made and subscript i stands for the experiment number. The ith estimate of D is given by

$$D_i = D(a_{1i}, a_{2i}...a_{mi}) \tag{1.37}$$

If we assume that the spread in values are small compared to the mean or the best values (this is what one would expect from a good experiment), the difference between the ith estimate and the best value may be written using a Taylor expansion around the best value as

$$\sigma_D^2 = \frac{1}{n-1} \sum_{i=1}^{n} \left[\frac{\partial D}{\partial a_1} \Delta a_{1i} + \frac{\partial D}{\partial a_2} \Delta a_{2i} + ... + \frac{\partial D}{\partial a_m} \Delta a_{mi} \right]^2 \tag{1.38}$$

where the partial derivatives are all evaluated at the best values for the a_i. The partial derivatives evaluated at the best values of a_i are also known as influence coefficients, usually represented as I_{a_i}. Note that only the first partial derivatives are retained in the above expansion. If a_i are all independent of one another then the errors in these are unrelated to one another and hence the cross terms $\sum_{i=1}^{N} \Delta a_{mi} \cdot \Delta a_{ki} = 0$ for $m \neq k$. Thus Equation 1.38 may be rewritten as

$$\sigma_D^2 = \frac{1}{n-1} \sum_{i=1}^{n} \left[\left(\frac{\partial D}{\partial a_1} \Delta a_{1i} \right)^2 + \left(\frac{\partial D}{\partial a_2} \Delta a_{2i} \right)^2 + ... + \left(\frac{\partial D}{\partial a_m} \Delta a_{mi} \right)^2 \right] \tag{1.39}$$

Noting that $\sum_{i=1}^{n} \frac{\Delta a_{ji}^2}{n-1} = \sigma_j^2$ we may recast Equation 1.39 in the form

$$\sigma_D^2 = \left[\left(\frac{\partial D}{\partial a_1} \sigma_1 \right)^2 + \left(\frac{\partial D}{\partial a_2} \sigma_2 \right)^2 + ... + \left(\frac{\partial D}{\partial a_m} \sigma_m \right)^2 \right] \tag{1.40}$$

Equation 1.40 is the error propagation formula. It may also be recast in the form

$$\sigma_D = \pm \sqrt{\left(\frac{\partial D}{\partial a_1}\sigma_1\right)^2 + \left(\frac{\partial D}{\partial a_2}\sigma_2\right)^2 + \ldots + \left(\frac{\partial D}{\partial a_m}\sigma_m\right)^2} \qquad (1.41)$$

or

$$\sigma_D = \pm\sqrt{(I_{a_1}\sigma_1)^2 + (I_{a_2}\sigma_2)^2 + \cdots + (I_{a_m}\sigma_m)^2} \qquad (1.42)$$

Example 1.18

A derived quantity D follows the relation $D = 0.023a_1^{0.8}a_2^{0.3}$ where a_1 and a_2 are measured quantities. In a certain case it has been determined that $a_1 = 20000 \pm 125$ and $a_2 = 5.5 \pm 0.2$. Determine the nominal value of D and specify a suitable uncertainty for the same. Which of the two quantities a_1 or a_2 has a bigger influence on the uncertainty in D?

Solution :

Step 1 Using the mean values we first estimate the best value for D.

$$D = 0.023 \times 20000^{0.8} \times 5.5^{0.3} = 105.8$$

Step 2 In this case it is possible to perform logarithmic differentiation to get the error propagation formula. Taking logarithms on both sides of the defining relation between D and the a's we have

$$\ln(D) = \ln(0.023) + 0.8\ln(a_1) + 0.3\ln(a_2)$$

Differentiating the above we get

$$\frac{dD}{D} = 0.8\frac{da_1}{a_1} + 0.3\frac{da_2}{a_2} \quad \text{or} \quad dD = 0.8\frac{Dda_1}{a_1} + 0.3\frac{Dda_2}{a_2}$$

We recognize the influence coefficients as $I_{a_1} = \dfrac{0.8D}{a_1}$ and $I_{a_2} = \dfrac{0.8D}{a_2}$. These may be used in Equation 1.42 to obtain the desired result.

Step 3 Calculate the influence coefficients now.

$$I_{a_1} = \frac{0.8 \times 105.8}{20000} = 4.234 \times 10^{-5}; \quad I_{a_2} = \frac{0.3 \times 105.8}{5.5} = 5.773$$

Step 4 Set $\sigma_{a_1} = 125$ and $\sigma_{a_2} = 0.2$, use Equation 1.42 to get

$$\sigma_D = \sqrt{(4.234 \times 10^{-5} \times 125)^2 + (5.773 \times 0.2)^2} = 1.270$$

Step 5 We also have the following:

$$I_{a_1}\sigma_{a_1} = 4.234 \times 10^{-5} \times 125 = 0.529; \quad I_{a_2}\sigma_{a_2} = 5.773 \times 0.2 = 1.155$$

The uncertainty in measured quantity a_2 has a bigger influence on the uncertainty of D.

Role of variances: In Example 1.18 we have seen that uncertainty in the measured quantity a_2 had a larger influence on the uncertainty in the derived quantity D. The error propagation formula simply states that the variance in the derived quantity is weighted sum of variances of the measured quantities. The weights are the respective squares of influence coefficients. The fractional contribution of the respective weighted variances to the variance of the derived quantity gives us the relative influences of the variances in the measured quantities. In Example 1.18 the fractional contributions to variance in D from a_1 is $\dfrac{(I_{a_1}\sigma_{a_1})^2}{\sigma_D^2} =$

$\dfrac{0.529^2}{1.270^2} = \dfrac{0.280}{1.613} = 0.174$ while that from a_2 is $\dfrac{(I_{a_2}\sigma_{a_2})^2}{\sigma_D^2} = \dfrac{1.155^2}{1.270^2} = \dfrac{1.333}{1.613} = 0.826.$

Example 1.19

Two resistances R_1 and R_2 are given as $1000\pm25\,\Omega$ and $500\pm10\,\Omega$. Determine the equivalent resistance when these two are connected in a) parallel and b) series. Also determine the uncertainties in these two cases.

Solution :

Given Data: $R_1 = 1000\,\Omega$, $R_2 = 500\,\Omega$, $\sigma_1 = 25\,\Omega$, $\sigma_2 = 10\,\Omega$,

Case a) Resistances connected in parallel:

Equivalent resistance is

$$R_p = \frac{R_1 \cdot R_2}{R_1 + R_2} = \frac{1000 \times 500}{1000 + 500} = 333.3\,\Omega$$

The influence coefficients are

$$I_1 = \frac{\partial R_p}{\partial R_1} = \frac{R_2}{R_1 + R2} - \frac{R_1 \times R_2}{(R_1 + R_2)^2} = \frac{500}{1000 + 500} - \frac{1000 \times 500}{(1000 + 500)^2} = 0.111$$

$$I_2 = \frac{\partial R_p}{\partial R_2} = \frac{R_1}{R_1 + R2} - \frac{R_1 \times R_2}{(R_1 + R_2)^2} = \frac{1000}{1000 + 500} - \frac{1000 \times 500}{(1000 + 500)^2} = 0.444$$

Hence the uncertainty in the equivalent resistance is

$$\sigma_s = \pm\sqrt{(I_1\sigma_1)^2 + (I_2\sigma_2)^2}$$
$$= \pm\sqrt{(0.111 \times 25)^2 + (0.444 \times 10)^2} = \pm5.24\,\Omega$$

Case b) Resistances connected in series:

Equivalent resistance is

$$R_s = R_1 + R_2 = 1000 + 500 = 1500\,\Omega$$

The influence coefficients are

$$I_1 = \frac{\partial R_s}{\partial R_1} = 1; \; I_2 = \frac{\partial R_s}{\partial R_2} = 1$$

Hence the uncertainty in the equivalent resistance is

$$\sigma_s = \pm\sqrt{(I_1\sigma_1)^2 + (I_2\sigma_2)^2} = \pm\sqrt{(1 \times 25)^2 + (1 \times 10)^2} = \pm26.93\,\Omega$$

1.5   Specifications of instruments and their performance

In this section we look at the limitations introduced by the instruments used for making measurements. In the past measuring instruments were mostly of the analog type with the reading displayed by a pointer moving past a scale. Resolution of such instruments were basically limited to the smallest scale division. Of course, in addition, the manufacturer would also specify the accuracy as a percentage of the full scale reading, based on calibration with reference to a standard.

In recent times most measuring instruments are digital in nature and the performance figures are specified somewhat differently. Take the example of meter that displays $4\frac{1}{2}$ digits[20]. The reading of the instrument may be anywhere between 0.0000 and 1.9999. The number of counts[21] is 20000. Accuracy specification is usually represented in the form ±(%of reading + counts). For example, typical specification of a DMM (Digital Multi Meter)) is of form ±(0.5 %of reading + 5 counts) when DC voltage is being measured. In a typical example, we may be measuring the voltage of a DC source whose nominal value is 1.5 V. This DMM would give a reading in between 1.492 and 1.508 V. We take an example below to show how the instrument specification affects the measurement.

Example 1.20

A resistor is picked up from a lot labeled 150 Ω with a precision of 1%. Its value is measured using a DMM which has a range of $0-600.0\ \Omega$, accuracy of ±(0.9 %of reading + 2 counts). What would be the expected outcome of the measurement?

Solution :

Step 1 The nominal value of the resistor is $R = 150\ \Omega$. Precision of 1% would mean that it may have a minimum value of $R_l = 148.5\ \Omega$ and a maximum value of $R_m = 151.5\ \Omega$.

Step 2 Let us assume that the actual value of the resistor is R_l. The meter will then give either of the readings shown in the last row below:

$R_l =$	148.5	Ω
0.9% of $R_l =$	1.3	Ω
Error due to count =	0.2	Ω

Reading is either 147.0 Ω or 150.0 Ω

Step 3 Let us assume that the actual value of the resistor is R_m. The meter will then give either of the readings shown in the last row below:

---

[20]Most significant digit can be either 0 or 1 while all other digits may have any value between 0 and 9.

[21]Number of levels equals the number of counts.

$R_m =$	151.5	Ω
0.9% of $R_m =$	1.4	Ω
Error due to count =	0.2	Ω

Reading is either 149.9 Ω or 153.1Ω

Step 4 Thus the actual value of the resistance indicated by the instrument may be anywhere between 147.0 and 153.1 Ω.

Example 1.21

In a wind tunnel air flow is maintained steady at a nominal speed of 15 *m/s*. A vane anemometer is used to measure the wind speed. Specify an error bar for the measurement if the resolution (4 digit display) of the anemometer is 0.01 *m/s* and the accuracy specification by the manufacturer of the anemometer is ±(3% reading + 0.20) *m/s*.

Solution :

The nominal value of the wind speed is $V = 15$ *m/s*. We assume that the reading of the anemometer is this value i.e. 15.00 *m/s*. Using the accuracy specification the uncertainty is calculated as follows:

1. 3% of reading $= \dfrac{3 \times 15.00}{100} = 0.45 \, m/s$
2. Total uncertainty is equal to $0.45 + 0.2 = 0.65 \, m/s$.

Hence the measured velocity is specified as $V = 15 \pm 0.65$ *m/s*. In terms of percentages the uncertainty in the wind speed is ±4.33%.

Concluding remarks

This chapter has set the tone for the rest of the book by presenting statistical principles that play important role in analysis and interpretation of experimental results. Properties of normal distribution are relevant in most measurements and hence have been discussed in detail. Test for normality and data rejection based on sound statistical principles have been presented. Other topics considered include sampling theory and error propagation.

Chapter 2

Regression analysis

This chapter deals with the presentation of experimental results or experimental data in the form of an algebraic expression relating the effect to cause or causes. Such an expression is useful as an aid for the designer who may use the experimental results in design of a complex engineering system. In a simple bivariate case (one cause and one effect) it is possible to use tabulated data, along with a suitable local interpolation function (linear, quadratic or polynomial), to derive the desired value of the effect at an arbitrary value of the cause, within the measurement range. For example, it is possible to use this procedure in evaluating the temperature when the measured output of a thermocouple is known. However, in multivariate case i.e. when there are many causes and one effect, interpolation of tabulated data is not easy. In such a case a global function representation of the data is useful. Regression analysis aims to determine the best global representation of experimental data using a chosen functional form.

2.1   Introduction to regression analysis

Regression analysis or curve fitting consists in arriving at a relationship that may exist between two or more variables. In the context of experiments, the variables represent cause(s) effect relationship, with a particular measured quantity depending on other measured quantities. For example, the height of an individual may be related to the age of the individual. Height may be measured using a scale while the age may be ascertained from the date of birth. We may treat age as the "cause" and height as the "effect" and look for a relationship between these two. Variability in the measured height may be due either to measurement errors or due to natural variations. These random fluctuations will affect the type of relationship as well as the quality of relationship.

Suitable plot of data will indicate the nature of the trend of data and hence will indicate the nature of the relation between the independent and the dependent variables. A few examples are shown in Figures 2.1(a),(b) and 2.2 where we have a single cause (x) and an effect (y).

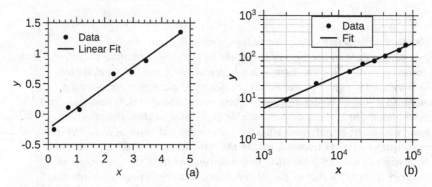

Figure 2.1: *(a) Linear relation between y and x (b)Linear relation between* $\log(y)$ *and* $\log(x)$

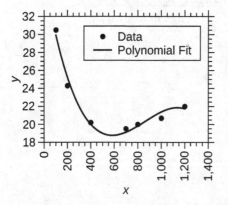

Figure 2.2: *Non-linear relation between y and x*

The linear graph shown in Figure 2.1(a) follows a relationship of the form $y = ax + b$ where a and b are parameters describing the fit. The linear relationship on the log-log plot shown in Figure 2.1(b) follows a relation of the form $y = ax^b$ (which may be recast in the linear form $\log(y) = \log(a) + bx$. The non-linear relationship shown in Figure 2.2 follows a polynomial relationship of the form $y = ax^3 + bx^2 + cx + d$. The parameters a, b, c, d are known as the fit parameters and need to be determined as a part of the analysis. The polynomial form is treated as a linear regression noting that it may be written in the linear form $y = ax_1 + bx_2 + cx + d$ where $x_1 = x^3$ and $x_2 = x^2$.

Linear fit is possible in all the cases shown in Table 2.1.

Table 2.1: *Regression laws covered by linear regression model*

Regression law	Type of fit	Comment
$y = ax + b$	Linear fit	Plots as a straight line on a linear graph sheet
$y = ax^b$	Power law fit	Plots as a straight line on a linear graph sheet
$y = ae^{bx}$	Exponential fit	Plots as a straight line on a log-log graph
$y = ax^2 + bx + c$	Quadratic (polynomial) fit	Multiple linear regression

2.2 Linear regression

2.2.1 Linear fit by least squares

Let $(x_1, y_1), (x_2, y_2), \ldots (x_n, y_n)$ be a set of ordered pairs of data. It is expected that there is a linear relation between y and x, as for example, by looking at a plot of the data using suitable graph paper. Thus, in the most obvious case, if we plot the data on a linear graph sheet as in Figure 2.1(a) the trend of the data should be well represented by a straight line. We notice that the straight line shown in Figure 2.1(a) *does not* pass through any of the data points shown by full symbols. There is a *deviation* between the data and the line and this deviation is sometimes positive, sometimes negative, sometimes large and sometimes small. If we look at the value given by the straight line as a local mean then the deviations are *assumed* to be normally distributed around the local mean. If all data have been obtained with equal care, one *may expect* the deviations at the various locations along the line to follow the same distribution with variance $\sigma_{y_i}^2 = \sigma^2$ for all i. The deviate data is said to be *homoscedastic*. The least square principle may be applied as under.[1]

$$\text{Minimise} : SSD = \sum_{i=1}^{n} [y_i - y_{fi}]^2 = \sum_{i=1}^{n} [y_i - (ax_i + b)]^2 \tag{2.1}$$

where $y_f = ax + b$ is the desired linear fit to data. We see that SSD is the sum of squares of deviations between the data and the fit and minimization of SSD will

---

[1]Least squares may also be interpreted as minimizing the square of the error vector $\sum_{i=1}^{n} (y_i - y_{f,i})^2$.

yield the proper choice of the mean line represented by the proper values of the parameters a and b. The minimization requires that

$$\frac{\partial SSD}{\partial a} = \sum_{i=1}^{n} [y_i - ax_i - b](-x_i) = 0 \tag{2.2}$$

$$\frac{\partial SSD}{\partial b} = \sum_{i=1}^{n} [y_i - ax_i - b](-1) = 0 \tag{2.3}$$

These equations may be rearranged as two simultaneous equations for a, b as given below:

$$\left(\sum_{i=1}^{n} x_i^2 \right) a + \left(\sum_{i=1}^{n} x_i \right) b = \sum_{i=1}^{n} x_i y_i \tag{2.4}$$

$$\left(\sum_{i=1}^{n} x_i \right) a + nb = \sum_{i=1}^{n} y_i \tag{2.5}$$

These are known as normal equations. The summation from $i = 1$ to n will not be indicated explicitly in what follows. These equations may be solved simultaneously to obtain a and b as:

$$a = \frac{n \sum x_i y_i - \sum x_i \sum y_i}{n \sum x_i^2 - (\sum x_i)^2} \tag{2.6}$$

$$b = \frac{\sum y_i \sum x_i^2 - \sum x_i \sum x_i y_i}{n \sum x_i^2 - (\sum x_i)^2} \tag{2.7}$$

We introduce the following definitions:

$$\bar{x} = \frac{1}{n} \sum x_i, \ \bar{y} = \frac{1}{n} \sum y_i, \ \sigma_x^2 = \frac{1}{n} \sum x_i^2 - \bar{x}^2,$$

$$\sigma_y^2 = \frac{1}{n} \sum y_i^2 - \bar{y}^2, \ \sigma_{xy} = \frac{1}{n} \sum x_i y_i - \overline{xy}$$

Alternately σ_x^2, σ_y^2 are also represented as σ_{xx} and σ_{yy}, respectively. These are the variances that are familiar to us. The last of the quantities (σ_{xy}) defined in the above is known as the covariance. With these definitions the slope of the line fit given by Equation 2.6 may be rewritten, after some manipulations as

$$a = \frac{\sigma_{xy}}{\sigma_x^2} \quad \text{or} \quad a = \frac{\sigma_{xy}}{\sigma_{xx}} \tag{2.8}$$

We get the intercept parameter b by the use of Equation 2.5 as

$$b = \bar{y} - a\bar{x} \tag{2.9}$$

Equation 2.9 indicates that the regression line passes through the point $\bar{x}, \bar{y}$. The fit line may hence be represented in the alternate form

$$Y_f = aX \tag{2.10}$$

where

$$Y_f = y_f - \bar{y} \tag{2.11}$$

$$X = x - \bar{x} \tag{2.12}$$

2.2.2 Uncertainties in the fit parameters

We have seen that the fit parameters a and b depend on measured y values that are subject to measurement errors. Hence it is clear that the parameters a and b should also have uncertainties due to the uncertainties in the measured y values. These uncertainties may be obtained by the use of the familiar error propagation formula. Differentiating expression 2.8 with respect to y_i we have

$$\frac{\partial a}{\partial y_i} = \frac{1}{\sigma_x^2} \frac{\partial \sigma_{xy}}{\partial y_i}$$

With $\sigma_{xy} = \frac{1}{n} \sum x_i y_i - \overline{xy} = \frac{1}{n} \sum x_i y_i - \overline{x} \frac{1}{n} \sum y_i$ we have $\frac{\partial \sigma_{xy}}{\partial y_i} = \frac{x_i}{n} - \frac{\overline{x}}{n}$. With this we get

$$\frac{\partial a}{\partial y_i} = \frac{1}{\sigma_x^2} \left[\frac{x_i}{n} - \frac{\overline{x}}{n} \right] = \frac{1}{\sigma_x^2} \frac{x_i - \overline{x}}{n}$$

According to error propagation formula $\sigma_a^2 = \sum \left(\frac{\partial a}{\partial y_i} \sigma_{y_i} \right)^2$ where $\sigma_{y_i}^2$ is the variance of y_i and hence the variance of slope parameter a is

$$\sigma_a^2 = \frac{1}{\sigma_x^4} \sum \left(\frac{x_i - \overline{x}}{n} \right)^2 \sigma^2$$

since $\sigma_{y_i}^2 = \sigma^2$ in the present case. Using the definition of variance it is easy to show that $\sum \left(\frac{x_i - \overline{x}}{n} \right)^2 = \frac{\sigma_x^2}{n}$. Hence the variance of a is given by

$$\sigma_a^2 = \frac{\sigma^2}{n \sigma_x^2} \tag{2.13}$$

Now consider the intercept parameter b which is given by $b = \overline{y} - a\overline{x}$. Variance in b is the sum of variance of $\overline{y}$ which is easily seen to be $\frac{\sigma^2}{n}$ and variance of $a\overline{x}$ which is obtained by multiplying the variance of a given in Equation 2.13 by $\overline{x}^2$. Thus we have

$$\sigma_b^2 = \frac{\sigma^2}{n} + \frac{\sigma^2 \overline{x}^2}{n \sigma_x^2} \tag{2.14}$$

Example 2.1

A temperature sensor is used to measure temperature between $20°C$ and $90°C$. The output of the sensor is conditioned by a suitable electronic circuit to yield output E in V. The following data has been collected in the laboratory by exposing the sensor to a suitable thermal environment whose temperature has been systematically varied and monitored by a standard sensor. Consider fitting the data to a linear relationship between E and temperature t. Obtain also the uncertainties associated with the slope and intercept parameters.

$t°C$	E, V	$t°C$	E, V	$t°C$	E, V
20	1.73	45	3.05	70	4.23
25	2.10	50	3.25	75	4.58
30	2.24	55	3.56	80	4.85
35	2.46	60	3.81	85	5.12
40	2.72	65	3.98	90	5.41

Solution :

Step 1 It is convenient to make a spreadsheet table as shown below. First column identifies the data number. The given temperature - sensor response data is in columns 2 and 3. The other quantities needed to calculate the fit parameters are in columns 4 - 6. Means are calculated column-wise and are shown in row 16.

#	$t°C$	E, V	t^2	E^2	tE
1	20	1.73	400	2.9929	34.6
2	25	2.10	625	4.4100	52.5
3	30	2.24	900	5.0176	67.2
4	35	2.46	1225	6.0516	86.1
5	40	2.72	1600	7.3984	108.8
6	45	3.05	2025	9.3025	137.25
7	50	3.25	2500	10.5625	162.5
8	55	3.56	3025	12.6736	195.8
9	60	3.81	3600	14.5161	228.6
10	65	3.98	4225	15.8404	258.7
11	70	4.23	4900	17.8929	296.1
12	75	4.58	5625	20.9764	343.5
13	80	4.85	6400	23.5225	388
14	85	5.12	7225	26.2144	435.2
15	90	5.41	8100	29.2681	486.9
Mean =	55	3.5393	3491.6667	13.7760	218.783

Step 2 We calculate the statistical parameters for the given data as follows.

$$\sigma_t^2 = \overline{t^2} - \overline{t}^2 = 3491.6667 - 55^2 = 466.7$$

$$\sigma_E^2 = \overline{E^2} - \overline{E}^2 = 13.7760 - 3.5393^2 = 1.249$$

$$\sigma_{tE} = \overline{tE} - \overline{t}\overline{E} = 218.783 - 55 \times 3.5393 = 24.12$$

Step 3 Slope and intercept parameters are obtained (using Equations 2.8 and 2.9) as

$$a = \frac{24.12}{466.7} = 0.0517; \quad b = 3.5393 - 0.0517 \times 55 = 0.6966$$

Step 4 The regression line is thus given by $E_f = 0.0517t + 0.6966$. The data and the fit are compared in the following table.

$t°C$	E, V	E_f, V	$(E - E_f)^2$	$t°C$	E, V	E_f, V	$(E - E_f)^2$
20	1.73	1.73	0.0000	60	3.81	3.80	0.0001
25	2.10	1.99	0.0121	65	3.98	4.06	0.0064
30	2.24	2.25	0.0001	70	4.23	4.32	0.0081
35	2.46	2.51	0.0025	75	4.58	4.57	0.0001
40	2.72	2.76	0.0016	80	4.85	4.83	0.0004
45	3.05	3.02	0.0009	85	5.12	5.09	0.0009
50	3.25	3.28	0.0009	90	5.41	5.35	0.0036
55	3.56	3.54	0.0004			$\sigma^2 =$	0.00254

That the fit is a good representation of the data is indicated by the proximity of the respective values in the second and third columns. The plot shown in Figure 2.3 is further proof of this.

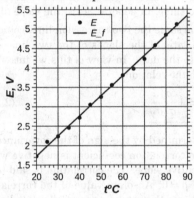

Figure 2.3: Comparison of data with the linear fit

Step 5 Uncertainties in the fit parameters are evaluated now. We need an estimate for the variance of the deviates between the data and the fit. This is obtained as $\sigma^2 = \sum\limits_{1=1}^{n} \dfrac{(E_i - E_{fi})^2}{n}$ which is indicated as 0.00254 in the table above.

Step 6 Uncertainty in the slope parameter is calculated using Equation 2.13 as

$$\sigma_a = \sqrt{\frac{0.00254}{15 \times 466.7}} = 0.0006$$

Step 7 Uncertainty in the intercept parameter is calculated using Equation 2.14 as

$$\sigma_b = \sqrt{\frac{0.00254}{15} + \frac{0.00254 \times 55^2}{15 \times 466.7}} = 0.0356$$

Step 8 Using the uncertainties in the parameters calculated above we may specify the fit line as

$$E_f = (0.0517 \pm 0.0006)t + (0.6966 \pm 0.0356)$$

2.2.3   Goodness of fit and the correlation coefficient

A measure of how good the regression line as a representation of the data is deduced now. In fact it is possible to fit two lines to data by (a) treating x as the independent variable and y as the dependent variable or by (b) treating y as the independent variable and x as the dependent variable. The former has been done above. The latter is described by a relation of the form $x = a'y + b'$. The procedure given earlier can be followed through to get the following (the reader is expected to show these results):

$$a' = \frac{\sigma_{xy}}{\sigma_y^2}; \quad b' = \bar{x} - a'\bar{y} \tag{2.15}$$

The second fit line may be recast in the form

$$y' = \frac{1}{a'}x - \frac{b'}{a'} \tag{2.16}$$

The slope of this line is $\frac{1}{a'}$ which is *not* the same, in general, as a, the slope of the first regression line. If the two slopes are the same the two regression lines coincide. Otherwise the two lines are distinct. The ratio of the slopes of the two lines is a measure of how good the form of the fit is to the data. In view of this we introduce the correlation coefficient ρ defined through the relation

$$\rho = \pm\sqrt{aa'} = \pm\frac{\sigma_{xy}}{\sigma_x\sigma_y} \tag{2.17}$$

The sign of the correlation coefficient is determined by the sign of the covariance. If the regression line has a negative slope the correlation coefficient is negative while it is positive if the regression line has a positive slope. The correlation is said to be perfect if $\rho = \pm 1$. The correlation is poor if $\rho \approx 0$. Absolute value of the correlation coefficient should be greater than 0.5 to indicate that y and x are at all related!

In Example 2.1 the correlation coefficient is positive. The pertinent parameters are $\sigma_t^2 = 466.7, \sigma_E^2 = 1.249$ and $\sigma_{tE} = 24.12$. With these the correlation coefficient is $\rho = \dfrac{24.12}{\sqrt{466.7 \times 1.249}} = 0.9990$. Since the correlation coefficient is close to unity the fit represents the data very closely (Figure 2.3 has already indicated this).

2.3   Polynomial regression

2.3.1   Method of least squares and normal equations

Sometimes the data may show a non-linear behavior that may be modeled by a polynomial relation. Consider a quadratic fit as an example. Let the fit equation be given by $y_f = ax^2 + bx + c$.[2] The variance of the data with respect to the fit is again minimized with respect to the three fit parameters a, b, c to get three normal equations. These are solved for the fit parameters. Thus we have

$$s^2 = \frac{1}{n}\sum(y_i - ax_i^2 - bx_i - c)^2 \tag{2.18}$$

---

[2]Polynomial fit is linear in the fit parameters and hence the regression model is linear.

Setting the partial derivatives of s^2 with respect to a, b, c to zero we get the following three normal equations.

$$a\sum x_i^4 + b\sum x_i^3 + c\sum x_i^2 = \sum x_i^2 y_i \qquad (2.19)$$

$$a\sum x_i^3 + b\sum x_i^2 + c\sum x_i = \sum x_i y_i \qquad (2.20)$$

$$a\sum x_i^2 + b\sum x_i + cn = \sum y_i \qquad (2.21)$$

Normal equations 2.19-2.21 are easily solved for the three fit parameters to complete the regression analysis.

2.3.2 Goodness of fit and the index of correlation or R^2

In the case of a non-linear fit we define a quantity known as the index of correlation to determine the goodness of fit. The fit is termed good if the variance of the deviates is much less than the variance of the y's. Thus we require the index of correlation R (spreadsheet programs usually display R^2 rather than R) defined below to be close to ± 1 for the fit to be considered good.

$$R^2 = 1 - \frac{s^2}{\sigma_y^2} = 1 - \frac{\sum(y - y_f)^2}{\sum(y - \overline{y})^2} \qquad (2.22)$$

Note that the subscript i has also been dropped in writing the above equation. The reader will realize that the summation is over i and it refers to the data point index. It can be shown that the R^2 is identical to the square of the correlation coefficient ρ^2 for a linear fit. The index of correlation compares the scatter of the data with respect to its own mean as compared to the scatter of the data with respect to the regression curve.

In practice we use the adjusted value of R_{Adj}^2 instead, taking into account the loss of degrees of freedom when regression is employed to represent the data. The degrees of freedom for calculating the variance of the y's with respect to its own mean is $n - 1$ as has been shown earlier. The degrees of df for calculating the variance with respect to the regression curve is $df = n - p - 1$ where p is the number of variables in the regression model. The -1 here accounts for the constant in the regression model. Thus we modify Equation 2.22 as

$$R_{Adj}^2 = 1 - \frac{(n-1)\sum(y - y_f)^2}{(n-p-1)\sum(y - \overline{y})^2} = 1 - (1 - R^2)\frac{(n-1)}{(n-p-1)} \qquad (2.23)$$

Example 2.2

The friction factor Reynolds number product fRe for laminar flow in a rectangular duct is a function of the aspect ratio $A = \dfrac{h}{w}$ where h is the height and w is the width of the rectangle. Use the following tabulated data to obtain a suitable fit.

A	0	0.05	0.1	0.125	0.167
fRe	96	89.81	84.68	82.34	78.81
A	0.25	0.4	0.5	0.75	1
fRe	72.93	65.47	62.19	57.87	56.91

Solution :

A plot of the given data indicates that a polynomial fit may be appropriate. For the purpose of the following we represent the aspect ratio as x and the fRe product as y. We seek a fit to data of the form $y_f = a_0 + a_1 x + a_2 x^2 + a_3 x^3$. The following tabulation helps in the regression analysis.

No.	x	y	x^2	x^3	x^4
	0.000	96.00	0.0000	0.0000	0.0000
2	0.050	89.81	0.0025	0.0001	0.0000
3	0.100	84.68	0.0100	0.0010	0.0001
4	0.125	82.34	0.0156	0.0020	0.0002
5	0.167	78.81	0.0279	0.0047	0.0008
6	0.250	72.93	0.0625	0.0156	0.0039
7	0.400	65.47	0.1600	0.0640	0.0256
8	0.500	62.19	0.2500	0.1250	0.0625
9	0.750	57.89	0.5625	0.4219	0.3164
10	1.000	56.91	1.0000	1.0000	1.0000
Sum =	*3.3420*	*747.0300*	*2.0910*	*1.6342*	*1.4095*

No.	x^5	x^6	xy	$x^2 y$	$x^3 y$
1	0.0000	0.0000	0.0000	0.0000	0.0000
2	0.0000	0.0000	4.4905	0.2245	0.0112
3	0.0000	0.0000	8.4680	0.8468	0.0847
4	0.0000	0.0000	10.2925	1.2866	0.1608
5	0.0001	0.0000	13.1613	2.1979	0.3671
6	0.0010	0.0002	18.2325	4.5581	1.1395
7	0.0102	0.0041	26.1880	10.4752	4.1901
8	0.0313	0.0156	31.0950	15.5475	7.7738
9	0.2373	0.1780	43.4175	32.5631	24.4223
10	1.0000	1.0000	56.9100	56.9100	56.9100
Sum =	*1.2799*	*1.1980*	*212.2553*	*124.6098*	*95.0595*

Note: Entries are rounded to 4 digits. However the calculations use available precision of the computer.

The column sums required for generating the normal equations are shown in *italics*. The four normal equations are then given in matrix form as

$$
\begin{pmatrix}
10 & 3.3420 & 2.0910 & 1.6342 \\
3.3420 & 2.0910 & 1.6342 & 1.4095 \\
1.6342 & 1.4095 & 1.2799 & 1.1980 \\
2.0910 & 1.6342 & 1.4095 & 1.2799
\end{pmatrix}
\begin{pmatrix}
a_0 \\ a_1 \\ a_2 \\ a_3
\end{pmatrix}
=
\begin{pmatrix}
747.0300 \\ 212.2553 \\ 124.6098 \\ 95.0595
\end{pmatrix}
$$

The normal equations are solved (use Gauss elimination method) to get the fit parameters:

$$a_0 = 95.67, \ a_1 = -121.00, \ a_2 = 132.38, \ a_3 = -50.23$$

The following table helps in comparing the data with the fit.

No.	x	y	y_f	$(y-y_f)^2$	$(y-\bar{y})^2$
1	0	96	95.6688	0.1097	453.5622
2	0.05	89.81	89.9434	0.0178	228.2214
3	0.1	84.68	84.8422	0.0263	99.5405
4	0.125	82.34	82.5139	0.0303	58.3238
5	0.167	78.81	78.9196	0.0120	16.8674
6	0.25	72.93	72.9074	0.0005	3.1435
7	0.4	65.47	65.2346	0.0554	85.2483
8	0.5	62.19	61.9850	0.0420	156.5752
9	0.75	57.89	58.1928	0.0917	282.6770
10	1	56.91	56.8223	0.0077	316.5908
Sum =	3.342	747.03	747.0300	0.3934	1700.7502

The column sums required for calculating the index of correlation are given in the last row of the table. The index of correlation uses the mean values of columns 5 and 6 given by $s^2 = 0.03934$ and $\sigma_y^2 = 170.0750$. The index of correlation is thus equal to $R = \sqrt{1 - \dfrac{0.03934}{170.0750}} = -0.9999$. Negative sign indicates that y decreases when x increases. The index of correlation is close to -1 and hence the fit represents the data very well. The adjusted R^2 is given by

$$R_{Adj}^2 = 1 - \frac{9 \times 0.3934}{6 \times 1700.7502} = 0.9997$$

R_{Adj}^2 also indicates a good fit. A plot of the data along with the fit given in Figure 2.4 also indicates this. The standard error of the fit is given by

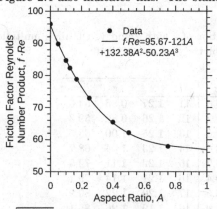

Figure 2.4: Comparison of data with cubic fit

$$\sigma = \sqrt{\frac{0.3934}{6}} = \pm 0.26,$$ rounded to two digits after decimals. Note that the number of degrees of freedom is used rather than the number of data points in calculating the error estimate.

2.3.3 Multiple linear regression

In the above we have considered cases that involved one independent variable and one dependent variable. Sometimes the dependent variable may be a function of more than one independent variable. For example, the relation of the form $Nu = aRe^b Pr^c$ is a common type of relationship between the Nusselt number (Nu, dependent variable) and Reynolds (Re) and Prandtl (Pr) numbers both of which are independent variables. By taking logarithms, we see that $\log Nu = \log a + b \log Re + c \log Pr$. It is thus seen that the relationship is linear when logarithms of the dependent and independent variables are used to describe the fit. Hence the relationship may be expressed in the form $z = ax + by + c$, where z is the dependent variable, x and y are independent variables and a, b, c are the fit parameters. The least square method may be used to determine the fit parameters.

Let the data be available in the form of a set of n - (x, y, z) values. The quantity to be minimized is given by

$$s^2 = \sum_i [z_i - ax_i - by_i - c]^2 \tag{2.24}$$

The normal equations are obtained by the usual process of setting the first partial derivatives with respect to the fit parameters to zero.

$$a \sum x_i^2 + b \sum x_i y_i + c \sum x_i = \sum x_i z_i \tag{2.25}$$

$$a \sum x_i y_i + b \sum y_i^2 + c \sum y_i = \sum y_i z_i \tag{2.26}$$

$$a \sum x_i + b \sum y_i + cn = \sum z_i \tag{2.27}$$

These equations are solved simultaneously to get the three fit parameters.

Example 2.3

The following table gives the variation of z with x and y. Obtain a multiple linear fit to the data and comment on the goodness of the fit.

i	x	y	z	i	x	y	z
1	1.40	0.21	6.3	11	1.27	0.86	54.0
2	1.39	0.26	9.2	12	1.26	0.89	59.3
3	1.37	0.36	19.1	13	1.25	1.00	63.5
4	1.36	0.39	23.6	14	1.23	1.04	68.7
5	1.35	0.46	26.7	15	1.22	1.11	72.5
6	1.33	0.52	29.4	16	1.21	1.14	79.1
7	1.32	0.57	36.5	17	1.20	1.22	80.9
8	1.31	0.64	41.9	18	1.19	1.26	85.1
9	1.29	0.73	45.5	19	1.18	1.27	89.5
10	1.28	0.77	50.4	20	1.17	1.39	94.5

Solution :

The calculation procedure follows that given previously. Several sums are required and these are tabulated below. We omit the columns corresponding to x, y and z in this table.

x^2	xy	y^2	xz	yz
1.9698	0.2898	0.0426	8.9119	1.3112
1.9290	0.3627	0.0682	12.7514	2.3973
1.8894	0.4886	0.1263	26.1931	6.7729
1.8511	0.5299	0.1517	32.1264	9.1962
1.8139	0.6238	0.2146	35.9415	12.3612
1.7778	0.6889	0.2669	39.2469	15.2082
1.7427	0.7523	0.3247	48.1429	20.7821
1.7087	0.8403	0.4132	54.8098	26.9527
1.6757	0.9392	0.5264	58.9557	33.0417
1.6437	0.9845	0.5897	64.6679	38.7361
1.6125	1.0923	0.7399	68.5852	46.4579
1.5822	1.1222	0.7959	74.5665	52.8863
1.5528	1.2510	1.0078	79.1782	63.7879
1.5242	1.2783	1.0721	84.8415	71.1566
1.4963	1.3538	1.2249	88.7040	80.2571
1.4692	1.3792	1.2948	95.8877	90.0146
1.4429	1.4640	1.4855	97.1927	98.6166
1.4172	1.4994	1.5864	101.3602	107.2391
1.3923	1.4986	1.6130	105.6190	113.6856
1.3679	1.6262	1.9332	110.4861	131.3459
32.8594	20.0650	15.4779	1288.1686	1022.2070

The last row contains all the sums required (along with $\sum x = 25.5965$, $\sum y = 16.0735$, and $\sum z = 1035.9123$)) and the normal equations are easily written down as under. In the table only four significant digits after the decimal point have been retained even though calculations have been made using available computer precision.

$$\begin{pmatrix} 32.8594 & 20.0650 & 25.5965 \\ 20.0650 & 5.4779 & 16.0735 \\ 25.5965 & 16.0735 & 20 \end{pmatrix} \begin{pmatrix} a \\ b \\ c \end{pmatrix} = \begin{pmatrix} 1288.1686 \\ 1022.2070 \\ 1035.9123 \end{pmatrix}$$

These equations are solved to get the fit parameters $a = -308.1596$, $b = 13.1425$ and $c = 435.6232$. The data and the fit may be compared by making

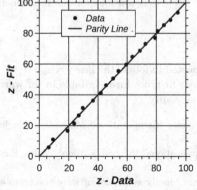

Figure 2.5: Parity plot vouching for quality of fit

a parity plot as shown in Figure 2.5. The parity plot is a plot of the fit (z_f)

along the abscissa and the given data (z) along the ordinate. The parity line is a line of equality between the two. The amount of departure of the data from the parity line is an indication of the quality of the fit. Figure 2.5 indicates that the fit is indeed very good. When the data is a function of more than one independent variable it is not possible to make plots between independent and dependent variables to evaluate the goodness of the fit. In such a case the parity plot is a way out.

We may also calculate the index of correlation as an indicator of the quality of the fit. This calculation is left to the reader!

2.4   General non-linear fit

The fit equation may sometimes have to be chosen as a non-linear relation that is not either a polynomial or in a form that may be reduced to the linear form. Parameter estimation problem becomes an optimization problem, generally involving the minimization of sum of squares of the residuals. Optimization (specifically multivariate optimization - of interest here), in general, is not amenable to calculus methods and requires the use of a search method to determine the best parameter set.

Following are some of the search methods that are available for parameter estimation.

- Steepest descent method
- Conjugate gradient or CG method
- Levenberg-Marquardt or LM method

Both SD and CG methods have been discussed in a companion book by the author.[3] The Levenberg-Marquardt method is described in the book by Rao.[4] The reader may consult more advanced texts on optimization to learn about other better parameter estimation methods.

Luckily for the interested user, it is possible to use software to make a non-linear fit. For example, **"QtiPlot"** is a plot program that is capable of doing non-linear fit using the Levenberg-Marquardt algorithm. We give below an example that has been solved using QtiPlot.

Example 2.4

Bivariate data shown in the table is expected to follow the law $y = a_0 + a_1 x^{a_2}$. Obtain the fit parameters using non-linear fit procedure available in software such as QtiPlot.[5] Test the deviates for normality.

---

[3]S.P. Venkateshan and Swaminathan Prasanna, "Computational Methods in Engineering", Ane Books, New Delhi, 2013

[4]S.S. Rao, "Optimization - Theory and Applications" (Second Edition), Wiley Eastern Limited, New Delhi, 1984

[5]Non-linear fit may also be made using other readily available software resources or by writing a program that uses library routines.

x	0.4	0.64	0.84	1.02	1.25	1.42
y_d	0.388	0.443	0.508	0.557	0.61	0.646
x	1.61	1.83	2.03	2.23	2.42	2.61
y_d	0.695	0.726	0.776	0.815	0.876	0.896

Solution :

Step 1 We enter the given data into two columns of table in QtiPlot environment.

Step 2 Choose these two columns and invoke **FitWizard** from drop down menu **Analyze**.

Step 3 Choose user defined fit law. In this case we enter A1+A2*x^A3 as the fit law. Choose initial parameters as A1=0.2, A2=0.5 and A3=1 to start the iteration process.

Step 4 Choose **Fitting Session**. Choose **Fit**. This action initiates the Levenberg-Marquardt search method to get the fit parameters. The following appears in the **Results log** window.

```
[24/05/13 4:53 PM Plot: "Graph2"]
Non-linear Fit of dataset: Table2, using function: A1+A2*x^A3
Weighting Method: No weighting
Scaled Levenberg-Marquardt algorithm with tolerance = 0.0001
From x = 4.0000000000000e-01 to x = 2.6100000000000e+00
A1 = 2.3170515212420e-01 +/- 2.9696319192181e-02
A2 = 3.1450537350840e-01 +/- 3.1119948578227e-02
A3 = 7.8495407088813e-01 +/- 6.2320813097002e-02


R^2 = 0.997926741301
Adjusted R^2 = 0.9971492692889
RMSE (Root Mean Squared Error) = 0.008352893958729
RSS (Residual Sum of Squares) = 0.000627937537372


Iterations = 5
Status = success
```

Thus the fit is given by $y_f = 0.2317 + 0.3145x^{0.7849}$. The data was devised by perturbing the relation $y = 0.25 + 0.3x^{0.8}$ by adding random errors to it. Figure 2.6(a) shows that the fit represents the data very well.

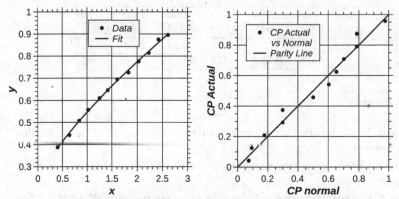

Figure 2.6: (a) Non-linear fit to data obtained using QtiPlot
(b) Test for normality of deviates in Example 2.4

Step 5 In order to check for normality of the deviates we may use any of the normality tests presented in Chapter 1.[6] The required data is generated and tabulated for the purpose of making a Q-Q plot.

i	d_i	$\dfrac{d_i}{\sigma_e}$	CP Norm	$\dfrac{i-0.5}{12}$
1	-0.011	-1.456	0.073	0.042
2	-0.010	-1.323	0.093	0.125
3	-0.007	-0.926	0.177	0.208
4	-0.004	-0.529	0.298	0.292
5	-0.004	-0.529	0.298	0.375
6	0.000	0.000	0.500	0.458
7	0.002	0.265	0.604	0.542
8	0.003	0.397	0.654	0.625
9	0.004	0.529	0.702	0.708
10	0.006	0.794	0.786	0.792
11	0.006	0.794	0.786	0.875
12	0.015	1.985	0.976	0.958
$\mu =$	0			
$\sigma_e =$	0.008			

Step 6 We make a plot of last two columns in the table as given in Figure 2.6(b). The deviates are close to the parity line and show no definite trend. Hence we conclude that the deviates follow a normal distribution with zero mean.

2.5 χ^2 test of goodness of fit

We have seen earlier that goodness of fit of a model to data may be judged by looking at either the correlation coefficient or the index of correlation. Another test that is usually made consists of the χ^2 test. This test is based on the following.

---

[6]The reader is encouraged to use other normality tests

Regression analysis is based on the argument that the errors with respect to the regression line are normally distributed. This is the basis for the least square analysis that is employed to estimate the fit parameters. Several examples have already been presented. The test statistic for comparing observed (y_d, data) and expected values (y_f, fit) is defined as χ^2 given by

$$\chi^2 = \sum_{i=1}^{n} \frac{[y_{d,i} - y_{f,i}]^2}{\sigma_i^2} \tag{2.28}$$

In the above n represents the number of x values for which the y values have been measured or obtained. σ_i^2 represents the variance of the measurement error at the data point i, obtained by repeated measurement. The number of degrees of freedom v is given by $n - p$ where p is the number of parameters. The Chi Square (χ^2) distribution is a one sided distribution that varies between 0 and ∞. The distribution also depends on v. The distribution is given by

$$f(\chi^2, v) = \frac{e^{-\left(\frac{\chi^2}{2}\right)}\left(\frac{\chi^2}{2}\right)^{\left(\frac{v}{2}-1\right)}}{2\Gamma\left(\frac{v}{2}\right)} \tag{2.29}$$

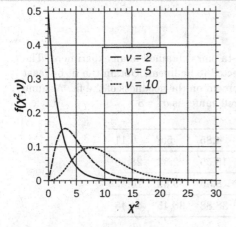

Figure 2.7: χ^2 distribution

Figure 2.7 shows the shape of the χ^2 distribution for three cases with different v values. We notice that the distribution becomes wider as v increases. In an application we are interested in finding the critical value of χ_α^2, say that which has only a chance of being exceeded 5% of the time ($\alpha = 0.05$). α is the area of the tail of the χ^2 distribution as indicated in Figure 2.8.

The χ^2 value is, in fact, determined by integrating Equation 2.29 and determining the "x" value that gives

$$\alpha = \int_x^\infty \frac{e^{-\frac{x}{2}}\left(\frac{x}{2}\right)^{\left(\frac{v}{2}-1\right)}}{2\Gamma\left(\frac{v}{2}\right)} dx \tag{2.30}$$

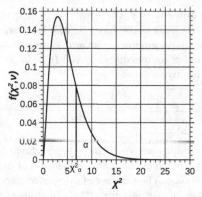

Figure 2.8: α as fraction of area under the χ^2 distribution

These values are available in the form of a table in most texts on Statistics. In spreadsheet program we invoke this by the function $\mathrm{CHIINV}(1 - \alpha, \nu)$. Table B.6 in Appendix B gives the critical χ^2 values for various degrees of freedom. In performing the χ^2 test one determines the χ^2 value using the tabulated experimental data for the appropriate ν. This value is compared with the critical value given in Table B.6. If the calculated χ^2 value is less than the critical value one concludes that the differences between the data and the fit are purely due to chance and that the fit is an acceptable representation of the data.

Example 2.5

The following table gives the (x, y) data pairs obtained in an experiment. The relationship between y and x is expected to be linear. Obtain the fit by least squares and perform a χ^2 test to comment on the goodness of the fit. Assume that the expected variance of measured values is $\sigma^2 = 8$.

x	3.25	4.22	4.85	5.5	6.11
y	16.56	21.82	18.55	22.33	28.98
x	6.78	7.32	8.56	9.82	10.42
y	26.98	35.02	38.88	36.45	44.45

Solution :

Step 1 We make use of a spreadsheet program to obtain the linear fit using the appropriate "trend line" option. Extract of spreadsheet is shown in Figure 2.9. The extract also shows the plot of data and the trend line using the plot routine available in the spreadsheet program.

Step 2 The trend line follows the law $y_f = 3.7633x + 3.1858$ and is shown plotted in Figure 2.9. The trend line has an R^2 value of 0.9062.

Step 3 We use the equation of the trend line to make the table given below, useful for calculating the χ^2 value. It is actually an extract of a spreadsheet redone in the form of a table. The entries in the second and third columns are the given data. The numbers in the fourth column represents the fit calculated

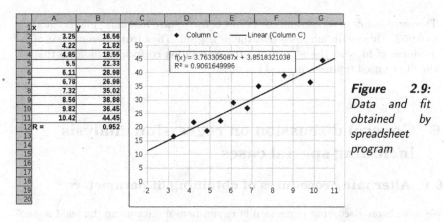

	A	B	C	D	E	F	G
1	x	y					
2	3.25	16.56					
3	4.22	21.82					
4	4.85	18.55					
5	5.5	22.33					
6	6.11	28.98					
7	6.78	26.98					
8	7.32	35.02					
9	8.56	38.88					
10	9.82	36.45					
11	10.42	44.45					
12	R =	0.952					

Figure 2.9: Data and fit obtained by spreadsheet program

using the previously given expression. Entries in the fifth column assists in the calculation of the χ^2 value.

No.	x	y_d	y_f	$\dfrac{(y_d - y_f)^2}{\sigma^2}$	$(y_d - y_f)^2$
1	3.25	16.56	16.08	0.02850	0.22798
2	4.22	21.82	19.73	0.54448	4.35588
3	4.85	18.55	22.10	1.57869	12.62953
4	5.5	22.33	24.55	0.61602	4.92818
5	6.11	28.98	26.85	0.56948	4.55582
6	6.78	26.98	29.37	0.71221	5.69764
7	7.32	35.02	31.40	1.63881	13.11051
8	8.56	38.88	36.07	0.99007	7.92058
9	9.82	36.45	40.81	2.37337	18.98699
10	10.42	44.45	43.07	0.23964	1.91716
$n =$	10		SUM =	9.2913	74.33027
$p =$	2			$\chi^2 =$	9.2913
$v =$	8			$\sigma_e =$	3.05

Extract of spreadsheet for Example 2.5

Step 4 The number of degrees of freedom is $v = 8$ in this case. The χ^2 value is obtained by the fifth column sum and is equal to 9.2913. The critical value of χ^2 with $\alpha = 0.05$ and $v = 8$ is 15.5073 as read off Table B.6. Since the χ^2 value for the fit is less than the critical value we may conclude that the linear fit is a good representation of the data and the departures from the fit are purely accidental.

Step 5 The error estimate uses the number of degrees of freedom as $v = n - p - 1 = 10 - 1 - 1 = 8$ (n=number of data, p=number of parameters excluding the constant) and yields

$$\sigma_e = \pm\sqrt{\frac{\text{Last column sum}}{v}} = \pm\sqrt{\frac{74.33027}{8}} = \pm 3.05$$

Step 6 The "R^2" value indicated in Figure 2.9 is internally calculated by the spreadsheet program and represents the square of the correlation coefficient.

The corresponding correlation coefficient is the square root of R^2 and is equal to 0.952. This again indicates that the *fit* is good. Thus two different tests of goodness of fit, viz. the χ^2 test and the correlation coefficient, indicate that the fit is a good representation of the data.

2.6 General discussion on regression analysis including special cases

2.6.1 Alternate procedures of obtaining fit parameters

We have been discussing regression of experimental data using the least square principle as the back bone. Even though it is based on the maximum likelihood principle when the errors are normally distributed, in general, least squares may be interpreted as indicating the shortest length of the error vector in n dimensions, where n represents the number of data points. The latter interpretation is the one that is the basis for least square regression using linearization of the data by taking logarithms, either of x or y alone or both x and y. We take a particular example to look at such a case here.

Example 2.6

Consider the data in Table below. It is expected to be represented well by a power law of form $y = ax^b$. Obtain the regression parameters using a power fit directly and compare with those obtained by linear fit to $\log(x)$ vs. $\log(y)$.

x	y	x	y	x	y
10000	2.656	410000	7.601	810000	8.718
35000	3.982	435000	7.788	835000	9.506
60000	4.599	460000	7.828	860000	8.983
85000	5.364	485000	8.085	885000	9.153
110000	5.193	510000	8.252	910000	9.539
135000	6.075	535000	8.003	935000	9.354
160000	5.847	560000	8.155	960000	9.727
185000	6.214	585000	8.33	985000	9.525
210000	6.594	610000	8.58	1010000	9.164
235000	6.632	635000	8.283	1040000	9.573
260000	6.835	660000	8.455	1060000	9.598
285000	7.179	685000	8.689	1090000	9.796
310000	7.137	710000	8.538	1110000	9.781
335000	6.834	735000	9.018	1140000	9.911
360000	7.162	760000	9.029	1160000	10.08
385000	7.214	785000	8.765		

Solution :

We make use of QtiPlot to obtain regression parameters in both cases. The data is entered as two columns in the spreadsheet program. Power law is specified in the form $f(x) = A_0 x^{A_1}$. Starting values for the parameters are specified as $A_0 = 0.4$ and $A_1 = 0.3$. Output from the program is as shown below.

[10/07/13 8:35 AM Plot: "Graph 5"]
Non-linear Fit of dataset: Table 1 y-Data, using function: A0*x^2A1
Weighting Method: No weighting
Scaled Levenberg-Marquardt algorithm with tolerance = 0.0001
From x = 1.0000000000000e+04 to x = 1.1600000000000e+06
A0 = 2.7698615691929e-01 +/- 1.9480131413622e-02
A1 = 2.5617382492432e-01 +/- 5.2763957766413e-03

R^2 = 0.9870283845753
Adjusted R^2 = 0.9864387656923
RMSE (Root Mean Squared Error) = 0.1945613017335
RSS (Residual Sum of Squares) = 1.703434505951

Iterations = 5
Status = success

Next we calculate the logarithms of each column of data and enter these in the columns of QtiPlot. The fit equation is specified as linear and the output of the program is as given below.

[10/07/13 8:48 AM Plot: "Graph 6"]
Linear Regression of dataset: Table2 LOGy-Data, using function: A*x+B
Weighting Method: No weighting
From x = 4.0000000000000e+00 to x = 6.0640000000000e+00
B (y-intercept) = -5.9526297891447e-01 +/- 2.3858716353470e-02
A (slope) = 2.6276386164443e-01 +/- 4.2231554363695e-03

R^2 = 0.9885095707425
Adjusted R^2 = 0.9879872785035
RMSE (Root Mean Squared Error) = 0.01241934251538
RSS (Residual Sum of Squares) = 0.006940803083143

In the latter case the power law parameters are given by

$$A_0 = 0.26276386164443, \quad A_1 = 10^{-0.59526297891447} = 0.2539434533$$

Rounding the fit parameters to 3 digits after decimals, we have the following:

	A_0	A_1
Power Fit	0.277	0.256
Linear Fit to log - log data	0.263	0.254

The two fits are now calculated using the regression parameters from the above table. The two fits are compared in Figure 2.10 by making a parity plot.

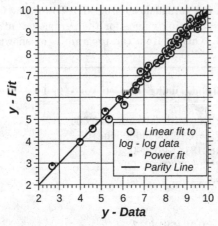

Figure 2.10: Parity plot comparing two ways of obtaining the fit

The two methods of fitting the data essentially give the same results. When it comes to determining the goodness of fit and the estimated errors of fit the calculations are done by using the data and the fits shown in the above figure. Appropriate quantities of interest are shown below:

	σ_e	R^2_{Adj}
Power Fit	0.193	0.9867
Linear Fit to log - log data	0.197	0.9861

It is thus clear that both procedures yield fits of comparable merit.

2.6.2   Segmented or piecewise regression

In all cases treated till now it was assumed that a single regression curve fit the data end to end. Many times this may not hold and we have different regression curves fitting the data in segments i.e. in different ranges of the independent variable. Such situations are common in heat transfer and fluid flow problems where different regimes of flow may occur. For example, when flow changes from laminar to turbulent, the data may show changes in the way it behaves and it may be necessary to divide the data into segments and perform regression individually, in these subranges.

When this is the case, we have to determine the parameters that represent the first segment, the parameters that represent the second segment and also the break point - the point at which the first segment ends and the second segment starts. It is normal to choose the segmental fits such that the function is continuous at the break point. This gives an additional condition that is to be satisfied. Thus, if the break point is at x_b we require that $y_{f,1}(x_b) = y_{f,2}(x_b)$ where $y_{f,1}$ and $y_{f,2}$ are the fit functions, respectively to the left and right of the break point.

A simple example with a single break point is presented below to demonstrate the procedure of segmented fit to data.

Example 2.7

Consider the bivariate data shown in the following table. Explore different possibilities of regressing the data. What is the best regression for this data? Base your answer on graphical representations and statistical parameters. If the standard deviation of data is uniform and equal to 0.4 perform χ^2 test and comment on the quality of the fits.

i	x_i	y_i	i	x_i	y_i	i	x_i	y_i
1	1	1.80	10	37	12.52	19	85	34.04
2	5	2.72	11	41	13.31	20	91	36.80
3	9	4.52	12	45	14.41	21	97	39.22
4	13	5.22	13	49	15.71	22	103	42.48
5	17	5.91	14	55	19.81	23	109	45.48
6	21	7.58	15	61	22.15	24	115	47.68
7	25	8.63	16	67	25.41	25	121	51.14
8	29	9.62	17	73	27.91	26	127	54.21
9	33	11.07	18	79	30.78	27	133	56.78

Solution :

Step 1 A plot of the data will show the options we should explore. A simple linear - linear plot is made as shown in Figure 2.11.

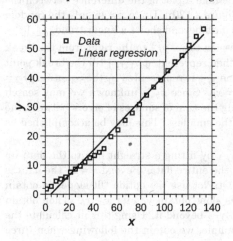

Figure 2.11: *Linear-Linear plot of data in Example 2.7 showing unsuitable linear fit*

Step 2 An examination of Figure 2.11 shows that the data is not linear, representable by a single line of regression. Two possibilities are explored. First one is a non-linear power fit to the data, end to end. Alternately we explore piecewise regression using two regression lines, with a break point around $x = 45$.

Step 3 Power fit to data is accomplished by using QtiPlot, choosing a regression model of form $y = A_0 + A_1 x^{A_2}$. We obtain the fit shown in Figure 2.12(a). Even though the fit seems to be much better than the linear fit shown in Figure 2.11, the data appears to show a recognizable pattern of deviations near the break point alluded to earlier.

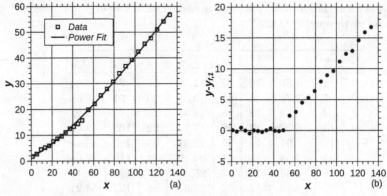

Figure 2.12: *(a) Linear-Linear plot of data in Example 2.7 showing a single power fit to data (b) Plot of difference between data and $y_{f,1}$*

Step 4 At this stage we are ready to look at the third alternative of fitting the data by two different linear segments with break point around $x = 45$. As a prelude to this we fit a regression line to the data between $x = 1$ and $x = 45$ using QtiPlot and label it as $y_{f,1}$. Make a plot of the difference between the given data and $y_{f,1}$ as shown in Figure 2.12(b). Up to $x = 45$ only the random noise in the data survives and the data beyond shows a linear trend.

Step 5 Fit a second line of regression to $y - y_{f,1}$ to the right from the break point and call it $y_{f,2}$. The data is then represented by $y_{f,1}$ till the break point and by $y_{f,1} + y_{f,2}$ beyond it. The line $y_{f,2}$ is represented by the equation $y_{f,2} = m(x - x_b)$ as it passes through $x = x_b$. Since x_b is unknown we may search for the best value of x_b by requiring the sum of squares of errors between the segmented fit and the data to be the smallest. This may be accomplished as given below.

Assume an x_b value and obtain $y_{f,1}$ by fitting a straight line to the data up to $x = x_b$. Now obtain $y - y_{f,1}$ for the entire data. Beyond $x = x_b$ the data is expected to be given by $y_{f,2}$. In QtiPlot use the option "fit slope" to obtain the slope of $y_{f,2}$, viz. s. With this slope obtain $y_{f,2}$ for $x \geq x_b$. Then obtain $y_f = y_{f,1}$ till $x = x_b$ and $y_f = y_{f,1} + y_{f,2}$ beyond it. Using this fit calculate the sum of squares of errors. For example, we obtain the following when three different values of x_b are used.

x_b	$\sum_{i=1}^{27}(y_i - y_{f,i})^2$
41	4.7737
45	2.1093
49	4.4466

It is apparent that the optimum value of x_b is close to 45. We may fit a quadratic to the above data, differentiate it with respect to x_b and set it equal to zero to obtain the optimum value. The reader may easily show that the optimum value is $x_{b,opt} = 45.13$.

We then obtain the fit with this value of x_b to obtain the best segmented fit to data. Figure 2.13 shows the data and the segmented fit that is continuous at $x = 45.13$.

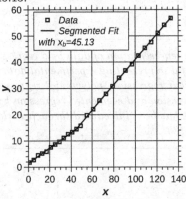

Figure 2.13: Linear-Linear plot of data in Example 2.7 showing segmented fit to data with break point at $x_b = 45.13$.

Step 6 Statistical parameters are now calculated to compare the three fits that have been presented above. In the case of single line of regression the number of parameters is 2 and hence the degrees of freedom is $v_1 = 27 - 2 = 25$. In the case of single power fit the number of degrees of freedom is $v_2 = 27 - 3 = 24$. In the case of the segmented fit there are five parameters, slope and intercept of first line segment, slope and intercept of the second line segment and the break point. However we have an extra condition that the regression be continuous at the break point. Hence the degrees of freedom is $v_3 = 27 - 5 + 1 = 23$. Standard deviations are estimated for the three fits. We also calculate R^2_{Adj} for each of the fits.

1. Single line fit end to end:
   $y_f = 0.4272x - 2.0054$; $\sigma_e = 1.76$; $R^2_{Adj} = 0.9903$
2. Single non-linear fit end to end:
   $y_f = 1.9716 + 0.1006x^{1.2914}$; $\sigma_e = 0.66$; $R^2_{Adj} = 0.9953$
3. Optimum segmented fit:
   $y_{f,1} = 0.2897x + 1.4449$; $y_{f,2} = 0.1924(x - x_b)$; $\sigma_e = 0.30$; $R^2_{Adj} = 0.9997$
   with $x_b = 45.13$.

Step 4 Improvement in the standard deviation as well as the R^2 value indicates that the segmented fit is the best bet.

Step 5 Using the various fits presented above, setting $\sigma = 0.4$ we may easily calculate the χ^2 for each of the fits. Critical values corresponding to the respective degrees of freedom may then be obtained by invoking the spreadsheet function CHISQINV$(0.9, v)$ corresponding to $\alpha = 0.1$. The results are tabulated and the appropriate comment in the last row speaks for the quality of the fit.

Fit	Single line	Single power fit	Segmented fit
v	25	24	23
χ^2	481.811	59.257	13.199
χ^2_{crit}	34.382	33.196	32.007
Remark	Unacceptable	Unacceptable	Acceptable

Concluding remarks

This chapter has considered in detail the representation of data by a regression line or curve. Such a representation is useful in presenting experimental data for applications in design and analysis. Linear regression, polynomial regression, non-linear regression, multiple linear regression and segmented regression have been presented in detail. Test of goodness has been introduced through various statistical parameters and also via the χ^2 test.

Chapter 3 Design of experiments

Chapter *3*

Design of experiments

This is a short chapter that deals with the concepts involved in the design of experiments (DOE). Goal of DOE is to obtain as much information as possible about a certain system using the smallest number of experiments. In this chapter we consider process related DOE as well as those that involve general physical systems that need to be characterized using experimental data.

3.1   Design of experiments

3.1.1   Goal of experiments

- Experiments help us in understanding the behavior of a (mechanical) system
- Data collected by systematic variation of influencing *factors* helps us to quantitatively describe the underlying phenomenon or phenomena

The goal of any experimental activity is to get the maximum information about a system with the minimum number of well designed experiments. An experimental program recognizes the major "*factors*" that affect the outcome of the experiment. The factors may be identified by looking at all the quantities that may affect the outcome of the experiment. The most important among these may be identified using a few exploratory experiments or from past experience or based on some underlying theory or hypothesis. The next thing one has to do is to choose the number of levels for each of the factors. The data will be gathered for these values of the factors by performing the experiments by maintaining the levels at these values.

Suppose we know that the phenomena being studied is affected by the pressure maintained within the apparatus during the experiment. We may identify the smallest and the largest possible values for the pressure based on experience, capability of the apparatus to withstand the pressure and so on. Even though the pressure may be varied "*continuously*" between these limits, it is seldom necessary to do so. One may choose a few discrete values within the identified range of the pressure. These will then be referred to as the levels.

Experiments repeated with a particular set of levels for all the factors constitute replicate experiments. Statistical validation and repeatability concerns are answered by such replicate data.

In summary an experimental program should address the following issues:

- Is it a single quantity that is being estimated or is it a trend involving more than one quantity that is being investigated?
- Is the trend linear or non-linear?
- How different are the influence coefficients?
- What does dimensional analysis indicate?
- Can we identify dimensionless groups that influence the quantity or quantities being measured?
- How many experiments do we need to perform?
- Do the factors have independent effect on the outcome of the experiment?
- Do the factors interact to produce a net effect on the behavior of the system?

3.1.2   Full factorial design

A full factorial design of experiments consists of the following:

- Vary one factor at a time
- Perform experiments for all levels of all factors
- Hence perform the large number of experiments that are needed!

- All interactions are captured (as will be shown later)

Consider a simple case:

Let the number of factors $= k$

Let the number of levels for the ith factor $= n_i$

The total number of experiments (n) that need to be performed is given by

$$n = \prod_{i=1}^{k} n_i \qquad (3.1)$$

If $k = 5$ and number of levels is 3 for each of the factors the total number of experiments to be performed in a full factorial design is $n = 3^5 = 243$! The goal of design of experiments (DOE) is to explore the possibility of reducing this large number of experiments to a smaller number and yet obtain all the information that we desire from the experiment.

3.1.3 2^k factorial design

Consider a simple example of a 2^k factorial design. Each of the k factors is assigned *only* two levels. The levels are usually designated as $High = 1$ and $Low = -1$. Such a scheme is useful as a preliminary experimental program before a more ambitious experimental study is undertaken. The outcome of the 2^k factorial experiment will help identify the relative importance of factors and also will offer some knowledge about the *interaction* effects. Let us take a simple case where the number of factors is also 2. Let these factors be x_A and x_B. The number of experiments that may be performed is 4 corresponding to the following combinations:

Experiment Number	x_A	x_B
1	1	1
2	-1	1
3	1	-1
4	-1	-1

Let us represent the outcome of each experiment to be a quantity y. Thus y_1 will represent the outcome of experiment number 1 with both factors having their "High" values, y_2 will represent the outcome of the experiment number 2 with the factor A having the "Low" value and the factor B having the "High" value and so on. The outcome of the experiments may be represented as the following "outcome" matrix:

$x_A \downarrow x_B \rightarrow$	+1	-1
+1	y_1	y_3
-1	y_2	y_4

A simple regression model that may be used to describe the relationship between y and the x's is one that can have up to four parameters. Thus we may represent the regression equation as

$$y = p_0 + p_A x_A + p_B x_B + p_{AB} x_A x_B \qquad (3.2)$$

where the p's are the regression parameters. In Equation 3.2 parameters p_A and p_B account for the "main" effects and the parameter p_{AB} accounts for the "interaction" effect due to the simultaneous effects of the two x's on the outcome of the experiment. The p's are the parameters that are determined by using the "outcome" matrix by the simultaneous solution of the following four equations:

$$p_0 + p_A + p_B + p_{AB} = y_1 \qquad (3.3)$$

$$p_0 - p_A + p_B - p_{AB} = y_2 \qquad (3.4)$$

$$p_0 + p_A - p_B - p_{AB} = y_3 \qquad (3.5)$$

$$p_0 - p_A - p_B + p_{AB} = y_4 \qquad (3.6)$$

The four Equations 3.3-3.6 are to be solved simultaneously to get the four parameters. If we add all the equations we get the value of parameter p_0 given by $p_0 = \dfrac{y_1 + y_2 + y_3 + y_4}{4}$ which is nothing but the mean of the y's. This is also obtained by the *substitution* of mean values for x_A and x_B given by the set $0,0$. In Example 3.1 we shall look at how the other parameters are obtained.

Equations 3.3-3.6 express the fact that the outcome may be interpreted as shown in Figure 3.1. It is seen that the values of $y - p_0$ at the corners of the square indicate the deviations from the mean value and hence the mean of the square of these deviations (we may divide the sum of the squares with the number of degrees of freedom = 3) is the variance of the sample data collected in the experiment. The influence of the factors may then be gaged by the contribution of each term to the variance, indicated by the magnitude of the p's. These ideas will be brought out by Example 3.1.

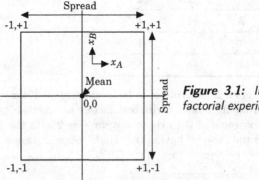

Figure 3.1: *Interpretation of 2^2 factorial experiment*

Example 3.1

A certain process of finishing a surface involves a machine. The machine has two speed levels x_A and the depth of cut x_B may also take on two values. The two values are assigned *+1* and *-1* as explained in the case of the 2^2 factorial experiment. The outcome of the process is the surface finish y that may have a value between *1* (the worst) to *10* (the best). A 2^2 factorial experiment was performed and the following matrix gives the results of the experiments:

$x_A \downarrow x_B \rightarrow$	+1	-1
+1	2	8.2
-1	1.5	3.5

Determine the regression parameters and comment on the results.

Solution :

The regression model given in Equation 3.2 is made use of. The four simultaneous equations for the regression parameters are given by

$$p_0 + p_A + p_B + p_{AB} = 2 \qquad (3.7)$$

$$p_0 - p_A + p_B - p_{AB} = 1.5 \qquad (3.8)$$

$$p_0 + p_A - p_B - p_{AB} = 8.2 \qquad (3.9)$$

$$p_0 - p_A - p_B + p_{AB} = 3.5 \qquad (3.10)$$

As mentioned earlier, the parameter p_0 is obtained as the mean of the y's given by

$$p_0 = \frac{2 + 1.5 + 8.2 + 3.5}{4} = \frac{15.2}{4} = 3.8$$

Equation 3.7-3.8+3.9-3.10 gives

$$p_A = \frac{2 - 1.5 + 8.2 - 3.5}{4} = \frac{5.2}{4} = 1.3$$

Equation 3.7+3.8-3.9-3.10 gives

$$p_B = \frac{2 + 1.5 - 8.2 - 3.5}{4} = \frac{-8.2}{4} = -2.05$$

Equation 3.7-3.8-3.9+3.10 gives

$$p_{AB} = \frac{2 - 1.5 - 8.2 + 3.5}{4} = \frac{-4.2}{4} = -1.05$$

Thus the regression equation based on the experiments is (based on Equation 3.2)

$$y = 3.8 + 1.3x_A - 2.05x_B - 1.05x_A x_B$$

The deviation d with respect to the mean is obviously given by

$$d = y - 3.8 = 1.3x_A - 2.05x_B - 1.05x_A x_B$$

It may be verified by the reader that the total sum of squares (SST) of the deviations is given by

$$SST = 4[p_A^2 + p_B^2 + p_{AB}^2] = 4[1.3^2 + 2.05^2 + 1.05^2] = 27.98$$

The sample variance is thus given by

$$s_y^2 = \frac{SST}{n-1} = \frac{27.98}{4-1} \approx 9.33$$

Contributions to the sample variance are given by 4 times the square of the respective parameters and hence we also have

$$
\begin{aligned}
SSA &= 4 \times 1.3^2 = 6.76 \\
SSB &= 4 \times 2.05^2 = 16.81 \\
SSAB &= 4 \times 1.05^2 = 4.41
\end{aligned}
$$

Here SSA means the sum of squares due to variation in level of x_A and so on. The relative contributions to the sample variance are represented as percentage contributions in the following table:

	Contribution	% Contribution
SST	27.98	100
SSA	6.76	24.16
SSB	16.81	60.08
SSAB	4.41	15.76

Thus the dominant factor is the depth of cut followed by the machine speed and lastly the interaction effect. In this example all these have significant effects and hence a full factorial experiment is justified.

3.1.4   More on full factorial design

We would like to generalize the ideas described above in what follows. Extension to larger number of factors as well as larger number of levels would then be straight forward. Let the High and Low levels be represented by + and - respectively (we have just dropped the numeral 1, assuming that that is what is implied). In the case of 2^2 factorial experiment design the following will hold:

	x_A	x_B	$x_A x_B$
Row vector 1	+	+	+
Row vector 2	+	-	-
Row vector 3	-	+	-
Row vector 4	-	-	+
Column sum	0	0	0
Column sum of squares	4	4	4

We note that the product of any two columns is zero. Also the column sums are zero. Hence the three columns may be considered as vectors that form an orthogonal set. In fact while calculating the sample variance earlier these properties were used without being explicitly spelt out!

Most of the time it is not possible to conduct a large number of experiments! Experiments consume time and money. The question that is asked is: "Can we reduce the number of experiments and yet get an adequate representation of the relationship between the outcome of the experiment and the variation of the factors?" The answer is in general "yes". We replace the full factorial design with a fractional factorial design. In the fractional factorial design only certain combinations of the levels of the factors are used to conduct the experiments. This ploy helps to reduce the number of experiments. The price to be paid is that *all* interactions will not be resolved.

Consider again the case of 2^2 factorial experiment. If the interaction term is small or *assumed* to be small only three regression coefficients $p_0, p_A,$ and p_B will be important. We require only three experiments such that the four equations reduce to only three equations corresponding to the three experimental data that will be available. They may be either:

$$p_0 + p_A + p_B = y_1 \qquad (3.11)$$

$$p_0 - p_A + p_B = y_2 \qquad (3.12)$$

$$p_0 + p_A - p_B = y_3 \qquad (3.13)$$

or

$$p_0 + p_A + p_B = y_1 \qquad (3.14)$$

$$p_0 - p_A + p_B = y_2 \qquad (3.15)$$

$$p_0 - p_A - p_B = y_4 \qquad (3.16)$$

depending on the choice of the row vectors in conducting the experiments.

In this simple case of two factors the economy of reducing the number of experiments by *one* may not be all that important. However it is very useful to go in for a fractional factorial design when the number of factors is large and when we expect some factors or interactions between some factors to be unimportant. Thus fractional factorial experiment design is useful when main effects dominate and interaction effects are of lower order. Also we may *at any time* do more experiments if necessitated by the observations made with a smaller set of experiments.

3.1.5 One half factorial design

For a system with k factors and 2 levels the number of experiments in a full factorial design will be 2^k. For example, if $k = 3$, this number works out to be $2^3 = 8$. The eight values of the levels would correspond to the corners of a cube as represented by Figure 3.2. A half factorial design would use 2^{k-1} experiments. With $k = 3$ this works out to be $2^2 = 4$. The half factorial design would cut the number of experiments by half. In the half factorial design we would have to choose half the number of experiments and they should correspond to four of the eight corners of the cube. We may choose the corners corresponding to a, b, c and abc as one possible set. This set will correspond to the following:

Point	x_A	x_B	x_C
a	+	-	-
b	-	+	-
c	-	-	+
abc	+	+	+

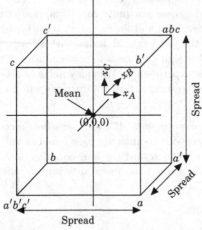

Figure 3.2: *Interpretation of 2^2 factorial experiment with $k = 3$*

We notice at once that the three column vectors are orthogonal. Also the points are obtained by requiring that the projection on to the left face of the cube gives a full factorial design of type 2^2 or 2^{k-1}. It is easily seen that we may also use the corners a', b', c' and $a'b'c'$ to get a second possible half factorial design. This is represented by the following:

Point	x_A	x_B	x_C
a'	+	+	-
b'	+	-	+
c'	-	+	+
$a'b'c'$	-	-	-

In each of these cases we need to perform only 4 experiments. Let us look at the consequence of this. For this purpose we make the following table corresponding to a, b, c, abc case.

Point	x_A	x_B	x_C	$x_A x_B$	$x_A x_C$	$x_B x_C$	$x_A x_B x_C$
a	+	-	-	-	-	+	+
b	-	+	-	-	+	-	+
c	-	-	+	+	-	-	+
abc	+	+	+	+	+	+	+
Column No.	1	2	3	4	5	6	7

Notice that Column vector 1 is identical to the Column vector 6. We say that these two are "aliases". Similarly the column vectors 2-5 and column vectors 3-4 form aliases. Let us look at the consequence of these.

The most general regression model that can be used to represent the outcome of the full factorial experiment would be

$$y = \boxed{\begin{aligned} p_0 + p_A x_A + p_B x_B + p_C x_C + p_{AB} x_A x_B \\ + p_{AC} x_A x_C + p_{BC} x_B x_C + p_{ABC} x_A x_B x_C \end{aligned}} \tag{3.17}$$

There are eight regression coefficients and the eight experiments in the full factorial design would yield all these coefficients. However we now have only four experiments and hence only four regression coefficients may be resolved. By looking at the procedure used earlier in solving the equations for the regression coefficients, it is clear that it is not possible to obtain the coefficients that form an *alias pair*. For example, it is not possible to resolve p_A and p_{BC}. These two are said to be "confounded". The student should verify that the following are also confounded: $p_B, p_{AC}; p_C, p_{AB}; p_0, p_{ABC}$. The consequence of this is that the best regression we may propose is

$$\boxed{y = p_0 + p_A x_A + p_B x_B + p_C x_C} \tag{3.18}$$

All interaction effects are unresolved and we have only the primary effects being accounted for. The regression model is a multiple linear model. The student may perform a similar analysis with the second possible half factorial design and arrive at a similar conclusion.

Generalization

The above ideas may be generalized as under:

Consider a 2^k full factorial design. The number of experiments will be 2^k. These experiments will resolve k main effects, ${}^kC_2$ - 2 factor interactions, ${}^kC_3$ - 3 factor interactions,......, ${}^kC_{k-1}$ - (k-1) factor interactions and 1 - k interactions. For example, if $k = 5$, these will be 5 main effects, 10 - 2 factor interactions, 10 - 3 factor interactions, 5 - 4 factor interactions and 1 - 5 factor interaction. The total of all these is equal to $2^5 - 1 = 31$).

In summary:

- The 2^k experiments will account for the intercept (or the mean) and all interaction effects and is referred to as Resolution k design. (With $k = 5$, we have full factorial design having resolution 5. Number of experiments is 32).
- Semi or half factorial design 2^{k-1} will be resolution $(k-1)$ design. (With $k = 5$, we have half factorial design having resolution 4. Number of experiments is 16).
- Quarter factorial design will have a resolution of $(k-2)$. (With $k = 5$, we have quarter factorial design having resolution 3. Number of experiments is 8). And so on...

The reader may determine all the aliases in these cases.

3.1.6   Other simple design

We now look at other ways of economizing on the number of experimental runs required to understand the behavior of systems. We take a simple example of characterizing the frictional pressure drop in a tube. A typical experimental set up will look like the one shown in Figure 3.3.

The factors that influence the pressure drop (measured with a differential pressure gage) between stations 1 and 2 may be written down as:

1. Properties of the fluid flowing in the tube: Density ρ, the fluid viscosity μ (2 factors)
2. Geometric parameters: Pipe diameter D, distance between the two stations L and the surface roughness parameter (zero for smooth pipe but non-zero positive number for a rough pipe). (2 or 3 factors)
3. Fluid velocity V (1 factor)

The total number of factors that may influence the pressure drop p is thus equal to 5 factors in the case of a smooth pipe and 6 factors in the case of a rough pipe. The reader may verify that a full factorial design with *just* two levels would require a total of 32 (smooth pipe) or 64 (rough pipe) experiments. In practice the velocity of the fluid in the pipe may vary over a wide range of values. The fluid properties may be varied by changing the fluid or by conducting the experiment at different pressure and temperature levels! The diameter and length of the pipe may indeed vary over a very wide range. If one were to go for a design of experiment based on several levels the number of experiments to be performed would be prohibitively large. The question

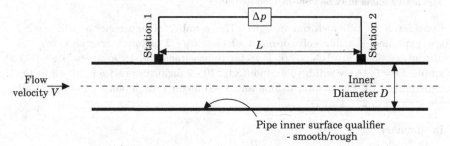

Figure 3.3: *Schematic of an experimental set up for friction factor measurement in pipe flow*

now is: "How are we going to design the experimental scheme?" How do we conduct a finite number of experiments and yet get enough information to understand the outcome of the measurement? We look for the answer in "dimensional analysis" (the student would have learnt this from his/her course in Fluid Mechanics), for this purpose. Dimensional analysis, in fact, indicates that the outcome of the experiment may be represented by a simple relationship of the form

$$\frac{\Delta p}{\rho V^2 \frac{L}{D}} = f\left(\frac{\rho V D}{\mu}, \frac{\epsilon}{D}\right) \tag{3.19}$$

The reader may verify that both the left hand side quantity and the quantities within the bracket on the right hand side are dimensionless. This means both are pure numbers! The "f" outside the bracketed term on the right hand side indicates a functional relationship. In Fluid Mechanics parlance $\dfrac{\Delta p}{\rho V^2}$ is known as the Euler number, $\dfrac{\rho V D}{\mu}$ is known as the Reynolds number and $\dfrac{\epsilon}{D}$ is non-dimensional roughness parameter, the ratio of pipe wall roughness to the pipe diameter. The fourth non-dimensional parameter that appears above is the ratio $\dfrac{L}{D}$. Essentially the Euler number is a function of only three factors, the Reynolds number, roughness parameter and the length to diameter ratio! The number of factors has been reduced from 6 to just 3! We may conduct the experiments with just a couple of fluids, a few values of the velocity, a couple of different diameter pipes of different lengths and wall roughnesses to identify the nature of the functional relationship indicated in Equation 3.19.

Friction factor for flow in a smooth pipe:

It is intended to verify the applicability of Blasius formula for friction factor variation with Reynolds number Re, for flow in a smooth pipe, given by $f = \dfrac{0.316}{Re^{0.25}}$. It is known from the literature that this formula is valid in the range $4 \times 10^3 \leq Re \leq 10^5$ and agrees with data with an error band of $\pm 5\,\%$. Device a suitable experimental program for this verification exercise.

Several decisions need to be taken while devising an experimental program for performing the verification exercise. These are listed below along with the decision.

1. **Fluid with which to perform the experiments:** Common fluids, with ready availability in any laboratory, are air and water. Pressure and temperature effects may be severe with air while they are mild in the case of water. Air may also require drying, especially on humid days. Hence we choose water as the working fluid.

2. **Selection of tube material:** Common tube materials used in the laboratory are PVC and drawn copper or brass tubes. Effect of corrosion is mild with these materials. Tubes made of mild steel is not suitable. In the present case we choose copper tubes. These have roughness value of $\epsilon = 1.5\ \mu m$.[1] For most purposes we may treat the pipe as smooth.

3. **Diameter and lengths of tubes to be used in the experiments:** Diameter and lengths need to be chosen keeping in mind the range of Reynolds numbers to be realized in the experiments. Length is determined by requiring the pressure drop to be large enough to be measured accurately. Length also should be such that entrance effects are not important and the flow is fully developed.[2] Choice of diameter and lengths will be made using a preliminary analysis to be given later.

---

[1] Obtained from http://www.enggcyclopedia.com/2011/09/absolute-roughness

[2] Refer to a book on Fluid Mechanics such as Frank M. White, Fluid Mechanics (SI units), 7th edition, McGraw-Hill, NY 2011

4. **Range of fluid velocities to be used in the experiments:** Velocity range may be selected based on the chosen pipe diameters and the required range of Reynolds number.

5. **Choice of instruments:** Calculations based on the Blasius formula for friction factor may be used to calculate the expected pressure drop over the range of velocities intended to be used in the experiments. Suitable differential pressure measuring device has to be chosen based on the expected pressure drop. Flow velocity needs to be measured using an instrument that is accurate in the chosen velocity range.

6. **Layout of the experimental setup:** Layout will have to be worked out to realize the goals of the experiment.

Selection of parameters:

Having decided on the tube material (drawn copper), diameters, velocities and lengths are chosen using preliminary analysis. Four different diameter tubes are chosen to cover the range of Reynolds numbers. They are 15.875, 19.05 22.225 and 41.275 mm OD tubes. The corresponding ID's are 14.453, 17.526, 20.599 and 38.786 mm. For the calculations we assume water to be available at a temperature of $27°C$. Corresponding water properties of interest are $T = 300\ K$; Dynamic Viscosity $\mu = 0.000852\ kg/ms$; Density $\rho = 996.59\ kg/m^3$ and Kinematic Viscosity $v = 8.549 \times 10^{-7}\ m^2/s$. We choose the flow velocity range of $0.25 \le V \le 2.25\ m/s$ and tabulate the relevant data as a table.

$D,\ mm$	$V,\ m/s$	Re	$\dfrac{L_e}{D}$
14.453	0.25	4226	11
14.453	0.80	13525	15
17.526	0.50	10250	14
17.526	1.00	22550	17
20.599	0.75	18071	16
20.599	2.00	48190	20
38.786	1.00	45368	20
38.786	2.25	102079	24

Thus we are able to cover the entire Reynolds number range over which the Blasius relation is valid. The flow is turbulent over the range of Reynolds number given in the exercise. Entry length for turbulent flow is approximately given by $\dfrac{L_e}{D} \approx$ $1.359Re^{0.25}$. Required entry lengths are shown in the last column of the table given above. We may choose an entry length of some 25 tube diameters in all cases. Pressure drop data is obtained by measuring the pressure drop down stream of the entry length. We may calculate the expected pressure drop per meter length of tube in each case and make another table as given below. Note that the Blasius formula for f has been used in these estimates.

D, mm	$V, m/s$	Re	$f_{Blasius}$	$\Delta p, Pa$	Head, mm water
14.453	0.25	4226	0.0392	84.5	8.6
14.453	0.80	13525	0.0293	646.6	66.1
17.526	0.50	10250	0.0314	223.2	22.8
17.526	1.10	22550	0.0258	887.1	90.7
20.599	0.75	18071	0.0273	370.9	37.9
20.599	2.00	48190	0.0213	2063.7	211.1
38.786	1.00	45368	0.0217	278.2	28.5
38.786	2.25	102079	0.0177	1149.8	117.6

Choice of instruments
Diaphragm type differential pressure instruments with different ranges may be used to measure the pressure drop. In order to increase the pressure differential we may choose different lengths of pipes over which the pressure drop is measured. For example, in the case of the smallest diameter tube and the smallest velocity we may use $L = 2.5\,m$ to get the minimum pressure differential of $2.5 \times 84.5 = 211.1\,Pa$. We may employ one differential pressure instrument with a range of $0 - 1000\,Pa$ and another of range $0 - 2500\,Pa$.

Flow velocity V is the mean velocity with which the fluid flows through the tube. It is convenient to estimate the flow velocity, by measuring the volumetric flow rate, using a suitable devise and divide it by the cross sectional area of the tube. Several options for measuring the volume flow rate are possible as will be discussed in a later chapter.

Control valves are used as necessary. A **centrifugal pump** of suitable capacity may be used for maintaining the flow in the apparatus. Choice of the pump is based on specifications of the manufacturer that match the requirements of flow and required head. **Schematic of the apparatus**

We are now ready to show a schematic of the test set up. It consists of the four tubes arranged horizontally between two headers as indicated in Figure 3.4. Headers are boxes to which the tubes may be attached, as indicated. Flow through each tube may be controlled by a suitable control valve as shown. During the experiments only one tube will be operational. Length between the outlet of the control valve and the first pressure tap is such that disturbances due to the valve are damped out and the flow has become fully developed. Differential pressure transmitter is connected between two static pressure taps. Water is drawn from a sump by a pump and passes through a flow meter before entering the test section. Water then flows through one of the four tubes and is sent back to the sump. The distance between the pressure taps are adjusted as previously discussed. Data is collected in the form of the mean velocity versus the pressure drop. The data thus collected may be analyzed to yield data in the form of friction factor vs Reynolds number.

The pressure gage attached to the header reads the mean pressure in the test section. A thermometer may also be inserted in to the header to measure the temperature of water.

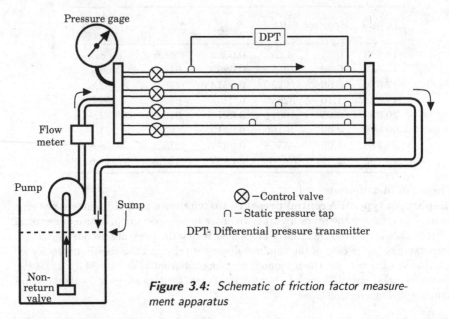

Figure 3.4: *Schematic of friction factor measurement apparatus*

Measurements have been made in an apparatus similar to that shown schematically here. Data was collected to cover the range of Re mentioned earlier. The data is compared with the Blasius friction factor in Figure 3.5.

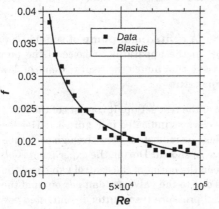

Figure 3.5: *Experimental verification of the Blasius friction factor formula for flow through a smooth tube*

The RMS error between the data and the Blasius friction factor formula obtained by taking the square root of the average of the sum of squares of the differences between the two is RMS Error = 0.0009. Hence the data may be expressed as $f = f_{Blasius} \pm 0.0009$.

It appears that the DOE schemes are rarely used in areas such as fluid mechanics and heat transfer. A possible reason for this is the inherently nonlinear nature of phenomena that involve fluid flow and heat transfer. We consider a simple example from convection heat transfer to discuss a few of the issues that impinge on how experiments are planned.

Consider turbulent flow of a fluid in a tube of circular cross section. Heat transfer from the fluid to the tube wall or vice versa is usually related by a relation of the form

$$Nu = aRe^b Pr^c \tag{3.20}$$

where Nu is the Nusselt number, Re is the Reynolds number and Pr is the Prandtl number of the fluid. The parameters a through c depend on the range of Reynolds and Prandtl numbers. Parameter c depends on whether the fluid or the wall is the hotter. We have two factors in this problem viz. the Reynolds number and the Prandtl number. It appears that the interaction effect alone is important and the main effects are not important. However, if we take logarithms on both sides of Equation 3.20, we get

$$\underbrace{\log(Nu)}_{Y} = \underbrace{\log(a)}_{A} + b\underbrace{\log(Re)}_{X_1} + c\underbrace{\log(Pr)}_{X_2} \tag{3.21}$$

It appears that the main effects are now important and the interaction effects are not!

Concluding remarks

This chapter has dealt with design of experiments - a task which every experimentalist needs to do before performing any measurements. Apart from deciding on the factors and levels, design of experiments involves the design of a suitable test set up and the choice of appropriate instruments for performing the measurements. When the reader is conversant with the material of the rest of the book this task will be easy to accomplish.

Exercise I

I.1 Errors and error distributions

Ex I.1: Find the mean and variance of the first ten terms of a geometric progression with 2 the first term and 1.05 the common ratio.

Ex I.2: Find the mean and variance of the first ten terms of an arithmetic progression with 2 the first term and 1.05 the common difference.

Ex I.3: A certain measurement gave the value of g, the acceleration due to gravity, as $9.804 \ m/s^2$. The precision of measurement is specified by the standard deviation given by $0.00055 \ m/s^2$. If the measurement is repeated what is the probability that the value is within $9.804 \pm 0.001 \ m/s^2$? You may assume that the probability distribution is normal.

Ex I.4: A certain measurement gave the value of specific heat of air as $1006 \ J/kg \ K$. The precision of measurement is specified by the standard deviation given by $35 \ J/kg \ K$. If the measurement is repeated what is the probability that the value is within $1006 \pm 45 \ J/kg \ K$? You may assume that the probability distribution is normal.

Ex I.5: Diameter of 360 shafts, in a production unit, were measured with a micrometer and the mean diameter was found to be $5.028 \ mm$. The standard deviation was determined as $\pm 0.002 \ mm$. How many of the measured values will lie within $\pm 0.002 \ mm$, $\pm 0.006 \ mm$, $\pm 0.009 \ mm$ of the mean value?
Is there a justification for assuming that the data is normally distributed? Explain.

Ex I.6: A certain length measurement is made 300 times and the mean of these is $2.253 \ m$. The standard deviation is determined as $\pm 2.3 \ mm$. How many of the measured values will lie within $\pm 0.0023 \ m$ of the mean value?

Ex I.7: A certain length measurement is made 100 times and the mean of these is $2.508 \ m$. The standard deviation is determined as ± 0.003 m. How many of the measured values will lie within $\pm 0.0015 \ m$, $\pm 0.006 \ m$, $\pm 0.009 \ m$ of the mean value? Is it reasonable to assume that the errors are distributed normally? Explain.

Ex I.8: Diameter of the spindle of a machine has been measured repeatedly and the data is given below:

Trial #	1	2	3	4	5
Diameter in mm	6.755	6.746	6.562	6.558	6.716

Determine the mean diameter of the spindle and the variance of the sample. Use t - statistics to specify an error bar with 95% confidence.

Ex I.9: In carefully collected data it is alright to discard data that are beyond 3 standard deviations of the mean. The following data has been obtained by repeated measurements of the melting point of an alloy:

Trial No.	Temperature $°C$	Trial No.	Temperature $°C$
1	454.69	6	453.63
2	454.62	7	453.71
3	453.56	8	455.51
4	454.31	9	453.79
5	454.49	10	453.74

It is known that the experiment has been carried out carefully and the standard deviation of the measured temperature is expected to be no more than $\pm 0.25K$. Obtain the best value of the melting point of the alloy and its standard deviation after discarding any data based on the criterion spelt out at the beginning of the exercise.

Ex I.10: An experimenter monitors the temperature of a constant temperture bath over a period of 30 minutes. Manufacturer has guaranteed that the the temperature of the bath will be stable with a standard deviation of $0.2°C$. The data collected by the researcher gave the following values for the deviate i.e. the actual reading minus the set point value in $°C$.

No.	1	2	3	4	5	6	7	8	9
Deviate	-0.07	0.34	-0.02	-0.02	0.03	0.26	0.17	0.34	-0.36

No.	10	11	12	13	14	15	16	17	18
Deviate	0.26	-0.01	0.45	0.17	0.03	-0.16	0.14	-0.02	0.13

Test the data for normality. Verify whether the manufacturer's claim is sustained.

Ex I.11: Two stes of data were collected by two researchers working in the same laboratory using the same experimental set up. The quantity being measured was the same, viz. the speed of sound in air at $298\,K$. From the measured data the deviates were calculated, with respect to the mean value and the deviates are tabulated below.

i	1	2	3	4	5	6	7
$d_{i,1}$	0.567	0.369	0.292	0.147	0.461	-0.632	0.153
$d_{i,2}$	-0.579	1.725	0.924	0.472	0.413	-0.099	0.052
i	8	9	10	11	12	13	14
$d_{i,1}$	0.778	-0.208	-0.258	-0.090	0.359	0.092	0.096
$d_{i,2}$	-0.213	-0.065	-0.662	-0.776	-0.160	0.357	-1.050

Verify whether the two sets of deviations are normally distributed. Is it necessary to discard any data treating them as outliers? Comment on the two sets of data based on their respective standard deviations.

Ex I.12: Specification given below is typical of a pressure transmitter.

Output	4 to 20 mA DC (2-wire)
Supply Voltage	12 to 32 V DC with V_{min} = 12V + (0.022A × loop resistance
Insulation Resistance	> 100 $M\Omega$
Adjustments	Zero: -10 to 110% FS
Rangeability	Span: -10 to 110% FS
Accuracy	±0.25% FS URL
Response Time	30 ms (adjustable)
Output Resolution	0.1% FS (URL)
Display	4 digit, 10 mm LCD with Red backlight
Display Accuracy	±0.25% FS (URL) + 1 digit

FS is Full Scale; URL is Upper Range Limit

Discuss accuracy and precision issues of this transmitter if the URL is 10 bar gauge.

Ex I.13: Repeated measurement of diameter of a shaft yields the following data. How would you specify the diameter and the corresponding uncertainty? Is the claim that the diameter has an uncertainty of less than ±0.05 cm valid?

No.	Diameter, cm	No.	Diameter, cm
1	2.37	9	2.46
2	2.48	10	2.38
3	2.42	11	2.42
4	2.37	12	2.40
5	2.37	13	2.46
6	2.42	14	2.39
7	2.48	15	2.38
8	2.41	16	2.55

Verify whether the deviates from the mean are normally distributed. Is it necesary to discard any data?

Ex I.14: It is claimed that the temperature is given by a certain thermometer with a standard deviation of 0.2°C. The following replicate data has been recorded by that thermometer. Does the data validate the claim? Is the deviation distribution normal? Verify using different tests discussed in the text.

No.	$t°C$	No.	$t°C$
1	34.92	9	34.82
2	35.13	10	34.78
3	35.00	11	34.93
4	34.91	12	35.00
5	34.78	13	34.96
6	35.09	14	35.19
7	35.00	15	35.07

Ex I.15: Replicate data obtained in a certain experiment is given as a table. The deviates have been calculated with respect to the mean so that the mean of the deviates is zero. Data is tabulated in ascending order. Make a box and whisker plot. Discard any data which you consider to be outliers. After discarding outliers make a fresh box and whisker plot. Comment based on the two box and whisker plots. Use a suitable test of normality on raw data and the data after discarding outliers.

Rank Order	Deviate	Rank Order	Deviate	Rank Order	Deviate
1	-1.004	13	-0.269	25	0.174
2	-0.969	14	-0.131	26	0.190
3	-0.853	15	-0.129	27	0.319
4	-0.809	16	-0.124	28	0.338
5	-0.539	17	-0.037	29	0.438
6	-0.467	18	0.040	30	0.552
7	-0.456	19	0.095	31	0.570
8	-0.393	20	0.096	32	0.571
9	-0.387	21	0.102	33	0.629
10	-0.330	22	0.104	34	0.886
11	-0.327	23	0.140	35	0.987
12	-0.322	24	0.150	36	1.167

I.2   Propagation of errors

Ex I.16: The radius of a circle and the radius of a sphere are measured subject to the same error. Which will have a larger fractional error, the area of the circle or the volume of the sphere? Explain.

Ex I.17: If both the radius and the height of a cylinder are measured with the same precision of 1% what is the specification for the precision of the estimated volume of the cylinder? What would you suggest to improve the precision of the estimated value of the volume of the cylinder?

Ex I.18: Volume of a cylinder is to be determined by using the measured values of its diameter and length. The measurements show that the diameter is 12.55±0.02 *mm* and the length is 199.5±0.05 *mm*. Determine the volume of the cylinder and specify an error bar for the same. Express the error bar also as a percentage.

Ex I.19: Two resistances R_1 and R_2 are given as $10000\pm125\Omega$ and $5000\pm75\Omega$. Determine the equivalent resistance when these two are connected in a) series and b) parallel. Also determine the uncertainties in these two cases.

Ex I.20: The diameter of a pipe is measured repeatedly and it is estimated to be $D = 0.025 \pm 0.001mm$. Air at $300K$ is flowing in the pipe with a velocity that has been measured as $V = 1.03 \pm 0.005m/s$. The viscosity of air obtained from the table may be expected to have an error of $\pm1\%$. The Reynolds number Re is calculated based on this data. What is the best value of Re and the uncertainty in it?

Ex I.21: Two resistances are connected in series and parallel. The values of the resistances are:$R_1 = 100.0 \pm 0.3\Omega$, $R_2 = 50 \pm 0.2\Omega$. Calculate the equivalent resistance and its uncertainty in each of these cases.

A $9\ V$ battery is connected across the two resistance arrangements. What are the currents and the uncertainties in each case.

Ex I.22: The diameter of a pipe is measured repeatedly and it is estimated to be $0.025m$. Helium at $305\ K$ is flowing in the pipe with a velocity that has been measured as $1.55m/s$. Estimate the Reynolds number based on this data. If the measured quantities are susceptible to respective uncertainties of 1%, 0.3% and 0.2% what is the error in the estimated value of the Reynolds number?

Ex I.23: A certain water heater operates at $110\ V$ and draws a current of $9.2\ A$. It is known that the voltage and current measurements are susceptible to errors of $\pm5\ V$ and $\pm0.15\ A$. What is the nominal power drawn by the heater and what is its uncertainty?

Ex I.24: The formula $y = Ae^{bx}$ relates the derived quantity y to the measured quantity x. If the values of A, b and x respectively are 10 ± 0.5, 0.5 ± 0.02 and 2 ± 0.05 what is the fractional error in the derived quantity y?

Ex I.25: Heat transfer from the walls of a tube to a flowing fluid is related by the formula $Nu = 0.023Re^{0.8}Pr^{0.3}$, where Nu is the Nusselt number based on diameter, Re is the Reynolds number based on diameter and Pr is the fluid Prandtl number. We are given the parameter values of Re=50000±500, $Pr =$ 4.71±0.25. Find the nominal value of Nusselt number and specify a suitable uncertainty for the same. Express the uncertainty also as a percentage.

Ex I.26: The inner and outer diameters of a cylindrical shell are measured and are found to be $ID = 55 \pm 0.3\ mm$, $OD = 85 \pm 0.3\ mm$. The thickness of the shell is independently measured and is found to be $t = 15 \pm 0.3\ mm$. What is the fractional error in the thickness determined from the first measurement involving the measurement of diameters and the second measurement that measures the shell thickness directly? The shell volume is determined from the measurements reported above. Estimate the shell volume and specify a suitable uncertainty on it.

Ex I.27: The efficiency of a uniform-area circular pin fin is given by, $\eta = \dfrac{\tanh(mL)}{mL}$

where $m = \sqrt{\dfrac{4h}{kD}}$. In the above, h is the heat transfer coefficient in $W/m^{2\circ}C$, k

is the thermal conductivity of the fin material in $W/m°C$, D is the diameter of the pin fin in m and L is the fin length in m. In a certain case $h = 67\pm3$; $k = 45\pm2$; $D = 0.01\pm0.0001$ and $L = 0.12\pm0.002$. Determine the fin efficiency along with its uncertainty.

Ex I.28: The temperature difference across a slab is estimated using the measured temperatures at its two boundaries. The two temperatures are $T_1 = 38°C$ and $T_2 = 24°C$. Each of these is subject to an error of $\pm 0.2°C$. What is the expected error in $T_1 - T_2$ in $°C$ and in %? Do you think it is better to measure the temperature difference directly? Explain.

Ex I.29: The area of a rectangle is estimated by measuring its length and breadth. These measurements are each subject to an uncertainty of $\pm0.5\%$. What is the % uncertainty in the estimated area of the rectangle?

Ex I.30: A triangle ABC has the following specifications:

$$\text{Length of side } AB \text{ is } c = 0.2 \pm 0.002 \ m$$
$$\text{Angle } B = 45 \pm 0.2° \ \text{Angle } A = 59 \pm 0.3°$$

Estimate the area of the triangle and also its uncertainty.

Ex I.31: Grashof Number is a non-dimensional parameter that occurs in free convection problems. It is given by $Gr = (g\beta\Delta TL)/v^2$, where g is the acceleration due to gravity, β is the isobaric compressibility of the medium, ΔT is a characteristic temperature difference in the problem, L is a characteristic length dimension and v is the kinematic viscosity of the medium. In a certain situation the following data is given: $g = 9.81 \ m/s^2$, $\beta = 0.00065K^{-1}$, $\Delta T = 298 \pm 0.5K$, $L = 0.05 \pm 0.001m$, $v = (14.1 \pm 0.3)10^{-6}m^2/s$. Determine the nominal value of Gr and its uncertainty.

Ex I.32: The length of a simple pendulum is measured as $23 \pm 0.2 \ cm$. The local acceleration due to gravity has been estimated as $9.65\pm0.03 \ m/s^2$. What is the period of oscillation of the pendulum and its uncertainty?

Ex I.33: The relation between distance traveled s in time t by a particle subjected to constant acceleration a is given by the well known formula $s = ut + at^2/2$ where u is the velocity at $t = 0$. From an observation the following data is given: $u = 1 \pm 0.01 \ m/s$, $a = 2.3\pm0.014 \ m/s^2$ and $t = 15\pm0.005 \ s$. What is the nominal distance traveled by the particle after $15 \ s$ and its uncertainty?

Ex I.34: The volume of a sphere is estimated by measuring its diameter by vernier calipers. In a certain case the diameter has been measured as $D = 0.0502 \pm 0.00005 \ m$. Determine the volume and specify a suitable uncertainty for the same.

I.3   Regression analysis

Ex I.35: The following data is expected to follow the linear law $y = mx + c$ where m and c are respectively the slope and intercept parameters.

x	1	2	3	4	5
y	5	12	11	14	20
x	6	7	8	9	10
y	24	27	26	32	33

Calculate the correlation coefficient. Based on this comment on the goodness of fit. Determine the fit parameters. Specify an error bar for the fit.

Ex I.36: The following data are expected to follow a linear relation of the form $y = ax + b$. Obtain the best fit by least squares. Estimate the value of y at $x = 3$ and specify an error bar for the same.

x	0.9	2.3	3.3	4.5	5.7	7.7
y	1.1	1.6	2.6	3.2	4	6

Make a plot to compare the data and the fit. Determine the correlation coefficient. Perform the χ^2 test and comment on the quality of the fit.

Ex I.37: Heat transfer data given below has been collected in an experiment involving turbulent tube flow of a fluid with a constant Prandtl number of $Pr = 4$. The data is expected to follow a law of form $Nu = aRe^b Pr^{0.3}$, where a and b are constants to be determined by the method of least squares.

Re(in thousands)	1	1.2	1.8	2.2	3
Nu	56.1	59.8	91.9	102.7	142.2
Re(in thousands)	4.2	5.1	6.3	7	9.4
Nu	185	192.9	236.1	271.3	390.6

Make suitable plots to assess the goodness of fit.

Ex I.38: Saturation temperature and saturation pressure of water are given in the following table:

$t,°C$	0	20	40	60	80	100	120
p,kPa	0.6108	2.337	7.375	19.92	47.36	101.33	198.54

Fit a curve of form $ln(p) = A + B/T$ (Arrhenius type relation), where T is in K, by the least square method and using EXCEL. Comment on the goodness of the fit

Ex I.39: The following data are expected to follow a linear relation of the form $y = ax + b$. Obtain the best fit by least squares. . Estimate the value of y at $x = 1.5$ and specify an error bar for the same.

x	0.4	0.6	0.8	1.0	1.2	1.4
y	-0.174	0.058	0.312	0.532	0.740	1.07
x	1.6	1.8	2.0	2.2	2.4	
y	1.306	1.584	1.91	2.084	2.337	

Ex I.40: The following data are expected to follow a linear relation of the form $y = ax + b$. Obtain the best fit by least squares. Calculate also the standard deviation of the predicted straight line variation from the data.

x	0.9	2.3	3.3	4.5	5.7	7.7
y	1.1	1.6	2.6	3.2	4.0	6.0

Make a plot to compare the data and the fit. Determine the correlation coefficient.

Ex I.41: The following data are expected to follow a functional form $Nu = 0.53Re^{0.51}$.

Re	12	20	30	40	100	300	400	1000	3000
Nu	1.8	2.7	3.0	3.5	5.1	10.5	10.9	18	32

Determine the index of correlation and comment on the goodness of the fit. Make two suitable plots showing the data and the fit. What is the expected error when the fit is made use of to estimate the Nusselt number Nu at a given Reynolds number Re?

Ex I.42: The following data follows a linear relation of form $z = ax + by + c$.

x	0.21	0.39	0.56	0.78	1.02	1.24	1.34	1.56	1.82
y	0.41	0.78	1.22	1.62	1.90	2.44	2.88	3.22	3.46
z	1.462	1.808	1.979	2.192	2.479	2.616	2.989	3.106	3.466

Make such a fit using the method of least squares. Determine the index of correlation and also make a parity plot. Can you specify a suitable error bar on z?

Ex I.43: The following data is expected to follow the law $y = ax^b$. Determine a and b by least squares and discuss the goodness of the fit by:
a. calculating the index of correlation and b. performing a χ^2 test.

x	1.21	1.35	2.4	2.75	4.5	5.1	7.1	8.1
y	1.18	1.65	4.90	8.90	20.10	31.30	56.00	81

Make different types of plots. Estimate the value of y when x= 3, 7.5. Specify errors for these.

Ex I.44: Thermal conductivity of Titanium varies with temperature as indicated below:

T, K	100	200	400	600	800	1000	1200	1500
$k, W/mK$	30.5	24.5	20.4	19.4	19.7	20.7	22.0	24.5

Use spreadsheet program of your choice and various possible trend line options to arrive at what you consider is the best one. Calculate the index of correlation and comment on the quality of the fit. Perform a Chi square test. Show the results by making different types of plots.

Ex I.45: Speed of sound in atmospheric air varies with temperature as shown in the table. Fit a suitable curve to the data after justifying the choice.

$t,°C$	-40	0	20	40	60	80
$a,m/s$	306.0	320.3	334.1	340.5	346.9	353.3

Ex I.46: x and y are expected to obey the law $y = a + cx^2$. Make a plot to see whether this is a reasonable expectation. Then determine a and c by the method of least squares. Discuss the goodness of fit by performing different tests. Do you think it is possible to use linear regression in this case? If yes use this to simplify your problem.

x	-1.0	-0.8	-0.6	-0.4	0	0.4	0.6	0.8	1.0
y	0.01	0.35	0.65	0.82	0.99	0.83	0.64	0.34	0.01

Ex I.47: The response of a thermocouple is represented by the relation $E = at + bt^2/2 + ct^3/3$ where E is in mV and t is in $°C$ and the reference junction is at the ice point. The data for Chromel-Alumel or K type thermocouple is given in the form of a table. What are the values of a, b and c for this thermocouple?

$t,°C$	37.8	93.3	148.9	204.4	260
E,mV	1.520	3.819	6.092	8.314	10.560
$t,°C$	371.1	537.8	815.6	1093.5	1371.1
E,mV	15.178	22.251	33.913	44.856	54.845

How good is the fit compared to the data? What will be the thermocouple output when $t = 300°C$? Use spreadsheet program of your choice for the curve fit using the built in "trend line" option.

Ex I.48: The following data is expected to follow the law $y = 1.439 + 0.0826x + 0.046x^2$.

x	1	2	3	4	5	6
y	1.57	1.79	2.10	2.52	3.02	3.61

Calculate the index of correlation. How good is the fit? What is the number of degrees of freedom? What is the standard error of fit with respect to data? What would be the correlation coefficient if the data were to be fit using a linear law? Compare this with the index of correlation calculated above and make a suitable comment.

Ex I.49: The variation of Prandtl number of Ethylene glycol with temperature is given in the table below:

T,K	273	280	290	300	310	320
Pr	617	400	236	151	103	73.5
T,K	330	340	350	360	370	373
Pr	55	42.8	34.6	28.6	23.7	22.4

It is observed that the logarithm of the Prandtl number is close to being a linear function of temperature. Fit a quadratic to $log(Pr)$ vs. T using EXCEL. Determine the error between the fit and the data. Calculate the index of correlation and comment on the fit. Make suitable plots using EXCEL. What would be the goodness of fit if linear relationship is assumed to exist between $log(Pr)$ and T?

Ex I.50: Density of water varies with temperature as shown in the table.

$T,°C$	0	20	40	60	80	100
$\rho, kg/m^3$	999.8	998.3	992.3	983.2	971.6	958.1
$T,°C$	120	140	160	180	200	
$\rho, kg/m^3$	942.9	925.8	907.3	886.9	864.7	

Make a plot to determine what will be a reasonably good fit to data. Determine the fit you propose. Discuss its merits.

Ex I.51: Following data consists of two sets of data obtained in a certain experiment in which the cause is x and the effect is y. The set y_1 was taken with instruments that were considered later to be not of good quality. The second set of data was taken after replacing the measuring instruments with those of good quality.

x	y_1	y_2	x	y_1	y_2
0.20	1.16	0.55	2.12	3.67	4.25
0.50	1.99	1.24	2.56	5.17	4.86
0.85	1.40	1.94	2.86	6.31	5.69
1.23	2.29	2.83	3.15	5.55	6.46
1.75	3.27	3.31	3.46	6.29	6.27

Is it reasonable to believe that the second set of data is a better representation of the relationship between the cause and the effect? Comment based on statistical parameters, assuming that the cause effect relationship is linear.

Ex I.52: Transient data of a first order system was taken in an experiment performed in the laboratory. Temperature readings have been converted by taking logarithm to base 10. Time t has units of s. It is believed that the system time constant might have changed at an intermediate time which remains unknown.

t	$\log(y)$	t	$\log(y)$	t	$\log(y)$
0	1.4783	40	1.7562	80	1.9095
5	1.5243	45	1.7884	85	1.9161
10	1.5640	50	1.8146	90	1.9282
15	1.5976	55	1.8360	95	1.9321
20	1.6297	60	1.8546	100	1.9392
25	1.6590	65	1.8698	105	1.9475
30	1.6823	70	1.8866	110	1.9524
35	1.7231	75	1.8965	115	1.9572

Obtain the best estimates for the time constant before and after the time constant changed and also the best estimate for the time at which the time constant change took place. Use a spreadsheet program for solving this problem.

I.4 Design of experiments

Ex I.53: A certain quantity y depends on three factors A, B and C. Four experiments were conducted as given in the table below, using two levels for each factor.

x_A	x_B	x_C	y
+	+	−	0.90
+	−	+	0.35
−	+	+	0.24
−	−	−	0.05

What would be a possible way of ralating the outcome to the factors? Obtain the parameters in such a relation. Compare the three factors based on their relative effects on the outsome.

Ex I.54: Friction factor for flow in a pipe is known to follow a relation of the type $f = aRe^b$ where Re is the Reynolds number and a, b are constants. The relation is valid only when the flow is fully turbulent, with the Reynolds number exceeding 10000. You are to design a suitable experimental scheme so that the values of a, b can be determined. Make use of the concepts of DOE for this purpose. Identify suitable instruments also for conducting the experiments.

Ex I.55: A certain physical system is governed by four factors. Discuss how you would use DOE principles to set up a fractional factorial scheme. What would be the deficiencies in your proposed scheme. Discuss how non-linear effects may be taken in to account using DOE.

Module **II**

Measurements of Temperature, Heat Flux and Heat Transfer Coefficient

---

This module consists of three chapters (4, 5 and 6) that deal with the measurement respectively of temperature, systematic errors in temperature measurement and heat flux. Measurement of temperature is common to all branches of science and engineering. It has also matured over a long period of more than two centuries. Several instruments based on different principles are available to measure temperature under various conditions. Since thermometry is a vast subject several books are available that deal primarily with thermometry. Keeping the needs of the reader in mind Chapter 4 presents the basic principles and also deals with issues such as measurement of temperature transients. Thermometric errors and their estimation is the topic considered in detail in Chapter 5. This is the longest chapter in this book. Even then the coverage is by no means complete. The reader should refer to the current literature for additional information. Typical applications of interest to the mechanical engineer have guided the choice and the order of presentation of the topics that fall under temperature measurement. Heat flux and heat transfer coefficient are quantities that are of primary interest in all thermal engineering applications. Techniques that are commonly employed are dealt with in Chapter 6.

Chapter 4

Measurements of
Temperature

Thermometry is the art and science of temperature measurement. Practical thermometry requires an understanding the basic principles of thermometry which are presented at the beginning of this chapter. This is followed by the description in detail of various methods of temperature measurement. These are based on thermoelectricity, resistance thermometry, radiation pyrometry and several other techniques. Towards the end we present in detail issues arising when temperature transients are to be measured.

4.1   Introduction

Temperature is a primary quantity that is commonly involved in mechanical systems and processes and needs to be measured with fair degree of precision and accuracy coupled with good repeatability. It is also necessary to measure variations of temperature with time and hence transient response of instruments that measure temperature are discussed in great detail. Systematic errors that may occur in measurement of temperature are also examined in sufficient detail.

4.2   Thermometry or the science and art of temperature measurement

4.2.1   Preliminaries

Temperature along with pressure is an important parameter that governs many physical phenomena. Hence the measurement of temperature is a very important activity in the laboratory as well as in industry. The lowest temperature that is encountered is very close to $0\,K$ and the highest temperature that may be measured is about $100,000\,K$. This represents a very large range and cannot be covered by a single measuring instrument. Hence temperature sensors are based on many different principles and the study of these is the material of several sections to follow.

We take recourse to thermodynamics to provide a definition for temperature of a system. Thermodynamics is the study of systems in equilibrium and temperature is an important intensive property of such systems. Temperature is defined via the so called "zeroth law of thermodynamics". A system is said to be in equilibrium if its properties remain invariant. Consider a certain volume of an ideal gas at a specified pressure. When the state of this volume of gas is disturbed it will eventually equilibrate in a new state that is described by two *new* values of volume and pressure. Even though we may not be able to describe the system as it is undergoing a change we may certainly describe the two end states as equilibrium states. Imagine two such systems that may interact through a wall that allows changes to take place in each of them. The change will manifest as changes in pressure and/or volume. If, however, there are no observable changes in pressure and volume of each one of them when they are allowed to interact as mentioned above, the two systems are said to be in equilibrium with each other and are assigned the *same temperature*. The *numerical value* that is assigned will have to follow some rule or convention as we shall see later.

The zeroth law of thermodynamics states that if a system C is in equilibrium separately with two thermodynamic systems A and B then A and B are also in equilibrium with each other. At once we may conclude that systems A and B are at the same temperature! Thermometry thus consists of using a thermometer (system C) to determine whether or not two systems (A and B) are at the same temperature.

Principle of a thermometer

Principle of a thermometer may be explained by referring to Figure 4.1. Consider a thermodynamic system whose state is fixed by two properties - coordinates - X and Y.

It is observed that several pairs of values of X, Y will, in general, be in equilibrium with a *second* system of fixed temperature (or a *fixed state*). These multiplicity of sates *must* all be characterized by the same temperature and hence represent or define an isotherm.

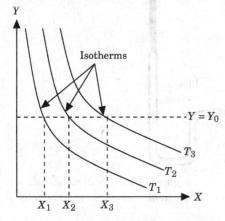

Figure 4.1: *Principle of thermometry explained*

Assume that one of the coordinates of the system (Y) is fixed at a value equal to Y_0. Then there is only one sate that will correspond to a given isotherm. If the system is allowed to equilibrate with a system characterized by different isotherms, the property X will change as indicated by the points of intersection $X_1, X_2, \cdots$ and so on. These will then correspond to the respective temperatures $T_1, T_2, \cdots$ and so on. We refer to X as the thermometric property and the system as a thermometer.

Table 4.1 shows several thermometers that are *actually* used in practice. The thermometric property as well as the symbol that is used to indicate it is also shown in the table.

Table 4.1: *Thermometers and thermometric properties*

Thermometer	Thermometric property	Symbol
Gas at constant volume	Pressure	P
Electric resistance under constant tension	Electrical resistance	R
Thermocouple	Thermal electromotive force	E
Saturated vapor of a pure substance	Pressure	P
Black body radiation	Spectral emissive power	$E_{b\lambda}$
Acoustic thermometer	Speed of sound	a

Constant volume gas thermometer

We look at the constant volume gas thermometer in some detail now. Schematic of such a thermometer is shown in Figure 4.2. It consists of a certain volume of a gas

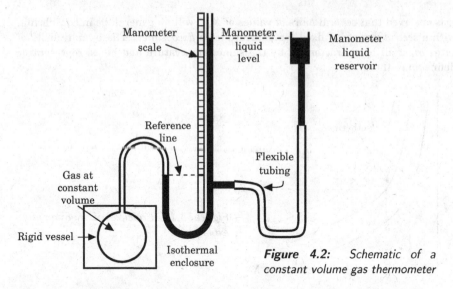

Figure 4.2: *Schematic of a constant volume gas thermometer*

contained in a rigid vessel. The vessel is connected to a "U" tube manometer with a side tube and height adjustable reservoir as shown. The volume is kept constant by making the meniscus in the left limb of the "U" tube always stay at the mark made on it. The right limb has a graduated scale attached to it as shown. The gas containing vessel is immersed in a constant temperature environment. The graduated scale helps in determining the pressure of the confined gas in terms of the manometer head (detailed discussion on pressure measurement using a manometer will be given in section 7.2).

The following experiment may be performed. Choose the pressure of the gas to have a definite value when the constant temperature environment corresponds to standard fixed state such as the triple point of water (or the ice point at one atmosphere pressure). Now move the thermometer into an environment at the steam point (boiling point of water at one atmosphere). The pressure of the gas has to be adjusted to a higher value than it was earlier by adjusting the height of the reservoir suitably so as to make the meniscus in the left limb of the "U" tube stay at the mark. The above experiment may be repeated by taking *less* and *less* gas to start with by making the pressure at the triple point of water to have a *smaller* and *smaller* value (the vessel volume is the same in all the cases). This may be done by evacuating the vessel by connecting it to a vacuum pump. The experiment may also be repeated with *different* gases in the vessel. The result of such an experiment gives a plot as shown in Figure 4.3. In the figure the ratio of the pressure inside the vessel at steam point to that at the triple point of water is plotted as a *function* of the initial gas pressure within the vessel. The ratio of the pressure of the gas p_{st} corresponding

to the steam point to that at the triple point of water p_{tp} tends to a unique number as $p_{tp} \to 0$ (the intercept on the pressure ratio axis) independent of the particular gas that has been used. This ratio has been determined very accurately and is given by $\dfrac{p_{st}}{p_{tp}} = 1.366049$. The gas thermometer temperature scale is defined based on this unique ratio and by *assigning* a numerical value of $T_{tp} = 273.16\,K$ or $t_{tp} = 0.01°C$.

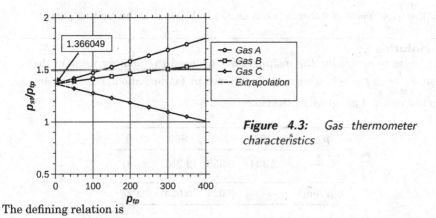

Figure 4.3: *Gas thermometer characteristics*

The defining relation is

$$\frac{T_{st}}{T_{tp}} = \frac{p_{st}}{p_{tp}} = 1.366049 \quad \text{or} \quad T_{st} = 1.366049 \times 273.16 = 373.149945 \approx 373.15\,K \quad (4.1)$$

The triple point of water that has been assigned the value of $273.16\,K$ is referred to as the *single fixed point* of thermometry. It is also referred to as the primary fixed point. At this temperature ice, liquid water and water vapor all coexist, if in addition, the pressure is maintained at $4.58\,mm$ Mercury column or $610.65\,Pa$. The ice point is at $273.15\,K$ or $0°C$, 1 atmosphere pressure and was used earlier as the primary fixed point in thermometry. In routine laboratory practice the ice point is easier to achieve and hence is commonly used.

Equation 4.1 may be generalized to define the constant volume gas thermometer temperature scale as

$$\boxed{\frac{T}{T_{tp}} = \lim_{p_{tp} \to 0} \frac{p}{p_{tp}} \quad \text{or} \quad T = 273.16 \lim_{p_{tp} \to 0} \frac{p}{p_{tp}}} \quad (4.2)$$

Thus the temperature ratio and pressure ratio are the same in the case of a constant volume gas thermometer. The latter is measured while the former is inferred. The message thus is clear! A measurable property that varies systematically with temperature is used to infer the temperature! The measured property is termed the thermometric property. The reason we take the pressure ratio as the pressure at the triple point goes to zero is because all gases behave ideally as the pressure of the gas tends to zero. The reader will recognize that the equality of temperature and pressure ratios for an ideal gas is Boyle's law, familiar to us from thermodynamics. The temperature scale defined through Equation 4.2 is referred to as the ideal gas scale of temperature.

Example 4.1

In determining the melting point of a certain alloy with a gas thermometer, an investigator finds the following values of the pressure p in mm mercury when the pressure p_{tp} at the triple point of water has the indicated value.

p_{tp}	100	200	300	400
p	233.4	471.6	714.7	962.9

If the triple point of water is taken as $273.16\,K$, what is the melting point of the alloy?

Solution :

In order to determine the melting point we need the limiting value of the ratio $\dfrac{p}{p_{tp}}$ as $p_{tp} \to 0$. This value is obtained by extrapolation. The ratios are calculated and are given by the following table.

p_{tp}	100	200	300	400
$\dfrac{p}{p_{tp}}$	2.334	2.358	2.382	2.407
Difference	$\cdots$	0.024	0.024	0.025

Since the common differences are constant we may extrapolate linearly to get

$$\lim_{p_{tp} \to 0} \frac{p}{p_{tp}} = 2.334 - 0.024 = 2.31$$

The melting point of the alloy on the ideal gas scale is then given by

$$T_{mp} = T_{tp} \lim_{p_{tp} \to 0} \frac{p}{p_{tp}} = 273.16 \times 2.31 = 630.99\,K$$

4.2.2   Practical thermometry

We have mentioned earlier that the range of temperatures encountered in practice is very wide. It has not been possible to device a single thermometer capable of measuring temperatures covering the entire range. Since all thermometers must be pegged with respect to the single fixed point viz. the temperature at the triple point of water, it is necessary to assign temperature values to as many reproducible states as possible, using the constant volume gas thermometer. Subsequently these may be used to calibrate other thermometers that may be used to cover the range of temperatures encountered in practical thermometry. Thermometers having overlap regions, where more than one is capable of being used, helps in covering the complete range of the temperature scale. These ideas are central to the introduction of International Temperature Scale 1990 (or ITS90, for short). The following is a brief description of ITS90.

Specification of ranges and corresponding thermometers according to ITS 90

- Between $0.65\,K$ and $5.0\,K$ T_{90} is defined in terms of the vapor-pressure temperature relations ^3He and ^4He.
- Between $3.0\,K$ and the triple point of neon - $24.5561\,K$ - T_{90} is defined by means of a helium gas thermometer calibrated at three experimentally realizable temperatures having assigned numerical values (defining secondary fixed points) and using specified interpolation procedures.

- Between the triple point of equilibrium hydrogen - 13.8033 K - and the freezing point of silver - 1234.93 K - T_{90} is defined by means of platinum resistance thermometer calibrated at specified sets of defining fixed points and using specified interpolation procedures
- Above the freezing point of silver - 1234.93 K - T_{90} is defined in terms of a defining fixed point and the Planck radiation law.

In the above T_{90} signifies temperature in K according to ITS90 and t_{90} signifies temperature in $°C$. It is noted that ITS 90 uses several "secondary fixed points" to define the temperature scale. These are shown in Table 4.2. Note that T is used to

Table 4.2: *Secondary fixed points specified in ITS90*

Equilibrium State	$T_{90}K$	$t_{90}°C$	Equilibrium State	$T_{90}K$	$t_{90}°C$
Triple point of H_2	13.8033	-259.3467	Triple point of Hg	234.3156	-38.8344
Boiling point of H_2 at 250 $mmHg$	17	-256.15	Triple point of H_2O	273.16	0.01
Boiling point of H_2 at 1 atm.	20.3	-252.85	Melting point of Ga	302.9146	29.7646
Triple point of Ne	24.5561	-248.5939	Freezing point of In	429.7483	156.5985
Triple point of O_2	54.3584	-218.7916	Freezing point of Sn	505.078	231.928
Triple Point of Ar	83.8058	-189.3442	Freezing point of Al	933.473	660.323
			Freezing point of Ag	1234.93	961.78

specify temperature in K and t in $°C$. Even though the ITS90 specifies only a few thermometers, in practice many other thermometers are also made use of. Many of these will also be discussed in this book.

Summary

Properties that vary systematically with temperature may be used as the basis of a thermometer. Several are listed here.

- Thermoelectric thermometer
  √ Based on thermo-electricity
  - Thermocouple thermometers using two wires of different materials

- Electric resistance
  √ Resistance thermometer using metallic materials like Platinum, Copper and Nickel etc.
  √ Thermistors consisting of semiconductor materials like Manganese-Nickel-cobalt oxide mixed with proper binders

- Thermal expansion
  √ Bimetallic thermometers
  √ Liquid in glass thermometer using mercury or other liquids
  √ Pressure thermometer

- Pyrometry and spectroscopic methods
  √ Radiation thermometry using a pyrometer
  √ Special methods like spectroscopic methods, laser based methods, interferometry etc.

4.3   Thermoelectric thermometry

Thermoelectric thermometry is based on thermoelectric effects or thermo-electricity discovered in the 19th century. They are:

- Seebeck effect discovered by Thomas Johann Seebeck in 1821[1]
- Peltier effect discovered by Jean Charles Peltier in 1824 [2]
- Thomson effect discovered by William Thomson (later Lord Kelvin) in 1847 [3]

The effects referred to above were all observed experimentally by the respective scientists. All these effects are *reversible* unlike heat diffusion (conduction of heat in the presence of temperature gradient) and Joule heating (due to passage of an electric current through a material of finite electrical resistance) which are *irreversible*. In discussing the three effects we shall ignore the above mentioned irreversible processes. It is now recognized that Seebeck - Peltier - Thomson effects are related to each other through the Kelvin relations (discussed later on).

4.3.1   Thermoelectric effects

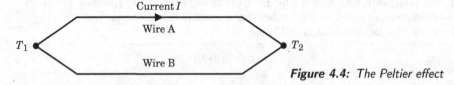

Figure 4.4: *The Peltier effect*

Consider two wires of dissimilar materials connected to form a circuit with two junctions as shown in Figure 4.4. Let the two junctions be maintained at different

---

[1] Thomas Johann Seebeck 1770 - 1831, German physicist
[2] Jean Charles Athanase Peltier 1785 - 1845, French physicist
[3] Kelvin, Lord William Thomson 1824 - 1907, Scottish physicist and engineer

temperatures as shown, by the application of heat at the two junctions. An electric current will flow in the circuit as indicated with heat absorption at one of the junctions and heat rejection at the other. This is referred to as the Peltier effect. The power absorbed or released at the junctions is given by $P = \dot{Q}_P = \pm \pi_{AB} I$ where π_{AB} is the Peltier voltage (this expression defines the Peltier emf), $\dot{Q}_P$ is rate at which the heat absorbed or rejected (the dot over Q indicates rate of change with respect to time). The direction of the current will decide whether heat is absorbed or rejected at the junction. For example, if the electrons move from a region of lower energy to a region of higher energy as they cross the junction, heat will be absorbed at the junction. This again depends on the nature of the two materials that form the junction. The subscript AB draws attention to this fact! The above relation may be written for the two junctions together as

$$\dot{Q}_P(net) = [\pi_{AB}|_{T_1} - \pi_{AB}|_{T_2}]I \qquad (4.3)$$

Note that the negative sign for the second term on the right hand side is a consequence of the fact that the electrons move from material A to material B (electrons flow in a direction opposite to the direction of current) at junction 1 and from material B to material A at junction 2.

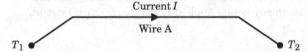

Current I

Wire A

T_1 T_2

Figure 4.5: The Thomson effect

Consider now a single conductor of homogeneous material (wire A alone of Figure 4.4) in which a temperature gradient exists (Figure 4.5). The current I is maintained by heat absorption or heat rejection along the length of the wire. Note that if the direction of the current is as shown the electrons move in the opposite direction. If $T_2 > T_1$, the electrons move from a region of higher temperature to that at a lower temperature. In this case heat will be rejected from the wire. The expression for heat rejected is $\dot{Q}_T = I \int_{T_1}^{T_2} \sigma_A dT$ where the Thomson heat is $\dot{Q}_T$ and σ_A is the Thomson coefficient (note that it has unit of emf/temperature) for the material. A similar expression may be written for the Thomson heat in conductor B. Thus, we have for the circuit shown in Figure 4.4

$$\dot{Q}_T(net) = I \int_{T_1}^{T_2} (\sigma_A - \sigma_B)dT \qquad (4.4)$$

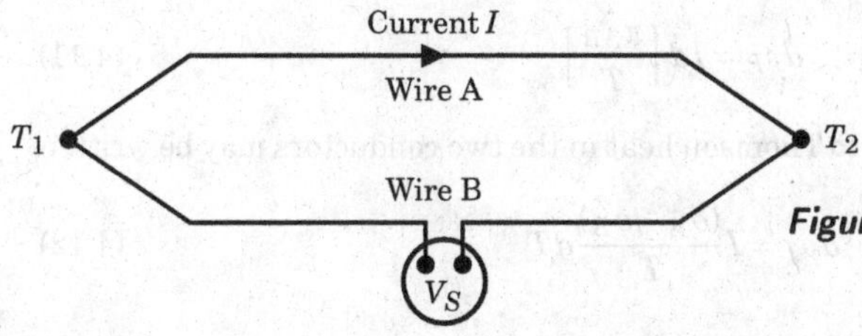

Current I

Wire A

T_1 T_2

Wire B

V_S

Figure 4.6: The Seebeck effect

If we cut conductor B (or A) as indicated in Figure 4.6 the Seebeck emf (V_S) appears across the cut. This emf is due to the combination of the Peltier (V_P) and Thomson (V_T) voltages. We may write the emf appearing across the cut as

$$V_S = V_P + V_T = \left[\pi_{AB}|_{T_1} - \pi_{AB}|_{T_2} \right] + \int_{T_1}^{T_2} (\sigma_A - \sigma_B) dT \tag{4.5}$$

We define the Seebeck coefficient (unit: V/K or $V/°C$) through the relation

$$\alpha_{AB} = \frac{dV_S}{dT} \tag{4.6}$$

If we take the temperature difference between T_2 and T_1 to tend to zero, Equation 4.5 may be rewritten in the differential form, using the definition of Seebeck coefficient given above as

$$\frac{dV_S}{dT} = \alpha_{AB} = \frac{d\pi_{AB}}{dT} + (\sigma_A - \sigma_B) \tag{4.7}$$

or

$$dV_S = \alpha_{AB} dT = d\pi_{AB} + (\sigma_A - \sigma_B) dT \tag{4.8}$$

Note that Equation 4.7 or 4.8 is a representation of conservation of energy and hence is nothing but the first law of thermodynamics applied to the thermocouple circuit.

Kelvin relations

Since the thermoelectric effects (Peltier and Thomson effects) are reversible in nature there is no net entropy change in the arrangement shown in Figure 4.4 (this is a consequence of the second law of thermodynamics). The entropy changes are due to heat interactions between the circuit and the surroundings due to Peltier and Thomson heats (strictly speaking, powers). The entropy change due to Peltier effect may be obtained as follows. We again assume that the temperature difference between the two junctions is a small differential quantity.

At junction 1 the entropy change is due to Peltier heat interaction and hence is given by

$$s_{P1} = \frac{\dot{Q}_{P1}}{T_1} = I \frac{\pi_{AB}(T_1)}{T_1} \tag{4.9}$$

Similarly at junction 2 the entropy change is

$$s_{P2} = \frac{\dot{Q}_{P2}}{T_2} = I \frac{\pi_{AB}(T_2)}{T_2} \tag{4.10}$$

Thus the net change in entropy due to Peltier heat interactions is

$$ds_P = Id\left[\frac{\pi_{AB}}{T} \right] \tag{4.11}$$

The net change in entropy due to Thomson heat in the two conductors may be written as

$$ds_T = I \frac{(\sigma_A - \sigma_B)}{T} dT \tag{4.12}$$

Combining these two the net entropy change is

$$ds_{net} = I\left\{d\left[\frac{\pi_{AB}}{T}\right] + \frac{(\sigma_A - \sigma_B)}{T}dT\right\} = 0 \tag{4.13}$$

The current I can have arbitrary value and hence the bracketed term in Equation 4.13 must be zero.

$$d\left[\frac{\pi_{AB}}{T}\right] + \frac{(\sigma_A - \sigma_B)}{T}dT = \frac{d\pi_{AB}}{T} - \pi_{AB}\frac{dT}{T^2} + \frac{(\sigma_A - \sigma_B)}{T}dT = 0 \tag{4.14}$$

Equation 4.8 may be rearranged as

$$(\sigma_A - \sigma_B)dT = dV_s - d\pi_{AB} \tag{4.15}$$

Introducing this in Equation 4.14 we get

$$\frac{d\pi_{AB}}{T} - \pi_{AB}\frac{dT}{T^2} + \frac{dV_s - d\pi_{AB}}{T} = 0 \tag{4.16}$$

or

$$\boxed{\pi_{AB} = T\frac{dV_s}{dT} = T\alpha_{AB} = T\alpha_S} \tag{4.17}$$

Here the Seebeck coefficient has been re designated as V_S, to be in tune with common practice. Equation 4.17 is known as the **First Kelvin relation**. Differentiate Equation 4.17 to get

$$d\pi_{AB} = \alpha_S dT + T d\alpha_S \tag{4.18}$$

Combine this with Equation 4.8 to arrive at the relation

$$\boxed{\sigma_A - \sigma_B = -T\frac{d\alpha_S}{dT} = -T\frac{d^2V_S}{dT^2}} \tag{4.19}$$

Equation 4.19 is known as the **Second Kelvin relation**.

Interpretation of the Kelvin relations

The Seebeck, Peltier and Thomson coefficients are normally obtained by experiments. For this purpose we use the arrangement shown in Figure 4.6 with the junction labeled 2 maintained at a suitable reference temperature, normally the ice point ($0°C$). The reference junction is also referred to as the cold junction. This is strictly correct if the measuring junction is above the reference temperature. In refrigeration applications the measuring junction may be at a temperature below the temperature of the reference junction. The junction labeled 1 will then be called the measuring junction. Data is gathered by maintaining the measuring junction at different temperatures and noting down the Seebeck voltage. If the measuring junction is also at the ice point the Seebeck voltage is identically equal to zero. The data is usually represented by a polynomial of suitable degree. For example, with Chromel (material A) and Alumel (material B) as the two wire materials, the expression is a fourth degree polynomial of form $V_S = a_1 t + a_2 t^2 + a_3 t^3 + a_4 t^4$ where the Seebeck voltage is in μV and the temperature is in $°C$. An inverse relation is also used in practice in the form $t = A_1 V_S + A_2 V_S^2 + A_3 V_S^3 + A_4 V_S^4$.

We shall see based on Example 4.2 that the coefficients in the polynomial are related to the three thermoelectric effects, on application of the Kelvin relations.

Example 4.2

The thermocouple response shown below (Copper - material A, Constantan - material B thermocouple with the reference (or cold) junction at the ice point) follows approximately the relation $V_S = at + bt^2$. Obtain the parameters a and b by least squares. Here t is in $°C$ and V_S is in mV.

t	37.8	93.3	148.9	204.4	260
V_S	1.510	3.067	6.647	9.523	12.572

Solution :

Since the fit follows a quadratic relation with zero intercept as specified above, it is equivalent to a linear relation between $E = V_S/t$ and t. Since V_S/t is a small number we shall work with $100 V_S/t$ and denote it as y. We shall denote the temperature as t. The following table helps in evolving the desired linear fit.

	t	y	t^2	y^2	ty
	37.8	4.015873	1428.84	16.12724	151.8
	93.3	4.251876	8704.89	18.07845	396.7
	148.9	4.46407	22171.21	19.92792	664.7
	204.4	4.659002	41779.36	21.7063	952.3
	260	4.835385	67600	23.38094	1257.2
Sum	744.4	22.22621	141684.3	99.22085	3422.7
Mean	148.88	4.445241	28336.86	19.84417	684.54

The statistical parameters are calculated and presented in the form of a table.

Variance of t	6171.606
Variance of y	0.084001
Covariance ty	22.73252
Slope of fit line	0.003683
Intercept of fit line	3.896856

Reverting back to t and V_S, we get the following relation:

$$V_S = 0.03897t + 3.683 \times 10^{-5} t^2$$

We compare the data with the fit in the table below:

t	V_S	$V_S(fit)$
37.8	1.518	1.526
93.3	3.967	3.956
148.9	6.647	6.619
204.4	9.523	9.504
260	12.572	12.622

Standard error of the fit with respect to data may be calculated as

$$\sigma_e = \sqrt{\frac{(V_S - V_S(fit))^2}{3}} = 0.035 mV$$

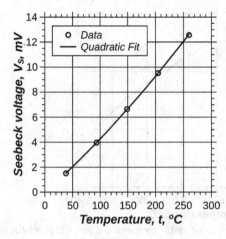

Figure 4.7: Comparison between data and the fit for a Copper Constantan thermocouple pair

The 3 in the denominator is the degrees of freedom. A plot is made to compare the data and the fit (Figure 4.7). It is clear that the thermocouple behavior is mildly nonlinear. The standard error of fit translates to approximately a temperature error of $1°C$!

Example 4.2 has shown that the thermoelectric data may be expressed in terms of a global polynomial to facilitate interpolation of data. A simple quadratic fit has been used to bring home this idea. In practice the appropriate interpolating polynomial may involve higher powers such that the standard error is much smaller than what was obtained in Example 4.2. As an example, the interpolating polynomial recommended for the K type thermocouple (Material A is Chromel and material B is Alumel) with the reference junction at the ice point is given as[4]:

$$V_S = 39.44386t + 5.895322 \times 10^{-5}t^2 - 4.2015132 \times 10^{-6}t^3 + 1.3917059 \times 10^{-10}t^4 \quad (4.20)$$

---

[4]ITS90 recommended relationship consists of a ninth degree polynomial plus an exponential term containing two more coefficients. Coefficients are tabulated in srdata.nist.gov/its90/type_j/jcoefficients_inverse.html

This is a fourth degree polynomial and passes through the origin. The Seebeck voltage is given in μV and the temperature is in $°C$. The appropriate parameters at the ice point are obtained as (note: $t = 0°C$ and $T = 273.15\,K$):

$$\alpha_S = \frac{dV_S}{dT} = \frac{dV_S}{dt} = 39.44\mu V/°C \tag{4.21}$$

$$\pi_{Chromel-Alumel} = T\alpha_S = 273.15 \times 39.444 = 10744.1\,\mu V \tag{4.22}$$

$$\sigma_{Chromel} - \sigma_{Alumel} = -T\frac{d\alpha_S}{dt} = -273.15 \times 2 \times 0.00005895 = -0.0322064\,\mu V/°C \tag{4.23}$$

Equation 4.21 follows from the definition while Equations 4.22 and 4.23 follow from Kelvin relations.

The variation of the Seebeck coefficient over the range of this thermocouple is given in Figure 4.8.

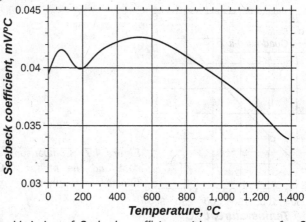

Figure 4.8: *Variation of Seebeck coefficient with temperature for Chromel-Alumel thermocouple pair*

A short excerpt from Table of Seebeck voltages is taken; the corresponding fit values are calculated using the above fourth degree polynomial and presented in Table 4.3. Note that the voltages in this Table are in mV. The maximum deviation is some $0.102\,mV$. The standard error is approximately $0.053\,mV$! This translates to roughly an error of $1.2°C$.

Table 4.3: *Short table of Seebeck voltages for Chromel Alumel thermocouple*

t	V_S	$V_S(fit)$	t	V_S	$V_S(fit)$
37.8	1.52	1.499	371.1	15.178	15.237
93.3	3.819	3.728	537.8	22.251	22.276
148.9	6.092	5.99	815.6	33.913	33.874
204.4	8.314	8.273	1093.3	44.856	44.879
260	10.56	10.581	1371.1	54.845	54.827

The above shows that the three effects are related to the various coefficients in the polynomial. The Seebeck coefficient and the Peltier coefficients are related to the first derivative of the polynomial with respect to temperature. The contributions to the first derivative from the higher degree terms are not too large and hence the Seebeck coefficient is a very mild function of temperature. In the case of K type thermocouple this variation is less than some 2% over the entire range of temperatures. The Thomson effect is related to the second derivative of the polynomial with respect to temperature.

4.3.2   On the use of thermocouple for temperature measurement

We shall be looking at basic theoretical aspects and practical aspects of measurement of temperatures using thermocouples. General ideas are explored first followed by important practical aspects. Simple or basic thermocouple circuit used for temperature measurement is shown in Figure 4.9.

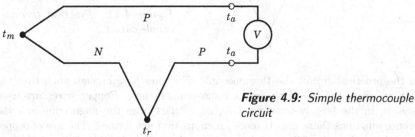

Figure 4.9: Simple thermocouple circuit

The basic thermocouple circuit consists of a wire of P type material (P stands for positive) and a wire of N type material (N stands for negative) forming a measuring junction and a reference junction as shown (more about P and N materials will be given later). The voltmeter is connected by making a break in the P type wire as shown. Thus two more junctions are formed between P type wires and binding posts of the voltmeter. As we shall see later the voltmeter will indicate the correct Seebeck voltage corresponding to t_m if the two extra junctions formed with the voltmeter are at the same temperature. If $t_m > t_r$, the voltmeter will indicate a positive voltage, the way it is connected.

We now look at the temperature variations along the wires that make up the simple circuit shown in Figure 4.9. It is likely that the thermocouple wires pass through a region of uniform temperature (eg. the laboratory space) a short distance away from the junctions that are maintained at the indicated temperatures. The state of affairs is schematically shown in Figure 4.10. It is clear that the temperature varies rapidly within a short distance from the measuring and reference junctions (the variation is determined by the thermal properties of the wire and the nature of the ambient) and come to the ambient temperature. Recall that the thermoelectric effects (Peltier and Thomson) are confined to either the junctions or to the region along the wires that have a temperature variation along them. Those regions along the wire that are isothermal *do not* give rise to any thermoelectric effects.

Hence it is possible to change the simple circuit shown in Figure 4.9 to a practical circuit shown in Figure 4.11.

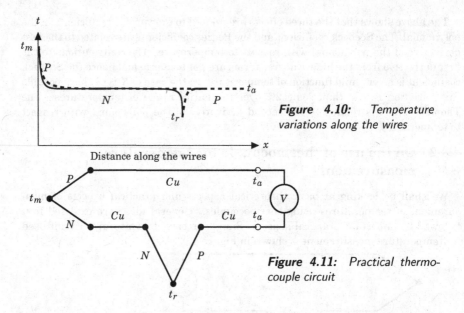

Figure 4.10: *Temperature variations along the wires*

Distance along the wires

Figure 4.11: *Practical thermocouple circuit*

In the practical circuit the thermocouple wires are long enough such that the ends away from the junctions are at room temperature. Copper wires are used as shown, in the largely isothermal region. Apart from the measuring and the reference junctions there are six more junctions that are formed! The use of copper wires reduces the cost of the installation since thermocouple wires are usually very expensive. To validate the use of copper lead wires we present below the three important laws of thermoelectric circuits.

Laws of thermoelectric circuits

1. Law of homogeneous materials

A thermoelectric current cannot be sustained in a circuit of a single homogeneous material however it varies in cross section by the application of heat alone.

2. Law of intermediate materials

The algebraic sum of the thermoelectric forces in a circuit composed of any number of dissimilar materials is zero if all the junctions are at the same temperature

The law is explained with reference to Figure 4.12. The Seebeck emf E developed is independent of the fact that a third material C forms two junctions with the + and - materials as shown in the first part of Figure 4.12. Since the material C is isothermal the situation is equivalent to a single measuring junction between the + and - materials as indicated in the latter part of the figure.

3. Law of successive or intermediate temperatures

This law is explained with reference to Figure 4.13. The Seebeck voltage is E_1 with the measuring junction at t_1 and the reference junction at t_2. The Seebeck

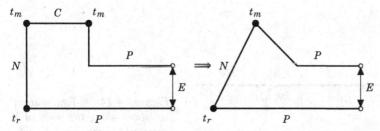

Figure 4.12: *Illustration of the law of intermediate materials*

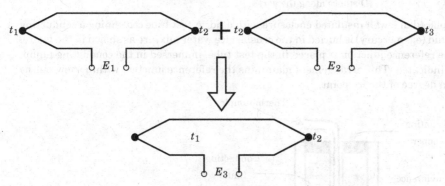

Figure 4.13: *Illustration of the law of intermediate temperatures. Note that $E_3 = E_1 + E_2$.*

voltage is E_2 with the measuring junction at t_2 and the reference junction at t_3. Then the Seebeck voltage is $E_3 = E_1 + E_2$ with the measuring junction at t_1 and the reference junction at t_3. Utility of this very important law will be brought out later.

Now we get back to the practical thermocouple circuit shown in Figure 4.11. The six extra junctions that are formed are all at a uniform temperature equal to the ambient temperature. The law 2 of thermo-electricity enunciated above guarantees that these do not have any effect on the Seebeck voltage developed by the thermocouple circuit. The copper wires are at room temperature (uniform temperature) and hence the law of intermediate materials asserts that the circuit is equivalent to one in which the copper wires are absent! In fact the temperature variations along the wires are as indicted in Figure 4.14. The reason for the use of copper lead wires is to cut down on the cost of expensive thermocouple wires. Sometimes compensating lead wires are made use of. These are made of the same material as the thermocouple wires but not of the same quality. They may also be made of cheaper alloys that have thermoelectric properties *closely* following the thermoelectric properties of the thermocouple wires themselves.

Ice point reference

The measurement of temperature by the use of a thermocouple requires a reference junction maintained at the ice point. This is achieved in the laboratory by the use of an arrangement shown schematically in Figure 4.15. Crushed ice water (as long as there is both ice and water the temperature remains fixed at the ice point)

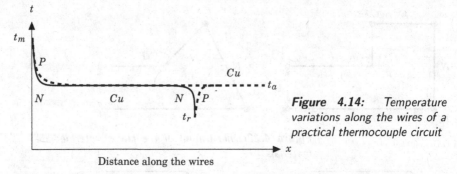

Figure 4.14: *Temperature variations along the wires of a practical thermocouple circuit*

is placed in a well insulated enclosure with a lid. A test tube containing a conducting liquid (eg. mercury) is buried in the crushed ice water mixture as shown in the figure. The reference junction is placed in the test tube, immersed in the conducting liquid, as indicated. This arrangement maintains the reference junction within a few tenths of a degree of the ice point.

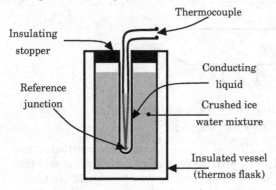

Figure 4.15: *Ice point reference junction*

It is seldom that individual ice point references are maintained while using several thermocouples for measuring temperatures at different points in an apparatus. The laws of thermo-electricity come to our help in designing a suitable arrangement with selector switch and a single ice point reference junction as shown in Figure 4.16.

The switch is generally a rotary switch with gold plated contacts that is maintained at a uniform temperature (for example the room temperature). Switches are available with a capacity of 8, 16 or 32 thermocouple connections. The switch is a double pole single throw rotary type that will connect each thermocouple pair to complete the circuit with the voltmeter and the reference junction.

4.3.3   Use of thermocouple tables and Practical aspects of thermoelectric thermometry

Even though, in principle, one can use any two dissimilar materials as candidates for constructing thermocouples thermometry demands that there be standardization so that one may use these with very little effort. Also no single thermoelectric thermometer can cover the wide range of temperatures met with in practice. The materials chosen must be available easily from manufacturers with guaranteed

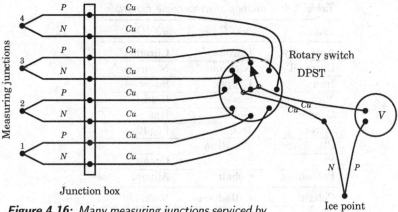

Figure 4.16: *Many measuring junctions serviced by a single reference junction (* DPST is a double pole single throw switch)

quality. In view of these only a few combinations of materials are made use of in day to day laboratory practice.

The common useful thermocouple materials are shown in Table 4.4. The entries in the table are such that all the materials that are below the one under consideration are negative with respect to it. This means that the Seebeck coefficient (or thermoelectric power) increases as the row count between two materials increases. The Seebeck voltages of materials are measured with respect to Platinum 67 (the platinum standard used by the National Institute of Standards and Technology - NIST - USA) as the standard second element. The columns correspond to the usable temperature ranges for the materials. Law of intermediate materials is invoked to combine the thermoelectric data of two materials that are individually measured with Platinum 67 as the reference material. Presumably Seebeck was experimenting with Antimony and Bismuth when he discovered thermo-electricity. It appears that he had hit upon materials with the largest thermoelectric power! If the candidate thermocouple material is positive with respect to Platinum 67 the Seebeck voltage will be positive across the candidate - Platinum 67 terminals. In general, if a candidate material is positive with respect to a second material the Seebeck voltage will be positive across the terminals.

Not all combinations of materials given in Table 4.4 are used in practice. A small number of them (Table 4.5) are used and are available from reputed manufacturers, in large quantities. The Seebeck coefficient in the last column indicates the sensitivity of the respective thermocouple combinations. Also the thermoelectric data are available in the form of tables for each of these thermocouple material combinations. As an example Table 4.6 shows an excerpt of the table appropriate for K type (Chromel is the positive element and Alumel is the negative element) thermocouple. The table assumes that the reference junction is maintained at the ice point. Note that the Seebeck voltage is specified with 3 digits after the decimal point. The last digit corresponds to $1\,\mu V$. Some interesting things may be noted by examining Table 4.6. The Seebeck coefficient for K type thermocouple is approximately $40\,\mu V/^\circ C$. If we use a voltmeter that can resolve $0.01\,mV$ or $10\,\mu V$ the

Table 4.4: Common thermocouple materials

$100°C$	$500°C$	$900°C$
Antimony	Chromel*	Chromel*
Chromel*	Nichrome*	Nichrome*
Iron	Copper	Silver
Nichrome*	Silver	Gold
Copper	Gold	Iron
Silver	Iron	$Pt_{90}Rh_{10}$
$Pt_{87}Rh_{13}$	$Pt_{90}Rh_{10}$	Pt
Pt	Pt	Cobalt
Palladium	Cobalt	Alumel
Cobalt	Palladium	Nickel
Alumel*	Alumel*	Palladium
Nickel	Nickel	Constantan
Constantan*	Constantan*	
Copel*	Copel*	
Bismuth		

* Materials are alloys sold under trade names

Table 4.5: Standard thermocouple types with color codes

Type	+ / - Wires	+/- Color	Seebeck coefficient $\mu V/°C^\star$
B	$Pt_{94}\%Rh_6\%/Pt$	Grey / Red	0.05
E	Chromel / Constantan	Purple / Red	60.9
J	Iron / Constantan	White / Red	51.7
K	Chromel / Alumel	Yellow / Red	40.5
R	$Pt87\%Rh13\%/Pt$	Black / Red	5.93
S	$Pt90\%Rh10\%/Pt$	Black / Red	6.02
T	Copper / Constantan	Blue / Red	40.7

*Representative values - vary with temperature

temperature resolution is about 0.25 $°C$! Even this is in not easily achieved in normal laboratory practice. Voltmeters capable of such high resolution are expensive and hence it is not possible to achieve sub degree resolution levels in ordinary laboratory practice. The accuracy limits that are possible to achieve are given in Table 4.7 for various thermocouple pairs. Note that the S type and Tungsten - Tungsten Rhenium thermocouples (special but expensive thermocouples - shown on the last two rows of Table 4.7) are useful for the measurement of very high temperatures normally inaccessible to other thermocouples. However the accuracy limits are not as good as for the other thermocouples.

The useful ranges of thermocouples are determined, (a) by the thermoelectric behavior and (b) by the physical properties of the wire materials, as the temperature is changed. At elevated temperatures the integrity of the thermocouple wire

Table 4.6: Typical thermocouple reference table

ITS90 Table for Type K type thermocouples									
°C	0	1	2	3	4	5	6	7	8	9
Seebeck voltage mV									
0	0.000	0.039	0.079	0.119	0.158	0.198	0.238	0.277	0.317	0.357
10	0.397	0.437	0.477	0.517	0.557	0.597	0.637	0.677	0.718	0.758
20	0.798	0.838	0.879	0.919	0.960	1.000	1.041	1.081	1.122	1.163
30	1.203	1.244	1.285	1.326	1.366	1.407	1.448	1.489	1.530	1.571
40	1.612	1.653	1.694	1.735	1.776	1.817	1.858	1.899	1.941	1.982
50	2.023	2.064	2.106	2.147	2.188	2.230	2.271	2.312	2.354	2.395
60	2.436	2.478	2.519	2.561	2.602	2.644	2.685	2.727	2.768	2.810
70	2.851	2.893	2.934	2.976	3.017	3.059	3.100	3.142	3.184	3.225
80	3.267	3.308	3.350	3.391	3.433	3.474	3.516	3.557	3.599	3.640
90	3.682	3.723	3.765	3.806	3.848	3.889	3.931	3.972	4.013	4.055
100	4.096	4.138	4.179	4.220	4.262	4.303	4.344	4.385	4.427	4.468
110	4.509	4.550	4.591	4.633	4.674	4.715	4.756	4.797	4.838	4.879
120	4.920	4.961	5.002	5.043	5.084	5.124	5.165	5.206	5.247	5.288
130	5.328	5.369	5.410	5.450	5.491	5.532	5.572	5.613	5.653	5.694
140	5.735	5.775	5.815	5.856	5.896	5.937	5.977	6.017	6.058	6.098

From: http://www.temperatures.com/tctables.html

Table 4.7: Accuracy and range values of some thermocouple sensors

Thermocouple	Full range °C	Accuracy °C or %	Range ISA* standard limits °C
Chromel Alumel,	-185 to 1371	±2°C	-18 to 277
K Type		±0.75%	277 to 1371
Iron - Constantan	-190 to 760	2±2°C	-18 to 277
J Type		±0.75%	277 to 760
Copper - Constantan,	-190 to 400	±2%	-190 to 60
T Type		0.8±0.8°C	-60 to 93
		±0.75%	93 to 370
$Pt_{90}Rh_{10}-Pt$	0 to 1760	±2.8°C	0 to 538
S Type		±0.5%	538 to 1482
$W-W_{74}Rh_{26}$	0 to 2870	±4.5°C	0 to 427
$W_{95}Rh_5-W_{74}Rh_{26}$		±1%	427 to 2870

*Instrument Society of America

materials as well as the junction is important. The materials of the wires are also prone to thermal fatigue when the junctions are subjected to thermal cycling during use.

In view of the fact that the materials close to the junctions experience these thermal cycles it is possible to discard a small length close to the junction and remake a junction for subsequent use. Materials also age during use and may have to be discarded if the thermoelectric properties change excessively during use.

To round off this discussion we present (Figure 4.17) the thermoelectric output of a K type thermocouple over the range of its usefulness. We notice that over an extended range the output *is* non-linear. Note that this pair of materials is far apart in Table 4.4. Hence the thermoelectric power is very large and next only to that for the J type (Iron - Constantan) thermocouple. The sensitivity also compares favorably with that of the T type (Copper - Constantan) thermocouple. Even though these three types have high sensitivities (a large value of Seebeck coefficient signifies high sensitivity) the ranges are different. K type has the widest range among these three types. Both K type and T type thermocouples are commonly used in laboratory and industrial applications, and have comparable sensitivities. Iron is prone to corrosion and hence the J type thermocouple is of limited utility.

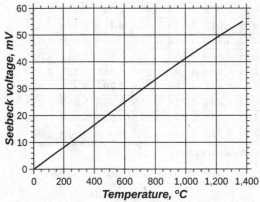

Figure 4.17: *Seebeck volts - temperature relationship for K type thermocouple*

Example 4.3

A K type thermocouple is used as shown in Figure 4.18(a) without a reference junction. The terminals of the voltmeter are at room temperature of $30°C$ while the measuring junction is at $100°C$. What is the voltmeter reading? What would have been the reading had it been connected as shown in Figure 4.18(b) with the reference junction at the ice point?

Solution :

Circuit as in (a):
The conditions are $t_m = 100°C$ and $t_{ref} = 30°C$
From the K type thermocouple table (Table 4.6) we read off the following:
Seebeck voltage with $t_m = 100°C$ and $t_{ref} = 0°C$ is $V_1 = 4.096\,mV$
Seebeck voltage with $t_m = 30°C$ and $t_{ref} = 0°C$ is $V_2 = 1.203\,mV$
Hence the voltmeter reading (using the law of successive temperatures) is

$$V = V_1 - V_2 = (4.096 - 1.203) = 2.893\,mV$$

Circuit as in (b):
The conditions are $t_m = 100°C$ and $t_{ref} = 0°C$
Hence the voltmeter reading is $V = V_1 = 4.096\,mV$

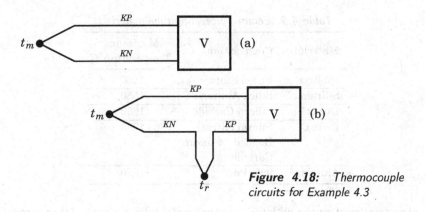

Figure 4.18: *Thermocouple circuits for Example 4.3*

Insulation systems

Thermocouples are made of wires of various diameters according to requirement. The P and N wires are expected to not contact each other (electrically) excepting at the junctions. Hence it is necessary to cover each wire with an electrical insulator. It is normal for the P and N wires to be individually covered with insulation, the two wires laid parallel to each other and covered with an outer sheath encasing both the wires. The insulation material is chosen with the temperature range in mind. A list of insulation materials along with the useful range of temperatures is given in Table 4.8. Sometimes the thermocouple is protected by a protective tube so as to protect

Table 4.8: *Thermocouple insulation systems*

Insulation	Temperature limits $°C$
Nylon	-40 to 160
PVC	-40 to 105
Enamel	Up to 107
Cotton over enamel	Up to 107
Silicone rubber over Fiberglass	-40 to 232
Teflon and Fiberglass	-120 to 250
Asbestos	-78 to 650
Tempered Fiberglass	Up to 650
Refrasil*	Up to 1083

*Trade name

the thermocouple junction from mechanical damage. The protective tube material is again chosen based on the temperature range, the nature of the process environment and the ruggedness desired (Table 4.9).

Some of the materials shown in Tables 4.8 and 4.9 are proprietary in nature and are identified by the trade name. Some times the protective tube material may be made of a metal or an alloy like stainless steel. Since metals are good conductors of electricity the insulation of the individual wires must be adequate to

Table 4.9: *Ceramic protecting tube materials*

Material	Composition	Maximum Temperature°C
Quatrtz	Fused Silica	1260
Siliramic*	Silica-Alumina	1650
Durax*	Silicon Carbide	1650
Refrax*	Silicon Nitride Bonded Silicon Carbide	1735
Alumina	99% Pure	1870

*Trade name

avoid any electrical contact with the protective metal tube. A sheathed thermocouple is typically like that shown in Figure 4.19.

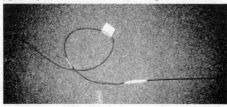

Figure 4.19: *Photograph of a K type thermocouple with stainless steel sheath*

Thermocouple junctions

Thermocouple junctions may be formed by various means. The most common method is to weld or fuse the two materials to form a junction. Welding is commonly made by twisting the two wires over a short length, passing a high current through the junction by discharging a capacitor charged to a high voltage. The momentary high current will heat the junction (due to contact resistance and the consequent arcing between the two wires) to a high temperature at which the two metals fuse together to form a nice bead. The entire process may be accomplished in an inert atmosphere to avoid any oxidation of the wires. The junctions may be of various types as shown schematically in Figures. 4.20.

> **Grounded junction:** Such a junction is used if one wants to avoid direct contact between the thermocouple bead and the process fluid. The protective tube provides a barrier between the junction and the process fluid. This type of arrangement increases the response time of the thermocouple when measuring temperature transients.
>
> **Exposed junction:** This arrangement is acceptable if the process space does not adversely affect the thermocouple materials. This arrangement decreases the response time of the thermocouple when measuring temperature transients.
>
> **Separated wire junction:** The two wires are allowed to float within the process space as shown. The process fluid (molten metal in metallurgical applications) provides the electrical connection between the P and N wires.

> **Button junction:** The P and N wires are attached to a copper (high conductivity material) button such that the thermal contact between the process fluids (may be gas like air) and the junction is enhanced. This is beneficial from the response time point of view as well as from the thermometric error point of view (as will be discussed later).

Junctions may be made by twisting the P and N wires together. Twisting action itself cold works the two materials to form good contact. In addition the two wires may also be welded as shown in Figure 4.21. The junction may also be formed by butt welding the two wires as shown in Figure 4.21.

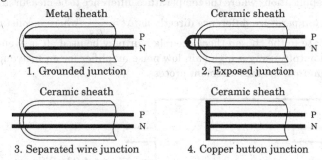

Figure 4.20: *Types of thermocouple junctions*

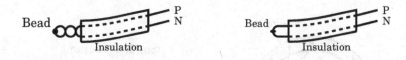

Figure 4.21: *More junction types*

Sometimes a thermocouple probe is constructed in the form of a bayonet or pencil as indicated in Figure 4.22. The figure shows a J type thermocouple probe with the protective tube of Iron (P type material) and a wire of Constantan (N type material) attached to the bottom of the tube with fiberglass insulation between the two. The connections to the external circuit are made through terminal blocks.

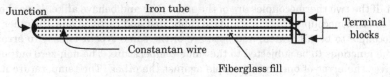

Figure 4.22: *Bayonet or pencil type of thermocouple probe*

Differential thermocouple

Thermocouples may be used in other ways than the ones described earlier. The temperature difference between two locations in a process may be obtained by measuring the two temperatures individually using two separate thermocouples and then taking the difference. The errors involved in each such measurement will propagate and give an overall error that may be unacceptable in practice. An alternative is to use the differential thermocouple shown in Figure 4.23 (this is just the basic thermocouple circuit with the hot and cold junctions at different temperatures!) and estimate the temperature difference directly once ΔV_S is measured. In applications where the temperature difference to be measured is small, we obtain the temperature difference directly as $\Delta t = \dfrac{\Delta V_S}{\alpha_S}$ where a constant value of α_S appropriate to the chosen thermocouple pair may be used. It is even possible to amplify the output using a high gain low noise amplifier (or an instrumentation amplifier) to improve the measurement process.

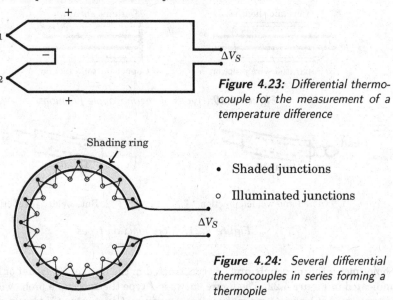

Figure 4.23: *Differential thermocouple for the measurement of a temperature difference*

Figure 4.24: *Several differential thermocouples in series forming a thermopile*

The differential thermocouple arrangement may also be used as a calibration arrangement. If $t_1 = t_2$, the differential thermocouple should give zero voltage output, if the two thermocouples are of the same type and behave alike. If we choose one of the thermocouples to be a standard calibrated and certified one, the other the thermocouple to be calibrated, and both are of the same type it is easy to arrange both the junctions to be subjected to the same temperature. The non-zero output, if any, gives the error of one thermocouple against the other. The temperature itself may be measured independently using a standard thermocouple so that we can have a calibration chart giving the thermocouple error as a function of temperature.

The amplification may also be accomplished by using several junctions in series as shown in Figure 4.24. This arrangement referred to as a thermopile, produces an output that is n times the value with a single differential thermocouple, where n is the total number of hot or cold junctions. In the example shown the thermopile is being used as a radiation flux sensor. The shading ring prevents illumination of the cold junctions and thus produces a number of differential thermocouples in series.

Thermocouples in parallel

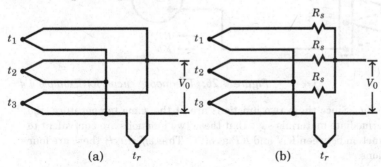

Figure 4.25: *Thermocouples connected in: (a) parallel (b) parallel with "swamp" resistors R_s*

In practice we may require the average temperature in a region where the temperature may vary. One way of doing this would be to measure all the temperatures individually with separate thermocouples and then take the mean. However, it is possible to measure the mean value directly using several thermocouples connected in parallel as shown in Figure 4.25(a). The thermoelectric output corresponding to the junctions are as indicated in the figure and these correspond to the respective measuring junction temperatures. Assuming that all the thermocouples are identical, the output voltage is given by

$$V_0 = \frac{V_1 + V_2 + V_3}{3} \tag{4.24}$$

Thus the inferred temperature is the mean of the measuring junction temperatures. The accurate averaging of the Seebeck voltages relies on thermocouple wire resistances being equal. If this is difficult to achieve one may use equal but large (large compared to the resistance of the thermocouple wires) "swamping" resistance in each circuit to alleviate the problem. This is indicated in Figure 4.25(b).

Example 4.4

Consider the thermopile arrangement shown in Figure 4.26. What will be the output voltage?

Solution :

In this example KP means positive wire of K type thermocouple and KN means negative wire of K type thermocouple. Cu refers to the copper lead wires.

Three materials are used in the circuit. KP and KN at the measuring temperature of 100°C form four junctions. There are three cold junctions between KN and KP at 30°C. There is one junction each between $KP - Cu$

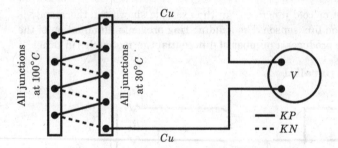

Figure 4.26: Thermopile circuit for Example 4.4

and $KN - Cu$. Since these two junctions are at the same temperature, the law of intermediate materials says that these two junctions are equivalent to a single junction between KN and KP at 30°C. Thus *effectively* there are four cold junctions.

Thus the voltage indicated will be four times that due to a measuring junction at 100°C and a reference junction at 30°C. By the law of intermediate temperatures, we thus have:

$$
\begin{aligned}
V &= 4 \times [\text{Seebeck voltage between } 100°C \text{ and } 30°C] \\
&= 4 \times [V_1 - V_2] \text{ as in Example 4.3} \\
&= 4 \times [4.095 - 1.203] = 11.568 \ mV
\end{aligned}
$$

Example 4.5

In an installation the extension wires of K type thermocouple were interchanged by oversight. The actual temperatures were: Measuring junction $t_m = 500°C$; Reference junction $t_{ref} = 0°C$; Junction between thermocouple wires and extension wires $t_j = 40°C$. What is the consequence of this error?

Solution :

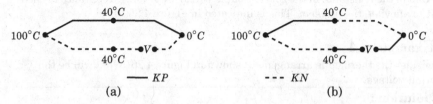

Figure 4.27: Thermocouple circuits for Example 4.5

Figure 4.27(a)-(b) respectively indicate the incorrect and the correct arrangements.

(a) Correct arrangement:
We indicate the Seebeck voltage developed when the measuring and reference

junctions are at $t°C$ and $0°C$ as $V_S(t)$. In this case the junctions between the thermocouple and lead wires do not play any role! The voltmeter reading will be (using K type thermocouple tables)

$$V = V_S(500) = 20.640mV$$

(b) Incorrect arrangement:
In this case there are effectively four junctions. The net voltage indicated is given by (we go round the circuit in the clockwise direction)

$$V = V_S(500) - V_S(40) + V_S(0) - V_S(40) = 20.640 - 1.611 - 0 - 1.611 = 17.418\ mV$$

If one were to convert this to a temperature based on the assumption that the measuring junction and reference junction are correctly connected, the temperature would be (using again the K type thermocouple tables) $424.2°C$. Thus the consequence of the mistake is that the temperature is underestimated by $74.8°C$!

This example shows the importance of connecting properly the thermocouple wires and lead wires.

4.4   Resistance thermometry

4.4.1   Basic ideas

Resistance thermometry depends on the unique relation that exists between resistance of an element and the temperature. The resistance thermometer is usually in the form of a wire and its resistance is a function of its temperature. Material of the wire is usually high purity Platinum. Other materials may also be used. The resistance variation of different materials is indicated by Table 4.10. Platinum resistance thermometer is also referred to as PRT (or PT) or Resistance

Table 4.10: *Resistance variation of different wire materials*

Material	Temperature Range $°C$	Element Resistance in Ω at $0°C$	Element Resistance in Ω at $100°C$
Nickel	-60 to 180	100	152
Copper	-30 to 220	100	139
Platinum	-200 to 850	100	136

Temperature Detector (RTD). Usually the resistance of the detector at the ice point is clubbed with it and the thermometer is referred to as, for example, PT100, if it has a resistance of 100 Ω at the ice point. The resistance of standard high purity Platinum varies systematically with temperature and it is given by the International standard calibration curve for wire wound Platinum elements (also referred to as Callendar-Van Dusen equation)[5]:

---

[5]Hugh Longbourne Callendar 1863 - 1930, British physicist

$$R_t = R_0[1 + K_1 t + K_2 t^2 + K_3(t - 100)t^3], \; -200°C < t < 0°C \qquad (4.25)$$

$$R_t = R_0[1 + K_1 t + K_2 t^2], \; 0°C < t < 661°C \qquad (4.26)$$

with R_t the resistance of the element at $t°C$, R_0 the resistance of the element at the ice point and the K's are given by the following.

$K_1 =$	$3.90802 \times 10^{-3} \, (°C)^{-1}$	(4.27a)
$K_2 =$	$-5.802 \times 10^{-7} \, (°C)^{-2}$	(4.27b)
$K_3 =$	$-4.2735 \times 10^{-12} \, (°C)^{-4}$	(4.27c)

The ratio $\dfrac{R_{100} - R_0}{100 R_0}$ is denoted by α, the temperature coefficient of resistance and is specified as $0.00385 \, (°C)^{-1}$. It is seen from Equations 4.25 and 4.26 that the resistance temperature relationship is non linear. The response of a Platinum resistance thermometer is usually plotted in the form of ratio of resistance at temperature t to that at the ice point, as a function of temperature, as shown in Figure 4.28. The platinum resistance sensor in Figure 4.29 is shown with a three wire arrangement. The resistance sensor is also available with four wire arrangement. The wire is wound round a bobbin made of an insulating material like mica. The wire is in the form of a "coiled coil" and wound round the bobbin (in bifilar arrangement to avoid any inductance) allowing for free expansion of the wire. The need for either three wire or four wire arrangements will be brought out later.

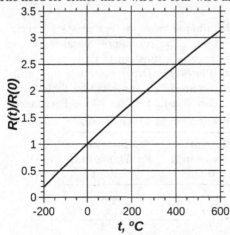

Figure 4.28: *Characteristics of a Platinum resistance thermometer*

4.4.2 Platinum resistance thermometer and the Callendar correction

As seen from Equations 4.25 and 4.26 the Platinum resistance thermometer has essentially a non-linear response with respect to temperature. However we define the linear Platinum resistance temperature scale (t_{Pt}) through the relation

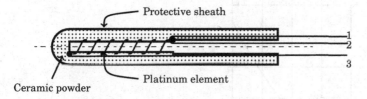

Figure 4.29: Schematic of a typical PRT sensor with three wire arrangement

$$t_{Pt} = \frac{R_t - R_0}{R_{100} - R0} \times 100 \tag{4.28}$$

Obviously the non-linearity will have to be taken into account to get the correct temperature from the linear scale defined by Equation 4.28. This is done by applying a correction to the Platinum resistance temperature as suggested by Callendar for temperatures above the ice point. We have, from Equation 4.25, with $t = 100°C$

$$\begin{aligned}
R_{100} &= R_0(1 + 100K_1 + 100^2K_2) \\
R_{100} - R_0 &= R_0(100K_1 + 100^2K_2) \\
R_t - R_0 &= R_0(K_1 t + K_2 t^2)
\end{aligned}$$

We may then write, for the Platinum resistance temperature

$$\begin{aligned}
t_{Pt} &= \frac{R_t - R_0}{R_{100} - R_0} \times 100 = 100 \times \frac{K_1 t + K_2 t^2}{100K_1 + 100^2 K_2} \\
&= \frac{K_1 t + K_2 t^2}{K_1 + 100K_2} = \frac{K_1 t + 100K_2 t - 100K_2 t + K_2 t^2}{K_1 + 100K_2} \\
&= t - \frac{100K_2 t - K_2 t^2}{K_1 + 100K_2} \approx t - \frac{100K_2 t - K_2 t^2}{K_1} \\
&= t + \frac{K_2}{K_1} \times 100^2 \left[\frac{t}{100} \left(\frac{t}{100} - 1 \right) \right]
\end{aligned}$$

The approximation used above is justified since $K_1 \gg 100K_2$. With the values of K_1 and K_2 given earlier by Equation 4.27, we have

$$\frac{K_2}{K_1} \times 100^2 = \frac{-5.802 \times 10^{-7}}{3.90802 \times 10^{-3}} \times 10^4 = -1.485 = -\delta \text{ (say)} \tag{4.29}$$

We then see that

$$t = t_{Pt} + \underbrace{\delta \left[\frac{t}{100} \left(\frac{t}{100} - 1 \right) \right]}_{\text{Correction}} \tag{4.30}$$

Equation 4.30 involves the temperature t on both sides. However, we note that the term shown with under-brace may be approximated by substituting t_{Pt} for t to get

$$t \approx t_{Pt} + \underbrace{\delta \left[\frac{t_{Pt}}{100} \left(\frac{t_{Pt}}{100} - 1 \right) \right]}_{\textbf{Callendar correction}} \tag{4.31}$$

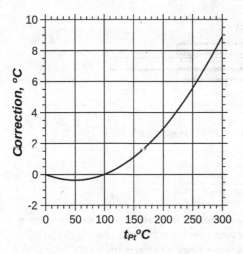

Figure 4.30: *Callendar correction as a function of t_{Pt}*

This is referred to as the Callendar equation and the second term (shown with under-brace) is the Callendar correction, represented as c. The Callendar correction is evidently zero at both the ice and steam points. The correction is non-zero at all other temperatures. Figure 4.30 shows the Callendar correction as a function of the Platinum resistance temperature over the useful range of the sensor.

Example 4.6

The resistance of a Platinum resistance sensor of $R_0 = 100\,\Omega$ was measured to be $119.4\,\Omega$. This sensor has α value of 0.00385 and $\delta = 1.485$. What is the corresponding temperature without and with Callendar correction?

Solution :

Using the given data, we get

$$R_{100} = R_0(1 + 100\alpha) = 100 \times (1 + 100 \times 0.00385) = 138.5\,\Omega$$

The measured resistance at the unknown temperature is $R_t = 119.4\,\Omega$. Hence the unknown temperature according to Platinum temperature scale is

$$t_{Pt} = \frac{R_t - R_0}{R_{100} - R_0} \times 100 = \frac{119.4 - 100}{138.5 - 100} \times 100 = 50.39°C$$

This is also the uncorrected value of the temperature. The Callendar correction is now obtained as

$$c = \delta\left(\frac{t_{Pt}}{100}\right)\left(\frac{t_{Pt}}{100} - 1\right) = 1.485 \times \left(\frac{50.39}{100}\right)\left(\frac{50.39}{100} - 1\right) = -0.37°C$$

The corrected temperature is thus given by

$$t = t_{Pt} + c = 50.39 - 0.37 = 50.02°C$$

4.4.3   RTD measurement circuits

The resistance of the RTD is determined by the use of a DC bridge circuit. As mentioned earlier there are two variants of RTD, viz. the three wire and the four wire systems. These are essentially used to eliminate the effect of the lead wire resistances that may adversely affect the measurement. There are two effects due to the lead wires: 1) they add to the resistance of the Platinum element and 2) the resistance of the lead wires also change with temperature. These two effects are reduced or even eliminated by either the three or four wire arrangements.

The lead wires are usually of larger diameter than the diameter of the sensor wire to reduce the lead wire resistance. In both the three and four wire arrangements, the wires run close to each other and pass through regions experiencing similar temperature fields (refer Figure 4.29). Hence the change in the resistance due to temperature affects all the lead wires by similar amounts. The resistances of the lead wires are compensated by a procedure that is described below.

Bridge circuit for resistance thermometry

Three wire arrangement for lead wire compensation

Figure 4.31 shows the bridge circuit that is used with RTD having three lead wires. The resistances R_1 and R_3 are chosen to be equal and the same as R_0 of the RTD. Two lead wires (labeled 2 and 3) are connected as indicated adding equal resistances to the two arms of the bridge. The third lead wire (labeled 1) is used to connect to the battery supply. Thus the bridge will indicate null (milli-ammeter will indicate zero) with $R_2 = R_0$ when the RTD is maintained at the ice point. During use, when the RTD is at temperature t, the resistance R_2 is adjusted to restore balance. If the lead wires have resistances equal to R_{s2} and $Rs3$, we have

$$R_t + R_{s3} = R_2 + R_{s2} \text{ or } R_t = R_2 + \underbrace{(R_{s2} - R_{s3})}_{\approx 0} \tag{4.32}$$

If the two lead wires are of the same size the term with under-brace should essentially be zero and hence the lead wire resistances have been compensated.

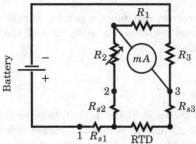

Figure 4.31: *Bridge circuit with lead wire compensation (three wire arrangement). $R_{s1} - R_{s3}$ are lead wire resistances.*

Four wire arrangement for lead wire compensation

The four wire arrangement also known as Kelvin sensing is a superior arrangement, with reference to lead wire compensation, as will be shown below. Figure 4.32

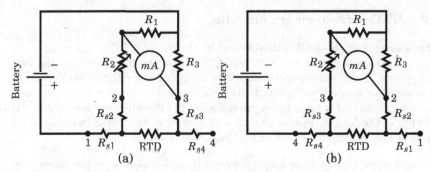

Figure 4.32: Bridge circuit with lead wire compensation (four wire arrangement). $R_{s1} - R_{s4}$ are the lead wire resistances.

shows the bridge arrangements that are used for this purpose. The choice of the resistances is made as given for the three wire arrangement. If the lead wires have resistances equal to $R_{s1} - R_{s4}$, we have the following.

Condition for bridge balance in arrangement shown in Figure 4.32(a):

$$R_t + R_{s3} = R_{2(a)} + R_{s2} \qquad (4.33)$$

where $R_{2(a)}$ is the resistance R_2 for which the bridge is under balance, in the arrangement (a).

Condition for bridge balance in arrangement shown in Figure 4.32(b):

$$R_t + R_{s2} = R_{2(b)} + R_{s3} \qquad (4.34)$$

where $R_{2(b)}$ is the resistance R_2 for which the bridge is under balance, in the arrangement (b).

We see that by addition of Equations 4.33 and 4.34, we get

$$R_t = \frac{R_{2(a)} + R_{2(b)}}{2} \qquad (4.35)$$

The lead wire resistances thus drop off and the correct resistance is nothing but the mean of the two measurements. Since the lead wire resistances actually drop off, the four wire scheme is superior to the three wire scheme.

Practical circuit with four wire RTD

An interesting way of using four wire RTD is shown in Figure 4.33. Two wires (1 and 4) are used to drive a constant current (~ mA) that gives rise to a measurable potential drop across the RTD. The other two lead wires (2 and 3) are connected to a voltmeter of high input impedance (digital volt meter or DVM) such that hardly any current is drawn by the voltmeter. In that case the voltmeter reading is not affected by the lead wires since they carry no current.

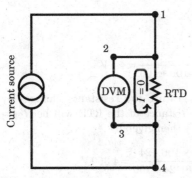

Figure 4.33: *Practical circuit with four wire RTD*

Example 4.7

An RTD has temperature coefficient of resistance at $20°C$ of $\alpha_{20} = 0.004 \, (°C)^{-1}$. If resistance at $20°C$ is $R_{20} = 106 \, \Omega$, determine the resistance at $25°C$. The above RTD is used in a bridge circuit with $R_1 = R_2 = R_3 = 100 \, \Omega$ and the supply voltage is $9 \, V$. Calculate the voltage the detector must be able to resolve in order to measure a $1°C$ change in temperature around $20°C$.

Solution :

Temperature coefficient of resistance α_t when the RTD is at temperature t represents the sensitivity of the RTD. We may define this quantity as

$$\alpha_t = \left[\frac{1}{R} \frac{dR}{dt} \right]\Big|_t = \frac{s_t}{R_t}$$

where s_t is the slope of the Resistance vs temperature curve for the RTD. With the given data the slope at $20°C$ may be determined as

$$s_{20} = \alpha_{20} \times R_{20} = 0.004 \times 106 = 0.424 \, \Omega/°C$$

Assuming the response of the sensor to be linear over small changes in temperature, the resistance of the sensor at $25°C$ may be determined as

$$R_{25} = R_{20} + s_{20} \times \Delta t = 106 + 0.424 \times (25 - 20) = 108.12 \, \Omega$$

Infer all voltages with reference to the negative terminal of the battery taken

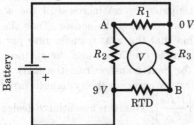

Figure 4.34: *Bridge circuit for Example 4.7*

as zero (ground)(see Figure 4.34). The voltmeter reads the potential difference between A and B. If there is change of temperature of $1°C$ the temperature of the RTD may either be $21°C$ or $19°C$.

Case (a): $t = 21°C$
The potentials are given by the following:

$$V_A = 9 - \frac{9R_2}{R_1 + R_2} = 9 - \frac{9 \times 100}{100 + 100} = 4.5\,V$$

If there is a change of $+1.0°C$ in temperature the resistance changes by $+0.424\,\Omega$ as given by the slope. The resistance of the RTD will be $R_{21} = 106.424\,\Omega$ in this case. The potential V_B is then given by

$$V_B = 9 - \frac{9R_{21}}{R_{21} + R_3} = 9 - \frac{9 \times 106.424}{106.424 + 100} = 4.360\,V$$

The voltmeter should read

$$V = V_A - V_B = 4.5 - 4.360 = 0.140\,V \text{ or } 140\,mV$$

Case (b): $t = 19°C$
There is no change in V_A in this case. If there is a change of $-1.0°C$ in temperature the resistance changes by $-0.424\,\Omega$ as given by the slope. The resistance of the RTD will be $R_{19} = 105.576\,\Omega$ in this case. The potential V_B is then given by

$$V_B = 9 - \frac{9R_{19}}{R_{19} + R_3} = 10 - \frac{9 \times 105.576}{105.576 + 100} = 4.378\,V$$

The voltmeter should read

$$V = V_A - V_B = 4.5 - 4.378 = 0.122\,V \text{ or } 122\,mV$$

The smaller of these or $0.140\,V$ or $122\,mV$ is the resolution of the voltmeter required for resolving $1°C$. Practically speaking we may choose a voltmeter with $100\,mV$ resolution for this purpose.

Effect of self heating of RTD

The bridge arrangement for measuring the resistance of the RTD involves the passage of a current through the sensor. Heat is generated by this current passing through the RTD. The heat has to be dissipated by an increase in the sensor temperature compared to the process space surrounding the sensor. Thus the self heating leads to a systematic error. Assume that the heat transfer rate per unit temperature difference - product of heat transfer coefficient and the surface area of the RTD or the dissipation constant - for heat transfer from the RTD to the surrounding medium is P_D W/K. The temperature excess or the systematic error due to self heating of RTD is then given by $\Delta t = \left[\dfrac{I^2 R_t}{P_D} \right]$. Thus it is essential to design the bridge circuit such that the current passing through the RTD is as small as possible. It also helps if the thermal conductance between the RTD and the process space is as high as possible.

Example 4.8

An RTD has $\alpha_{20} = 0.005/°C$, $R_{20} = 500\,\Omega$ and a dissipation constant of $P_D = 30\,mW/°C$ at $20°C$. The RTD is used in a bridge circuit with $R_1 = R_3 = 500\,\Omega$ and R_2 is a variable resistor used to null the bridge. If the supply voltage is $10\,V$ and the RTD is placed in a bath at the ice point, find the value of R_2 to null the bridge. Take the effect of self heating into account.

Solution :

Figure 4.34 in Example 4.7 is appropriate for this case also. Since $R_1 = R_3 = 500\,\Omega$, at null $R_t = R_2$. Hence the current through the RTD is given by $I = \dfrac{V_s}{R_2 + 500}$ where V_s is the battery supply voltage. The heat dissipation in the RTD is thus given by

$$\dot{Q} = I^2 R_2 = \left[\frac{V_s}{R_2 + 500}\right]^2 R_2$$

This self heating gives rise to a temperature excess of

$$\Delta t = \frac{\dot{Q}}{P_D} = \left[\frac{V_s}{R_2 + 500}\right]^2 \frac{R_2}{P_D}$$

over and above that prevailing in the process space. Since the process space is at the ice point (i.e. $0°C$), Δt is also the temperature of the RTD. Assume linear variation of resistance of RTD with temperature. Then we have

$$R_2 = R_{20}\left[1 - \alpha_{20}\left(20 - \left\{\frac{V_s}{R_2 + 500}\right\}^2 \frac{R_2}{P_D}\right)\right]$$

Introducing the numerics we have

$$R_2 = 500\left[1 - 0.005\left(20 - \left\{\frac{10}{R_2 + 500}\right\}^2 \frac{R_2}{0.03}\right)\right] \qquad (4.36)$$

This is a non-linear equation that needs to be solved for R_2. Either a trial and error solution or Newton Raphson method may be used for this purpose. The former starts with the trial value $R_2 = R_2^0 = R_0$ calculated at the ice point as the starting point. Thus

$$R_2^0 = R_{20} \times (1 + \alpha[0 - 20]) = 500 \times (1 - 0.005 \times 20) = 450\,\Omega$$

Substitute this for R_2 on the right hand side of Equation 4.36 to get a better value as

$$R_2^1 = 500\left[1 - 0.005\left(20 - \left\{\frac{10}{450 + 500}\right\}^2 \frac{450}{0.03}\right)\right] = 454.2\,\Omega \qquad (4.37)$$

It so happens that we may stop after just one iteration! The resistance required to null the bridge is thus equal to $454.2\,\Omega$. The temperature error in this case is calculated as $\Delta t = \left\{\dfrac{10}{454.2 + 500}\right\}^2 \left(\dfrac{454.2}{0.03}\right) = 1.66°C$. This is a fairly large error and is unacceptable in laboratory practice.

4.4.4 Thermistors

Resistance thermometry may be performed using thermistors. Thermistors are many times more sensitive than RTD and hence are useful over small ranges of temperature. They are small pieces of ceramic material made by sintering mixtures of metallic oxides of Manganese, Nickel, Cobalt, Copper, Iron etc. They come in different shapes as shown in Figure 4.35. These are known as NTC - Negative Resistance Coefficient - thermistors since resistance of these thermistors decrease non-linearly with temperature.[6]

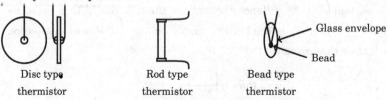

Disc type   Rod type   Bead type
thermistor   thermistor   thermistor

Figure 4.35: *Typical thermistor types*

Thermistors are extremely sensitive but over a narrow range of temperatures. The resistance temperature relation is known to follow the exponential law

$$R_t = R_0 e^{\beta\left[\frac{1}{T} - \frac{1}{T_0}\right]} \qquad (4.38)$$

This equation is based on physics of the solid state of semi-conductors. Here β is a constant (very nearly so, but, in general a function of T) and all temperatures are in Kelvin, T_0 is the ice point temperature and R_0 the corresponding resistance of the thermistor. A typical thermistor has the following specifications:

$$R_{25} = 2000\ \Omega; \quad \frac{R_0}{R_{70}} = 18.64 \qquad (4.39)$$

As usual the subscript on R represents the temperature in $^\circ C$. This thermistor has $\beta = 3917\,K$ and a resistance of 6656 Ω at $0^\circ C$. We may compare the corresponding numbers for a standard Platinum resistance element. For PT100, we have

$$R_0 = 100\ \Omega; \ R_{70} = R_0 \times (1 + 0.00385 \times 70) = 1.27 R_0 = 127\ \Omega$$

Hence we have, for PT100

$$\frac{R_0}{R_{70}} = \frac{1}{1.27} = 0.79 \qquad (4.40)$$

This clearly shows that the thermistor is extremely sensitive in comparison with a RTD. However the highly non-linear response of the thermistor is an undesirable quality and makes it useful over small ranges.

The resistance change is thus very mild in the case of a PRT as compared to a thermistor. The variation of the resistance of the thermistor described above is shown plotted in Figure 4.36. Resistance variation of a thermistor given by Equation

---

[6]We will not consider PTC - Positive Temperature Coefficient - thermistors here. PTC thermistors are ceramic components whose electrical resistance rapidly increases when a certain temperature is exceeded.

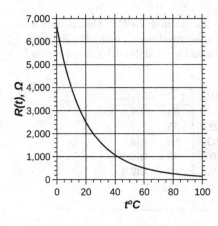

Figure 4.36: *Variation of resistance of thermistor with temperature*

4.38 may be recast, by taking natural logarithms, as

$$\frac{1}{T} = \frac{\ln(R)}{\beta} + \frac{1}{T_0} - \frac{\ln(R_0)}{\beta} \tag{4.41}$$

The above may be recast in the form

$$\frac{1}{T} = A + B\ln(R) \tag{4.42}$$

where A and B are constants that are related to R_0 and β. This equation is not good enough in practice and hence it is modified as

$$\boxed{\frac{1}{T} = A + B\ln(R) + C\{\ln(R)\}^3} \tag{4.43}$$

which is based on simple curve fit. The right side of above equation is a polynomial in $\ln(R)$ with the square term missing. This equation is known as Steinhart-Hart[7] thermistor Equation. The reader may refer to an interesting article "Negative Temperature Coefficient Thermistors Part I: Characteristics, Materials, and Configurations" by Gregg Lavenuta [8] that has appeared in SENSORS on May 1, 1997, for interesting details about thermistors.

The constants appearing in Equation 4.43 may be determined by measuring the resistance of the thermistor at three different temperatures. The temperatures are usually chosen to cover the desired range of temperatures over which the thermistor is to be used. This procedure is demonstrated using an example that follows.

Example 4.9

Data for a certain thermistor is available between −20 and 70°C and is given below. Obtain a representation of the data in terms of Steinhart-Hart

---

[7]J.S. Steinhart and S.R. Hart, 1968, "Calibration Curves for Thermistors," Deep Sea Research and Oceanographic Abstracts, Vol.15, pp. 497-503.

[8]http://www.sensorsmag.com/sensors/temperature/negative-temperature-coefficient-thermistors-part-i-characte-811

model using the data corresponding to −20, 25 and 70°C. Also obtain a representation of the data in terms of an exponential model using the data at the two extremes. Compare the results of these two models with the tabulated data.

$t°C$	$R(t)\Omega$	$t°C$	$R(t)\Omega$	$t°C$	$R(t)\Omega$
- 20	12540	15	1673	50	307.4
- 15	9205	20	1289	55	247.0
- 10	6806	25	1000	60	199.6
-5	5069	30	780.5	65	162.1
0	3803	35	612.9	70	132.3
5	2873	40	484.2		
10	2185	45	384.7		

Solution :

Step 1 Steinhart-Hart model:

We choose the three data points as suggested in the statement of the problem and calculate the required quantities to formulate the equations for the three constants in the Steinhart-Hart model.

	$t°C$	$R(t)\Omega$	$T\,K$	$\ln(R)$	$\{\ln(R)\}^3$	$1/T$
Low	-20	12540	253.15	9.4367	840.3448	3.9502×10^{-3}
Mid	25	1000	298.15	6.9078	329.6179	3.3540×10^{-3}
High	70	132.3	343.15	4.8851	116.5770	2.9142×10^{-3}

Step 2 Equations governing the constants $A - C$ are then given by

$$A + 9.4367B + 840.3448C = 3.9502 \times 10^{-3}$$
$$A + 6.9078B + 329.6179C = 3.3540 \times 10^{-3}$$
$$A + 4.8851B + 116.5770C = 2.9142 \times 10^{-3}$$

Step 3 These equations are solved easily to get $A = 1.9273 \times 10^{-3}$, $B = 1.9750 \times 10^{-3}$ and $C = 1.8942 \times 10^{-7}$.

Step 1 Exponential model:

We choose the data at −20 and 70°C to evaluate β and R_0 in the exponential model. We have the following two equations:

$$\frac{1}{253.15} = \frac{9.4367}{\beta} + \frac{1}{273.15} - \frac{\ln(R_0)}{\beta}$$
$$\frac{1}{343.15} = \frac{4.8851}{\beta} + \frac{1}{273.15} - \frac{\ln(R_0)}{\beta}$$

Step 2 We solve for β to get

$$\beta = \frac{9.4367 - 4.8851}{\frac{1}{343.15} - \frac{1}{253.15}} = 4393.2$$

Step 3 Using this β in the first of the previous equation we get

$$R_0 = \frac{12540}{e^{4393.2 \times \left(\frac{1}{253.15} - \frac{1}{273.15}\right)}} = 3519.2 \, \Omega$$

Step 4 We calculate the resistance of the thermistor at $25°C$ using the exponential model and obtain

$$R_{25} = 3519.2 e^{4393.2 \times \left(\frac{1}{298.15} - \frac{1}{273.15}\right)} = 913.6 \, \Omega$$

Thus the exponential model has an error of $\Delta T = 913.6 - 1000 = -86.4 \, \Omega$ at the third data point available at $t = 25°C$. Note that the Steinhart-Hart model has zero error at this point.

Step 5 Comparison of models with data:
For purposes of comparison we need to calculate the resistance vs. temperature values for the two models. In the case of the exponential model this is easily done by using Equation 4.38. In the case of Steinhart-Hart model we need to invert the relation between T and $\ln(R)$. In order to use the same constants as were determined above, the inverse relation is given by

$$X = \frac{A - \frac{1}{T}}{C}; Y = \sqrt{\left(\frac{B}{3C}\right)^3 + \left(\frac{X}{2}\right)^2}$$

$$R = e^{\left\{\left(Y - \frac{X}{2}\right)^{1/3} - \left(Y + \frac{X}{2}\right)^{1/3}\right\}}$$

Step 6 We calculate the resistance vs. temperature by the two models and make a plot as shown in Figure 4.37(a). The data and the Steinhart-Hart model do not resolve as separate curves in this figure. Since the exponential model has larger errors, it resolves as the dashed curve shown therein.

Step 7 In order to look at the errors between the data and the two models we plot the respective errors with respect to temperature as shown in Figure 4.37(b). It is seen that the Steinhart-Hart model is superior to the exponential model. The maximum error in the former is approximately $-13 \, \Omega$ while it is as much as $-320 \, \Omega$ in the case of the latter.

Step 8 Sensitivity at $t = 25°C$:
Sensitivity, defined as $s_t = \frac{1}{R}\frac{dR}{dt}$, is calculated using the two models. Since the calculation of derivative using the Steinhart-Hart model is difficult we make use of finite differences to obtain the same. We calculate the resistance of the thermistor at $t = 24°C$ and $t = 26°C$ and obtain the derivative by central differences. The derivative is easily obtained using the exponential model as $\frac{dR}{dt} = \frac{dR}{dT} = -\frac{\beta R}{T^2}$ or $s_t = -\frac{\beta}{T^2}$. The calculations are shown in the form of a short table.

$t°C$	R, Ω	$\dfrac{dR}{dt}, \Omega/°C$	$°C^{-1}$
Steinhart-Hart model			
24	960.0		
25	1000.0	-36.8	-0.0452
26	869.7		
Exponential model			
25	913.6	-35.1	-0.0494

Both models yield roughly the same sensitivity for the thermistor.

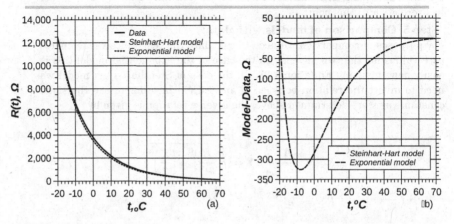

Figure 4.37: *(a) Variation of resistance vs. temperature for the thermistor of Example 4.9. (b) Errors due to Steinhart-Hart and exponential models*

Sensitivity and temperature errors

In Example 4.9 we have seen that the model equations have some errors with reference to the data. Because the three parameter Steinhart-Hart model is better we shall look at the temperature errors that occur because of its use. Calculation shows that the maximum difference between the model and the data occurs at $-10°C$ and is equal to $-12.7\,\Omega$. The sensitivity at this temperature may be obtained using the procedure adopted in the above example and is equal to -0.0597. Hence the change in resistance per $°C$ is given by $-0.0597 \times 6806 = -406.1\,\Omega$. Hence the error in resistance because of the model translates to an error in temperature of $\dfrac{-12.7}{-406.1} = 0.03°C$.

Thus there is practically no error because of the use of the model in place of the data. If we assume that the resistance is measured within $\pm10\,\Omega$ of the true value, the corresponding temperature error is only $\dfrac{10}{406.1} = 0.025 \approx 0.03°C$. Thus we see that the thermistor is a precision thermometer. We note in passing that lead wire compensation discussed in the case of RTD is also valid here even though the effect of lead wire resistances are not as severe as in the case of RTD. Four wire Kelvin sensing may be made use of in the case of thermistor measuring circuits also.

Typical thermistor circuit for temperature measurement

Thermistor temperature sensing involves essentially the measurement of the resistance of the thermistor at its temperature. This is invariably done by converting the resistance to a voltage and measuring it. Two circuits are discussed below.

The first one is a full bridge circuit shown in Figure 4.38(a).[9] The bridge consists of three equal resistances $R_1 - R_3$ in the three arms of the bridge and the thermistor (R_t) in the fourth arm (note the symbol used for a thermistor). The bridge is in balance at a particular temperature when $R_t = R_3$. This is usually chosen as the mid value such as $t = 25°C$. At any other temperature the bridge goes out of balance, develops a voltage which is measured by the voltmeter connected as shown in the figure.

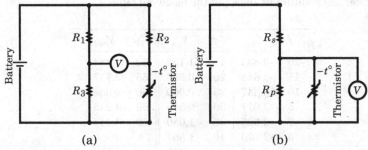

(a)   (b)

Figure 4.38: *(a) Thermistor bridge circuit (b) Alternate measuring circuit, essentially a voltage divider*

The second circuit which may be used is shown in Figure 4.38(b). It essentially is a voltage divider circuit (or half bridge circuit) with an additional parallel resistance R_p chosen carefully so as to make the equivalent resistance almost linear with temperature of the thermistor. Two examples are given below to indicate how these two circuits may be used for a typical thermistor.

Example 4.10

The thermistor of Example 4.9 is connected in a bridge circuit as shown in Figure 4.38(a). Resistances $R_1 - R_3$ are all equal and fixed at $1000\ \Omega$ while the resistance of thermistor alone varies with temperature. $9\ V$ is applied across the bridge and the imbalance voltage is measured using a voltmeter capable of resolving $10\ mV$. Determine output voltage as a function of process temperature to which the thermistor is exposed. Discuss issues regarding linearity and sensitivity. You may use the Steinhart-Hart model to evaluate the resistance of the thermistor as a function of temperature.

Solution :

---

[9]Note: Symbol used in this figure represents a NTC thermistor. The bent line indicates non-linear resistance temperature relationship. The label $-t°$ indicates that resistance decreases with temperature

Analysis of the bridge circuit is fairly simple and straightforward. The voltmeter reads the imbalance voltage given by

$$V_{out} = V_b \left[\frac{R_3}{R_1 + R_3} - \frac{R_t}{R_2 + R_t} \right]$$

where V_b is the battery voltage, V_{out} is the voltmeter reading and R_t is the thermistor resistance at its temperature t. It is obvious that the voltmeter will read zero when the thermistor is at $25°C$ since all the resistances will be the same and the bridge is in balance. For $t < 25°C$ the reading is negative while it is positive for $t > 25°C$.

With $V_b = 9\,V$, $R_1 = R_2 = R_3 = 1000\,\Omega$, the voltmeter reading varies with thermistor temperature as shown in the following table.

t	V_{out}	t	V_{out}	t	V_{out}
-20	-3.835	15	-1.133	50	2.384
-15	-3.618	20	-0.568	55	2.717
-10	-3.347	25	0.000	60	3.003
-5	-3.017	30	0.555	65	3.245
0	-2.626	35	1.080	70	3.448
5	-2.176	40	1.564		
10	-1.674	45	2.000		

This data is also shown plotted in Figure 4.39.

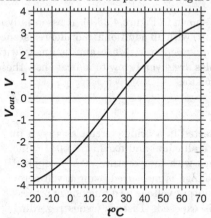

Figure 4.39: *Voltmeter reading variation with thermistor temperature*

The voltmeter reading varies approximately linearly with temperature between $0°C$ and $50°C$.[10] There is some nonlinearity outside this band.

---

[10]Linearity is not an issue when the voltage is measured by a digital recorder that may be manipulated by a computer. Voltage reading may easily be converted to temperature using software.

Making the linear assumption observed above, the voltage sensitivity is given by

$$\text{Voltage sensitivity} \quad = \quad \text{Slope of voltage vs.} \, R_t = \frac{2.384 - (-2.626)}{50 - 0}$$
$$= \quad 0.1 \, V/^{\circ}C \quad \text{or} \quad 100 \, mV/^{\circ}C$$

where the voltage values have been taken from the table. It is mentioned in the problem statement that the voltmeter can resolve 10 mV. Hence the thermistor bridge circuit will be able to resolve $\frac{10 \, mV}{100 \, mv/^{\circ}C} = 0.1^{\circ}C$. This is a very good situation!

Example 4.11

Thermistor considered in Example 4.10 is connected as shown in Figure 4.38(b). The parallel resistance is taken as $R_p = 1000 \, \Omega$. The series resistance R_s is also chosen to be 1000 Ω. Determine the variation of voltmeter reading with thermistor temperature if the battery supply voltage is 9 V.

Solution :

The equivalent resistance of R_t in parallel with R_p is easily seen to be $R_{eq} = \frac{R_t R_p}{R_t + R_p}$. The voltmeter reading is given by $V_{out} = V_b \frac{R_{eq}}{R_{eq} + R_s}$. In the present case $V_b = 9 \, V$, $R_p = R_s = 1000 \, \Omega$ and hence we have

$$R_{eq} \quad = \quad \frac{1000 R_t}{R_t + 1000}$$
$$V_{out} \quad = \quad 9 \frac{R_{eq}}{R_{eq} + 1000}$$

The output read by the voltmeter is shown as a function of thermistor temperature in the following table.

t	V_{out}	t	V_{out}	t	V_{out}
-20	4.327	15	3.448	50	1.699
-15	4.265	20	3.232	55	1.476
-10	4.185	25	3.000	60	1.276
-5	4.084	30	2.736	65	1.098
0	3.961	35	2.466	70	0.942
5	3.814	40	2.199		
10	3.643	45	1.941		

A plot of this data is also shown in Figure 4.40. The output is a decreasing function of temperature of the thermistor. It appears to be linear between $0^{\circ}C$ and $60^{\circ}C$. Nonlinearity is found outside this band.

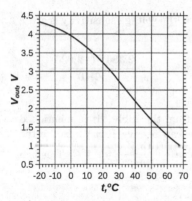

Figure 4.40: *Variation of voltmeter reading with temperature*

The voltage sensitivity of the thermistor may again be estimated using the linear portion of the output. We have

$$\text{Voltage sensitivity} = \text{Slope of voltage vs. } R_t = \frac{1.276 - 3.961}{60 - 0}$$

$$= -0.054 \, V/^{\circ}C \quad \text{or} \quad -54 \, mV/^{\circ}C$$

Here the voltage values have been taken from the table. Sensitivity of the thermistor in this arrangement is roughly 50% of the value achieved with the bridge circuit. In general, linearization of output, leads to a reduction of sensitivity.

As in the case of RTD a thermistor is also subject to self heating since a current invariably passes through it while in operation. The dissipation factor determines the temperature increase due to heat generated in the thermistor. The following example demonstrates how the measuring circuit may be designed to keep the effect of self heating within specified limit.

Example 4.12

Manufacturer of NTC thermistor has specified a resistance of $1.0 \, k\Omega$ at $25^{\circ}C$ with $\beta = 3100 \, K$. Both have an uncertainty of 1%. Dissipation constant for the thermistor for operation in air is given as $P_D = 2 \, mW/^{\circ}C$. The operating temperature range of the thermistor is from -50 to $90^{\circ}C$. This thermistor is used in an application where the temperature ranges between 30 and 60 $^{\circ}C$. Design a suitable circuit such that the thermistor gives temperature within $0.2^{\circ}C$ of the true value.

Solution :

Step 1 Assumptions: We propose to make use of a circuit similar to the one shown in Figure 4.38(b). We choose the series and parallel resistances to be equal i.e. $R_s = R_p$. We shall assume that the power supply is a regulated DC supply with $V_b = 6 \, V$.

Step 2 Power dissipation: The equivalent resistance of thermistor in parallel with R_p is given by $R_{eq} = \dfrac{R_p R_t}{R_p + R_t}$ where R_t is the resistance of thermistor at its temperature t. Voltage across this equivalent resistance is given by $V_t = \dfrac{V_b R_{eq}}{R_s + R_{eq}}$. With these the current through the thermistor may be evaluated as

$$I_t = \frac{V_t}{R_t} = \frac{V_b R_{eq}}{R_t(R_s + R_{eq})}$$

The power dissipation by the thermistor due to self heating is then given by

$$P_t = I_t^2 R_t$$

Step 3 Temperature increase due to power dissipation: It is given by dividing the thermistor power dissipation due to self heating by the thermistor by the dissipation constant i.e. $\delta t = \dfrac{P_t}{P_D}$.

Step 4 Maximum power dissipation: The maximum power is dissipated by the thermistor when it is at the lowest temperature in the range of interest viz. $t = 30°C$. Resistance of the thermistor is calculated at this temperature as

$$R_{30} = R_0 e^{\beta\left(\frac{1}{T} - \frac{1}{T_0}\right)} = 1000 e^{3100\left(\frac{1}{30+273.15} - \frac{1}{25+273.15}\right)} = 842.4\,\Omega$$

We choose a range of values of R_p or R_s to find out the power dissipated by the thermistor and the corresponding temperature increase due to self heating. This is best shown as a table.

R_p or R_s, Ω	R_{eq}, Ω	V_t, V	I_t, A	P_t, W	Δt, $°C$
10000	777.0	0.433	0.00051	0.00022	0.11
7500	757.3	0.550	0.00065	0.00036	0.18
5000	720.9	0.756	0.00090	0.00068	0.34

Step 5 Choice of R_s or R_p: It is thus clear that a resistance value of $R_s = R_p = 7500\,\Omega$ will meet the requirement of guaranteeing no more than $0.2°C$ change in the temperature of the thermistor due to self heating.

Step 6 Effect of uncertainty in β and R_0: We may use error propagation formula and obtain the $\pm\sigma$ limits on R_{30} as $\pm 8.5\,\Omega$. Corresponding 95% confidence limits are $R_{30} = 859.5\,\Omega$ and $R_{30} = 825.3\,\Omega$. The temperature increase due to self heating changes marginally and the design is still alright as far as the requirement that it be less than $0.2°C$ is concerned.

4.5 Pyrometry

Pyrometry is the art and science of measurement of high temperatures. According to the International Practical Temperature Scale 1968 (IPTS68) that preceded ITS90, Pyrometry was specified as the method of temperature measurement above the gold point $1064.43°C$ ($1337.58\,K$). Pyrometry makes use of radiation emitted by a surface (usually in the visible part of the spectrum but it is possible to use other parts of the electromagnetic spectrum also) to determine its temperature. Pyrometry is thus a non-contact method of temperature measurement. We shall introduce basic concepts from radiation theory, for the sake of completeness, before discussing Pyrometry in detail.

4.5.1   Radiation fundamentals

Black body radiation

Black body radiation exists inside an evacuated enclosure whose walls are maintained at a uniform temperature. The walls of the enclosure are assumed to be impervious to transfer of heat and mass. Black body radiation is a function of the wavelength λ and the temperature T of the walls of the enclosure. The amount of radiation heat flux leaving the surface of the enclosure, in a narrow band $d\lambda$ around λ, is called the spectral emissive power and is given by the Planck distribution function

$$E_{b\lambda}(T) = \frac{C_1}{\lambda^5}\frac{1}{e^{\left(\frac{C_2}{\lambda T}\right)} - 1} \tag{4.44}$$

where $C_1 = 3.742 \times 10^8\ W\mu m^4/m^2$ is the first radiation constant and $C_2 = 14390\ \mu m\ K$ is the second radiation constant. There is no net heat transfer from the surface of an enclosure and hence it receives the same flux as it emits. Also it may be seen that the -1 in the denominator of Equation 4.44 is much smaller than the exponential term as long as $\lambda T \ll C_2$. This is indeed true in Pyrometry applications where the wavelength chosen is around 0.66 μm and the measured temperature may not be greater than 5000 K. It is then acceptable to approximate the Planck distribution function by the Wein's approximation given by

$$E_{b\lambda}(T) = \frac{C_1}{\lambda^5}e^{\left(-\frac{C_2}{\lambda T}\right)} \tag{4.45}$$

It is clear from Figure 4.41 that the error in using the Wein's approximation in lieu of the Planck function is around 1.2% even at a temperature as high as 5000 K. The spectral black body emissive power has strong temperature dependence. This is

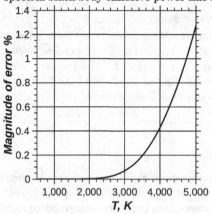

Figure 4.41:   *Error in using Wein's approximation instead of the Planck function*

basically the reason for its use in thermometry. In fact the emissive power peaks at a wavelength temperature product of

$$\lambda_{max}T = 2898\mu m K \tag{4.46}$$

This is referred to as Wein's displacement law. We show spectral black body emissive power plots for various temperatures of the black body in Figure 4.42. The peaks of the individual curves do, in fact, satisfy the Wein's displacement law.

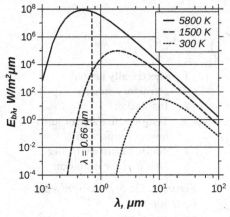

Figure 4.42: *Black body characteristics and the Wein's displacement law (wavelength of 0.66 μm is normally used in optical Pyrometry)*

If we imagine keeping the wavelength fixed at say 0.66 μm (in the visible part of the electromagnetic spectrum - dashed vertical line in Figure 4.42), we see that the ordinate is a strong function of temperature! Higher the temperature the brighter the surface as viewed by the eye. This is basically the idea central to Pyrometry. Actual surfaces, however, are not black bodies and hence they emit less radiation than a black surface at the same temperature. We define the spectral emissivity ε_λ as the ratio of the emissive power of the actual surface $E_{a\lambda}(T)$ to that from a black surface $E_{b\lambda}(T)$ at the same temperature and wavelength.

$$\varepsilon_\lambda = \frac{E_{a\lambda}(T)}{E_{b\lambda}(T)} \tag{4.47}$$

Black body reference - Cavity radiator

Radiation measurements, in general, and Pyrometry, in particular requires a reference black body source that may be maintained at the desired temperature. This is achieved by making use of a black body cavity shown schematically in Figure 4.43. The black body cavity consists of an electrically heated refractory sphere with a small opening as shown. The surface area of the sphere is much larger than the area of the opening through which radiation escapes to the outside. The radiation leaving through the opening is very close to being black body radiation at a temperature corresponding to the temperature of the inside surface of the sphere. Figure 4.44 shows that the effective emissivity of the opening is close to unity. The emissivity of the surface of the sphere is already high, 0.96, for the case shown in the figure). The area ratio for a typical cavity with diameter of sphere of 0.3 m and diameter of port equal to 0.05 m is

$$\frac{A_{port}}{A_{sphere}} = \frac{1}{4}\left(\frac{0.05}{0.3}\right)^2 \approx 0.007 \tag{4.48}$$

The effective emissivity for this area ratio is about 0.995! Many a time the cavity radiator has the sphere surrounded on the outside by a material undergoing phase change (solid to liquid). The melting point of the material will decide the exact temperature of the black body radiation leaving through the opening, as long as the material is a mixture of solid and liquid phases.

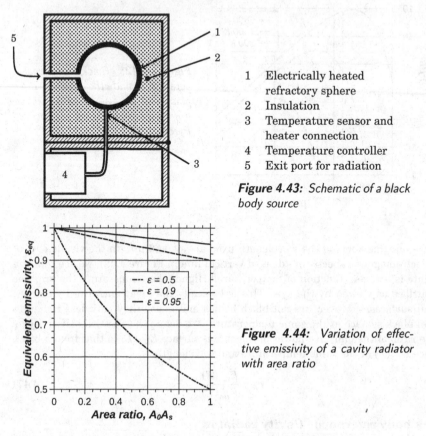

1   Electrically heated
    refractory sphere
2   Insulation
3   Temperature sensor and
    heater connection
4   Temperature controller
5   Exit port for radiation

Figure 4.43: *Schematic of a black body source*

Figure 4.44: *Variation of effective emissivity of a cavity radiator with area ratio*

4.5.2   Brightness temperature and the vanishing filament pyrometer

Brightness temperature

It is defined such that the spectral emissive power of the actual surface is the same as that of a hypothetical black body at the brightness temperature T_B. Thus:

$$E_{a\lambda}(T) = E_{b\lambda}(T_B) \tag{4.49}$$

If the emissivity of the surface (referred to as the target) is ε_λ, we use Equation 4.47 to write

$$E_{a\lambda}(T) = \varepsilon_\lambda E_{b\lambda}(T) = E_{b\lambda}(T_B)$$

Using Wein's approximation this may be re written as

$$\varepsilon_\lambda \frac{C_1}{\lambda^5} e^{\left(-\frac{C_2}{\lambda T}\right)} = \frac{C_1}{\lambda^5} e^{\left(-\frac{C_2}{\lambda T_B}\right)}$$

We may cancel the common factor $\dfrac{C_1}{\lambda^5}$ on the two sides, take natural logarithms, and rearrange to get

$$\frac{1}{T} - \frac{1}{T_B} = \frac{\lambda}{C_2}\ln(\varepsilon_\lambda) \qquad (4.50)$$

Equation 4.50 is referred to as the ideal pyrometer equation. This equation relates the actual temperature of the target to the brightness temperature of the target. The brightness temperature itself may be measured using a vanishing filament pyrometer. The brightness temperature of a target is less than or equal to the actual temperature of a target since surface emissivity is between zero and one. In measurement using a pyrometer the intervening optics may introduce attenuation due to reflection of the radiation gathered from the target. It is also possible that one introduces attenuation intentionally as we shall see later. We account for the attenuation by multiplying the emissivity of the surface by a transmission factor τ_λ (which is again between zero and one) to get

$$\frac{1}{T} - \frac{1}{T_B} = \frac{\lambda}{C_2}\ln(\varepsilon_\lambda \tau_\lambda) \qquad (4.51)$$

We refer to Equation 4.51 as the practical pyrometer equation. We infer from the above that the brightness temperature of a surface depends primarily on the surface emissivity, for a given or specified actual temperature. For a black body, of course, the brightness temperature is the same as the actual temperature. Therefore, if a surface is as bright as a black body at the gold point (melting temperature of gold =1337.58 K), the actual temperature should vary with spectral emissivity of the surface as indicated in Figure 4.45. Since no surface has zero emissivity we allow it to vary from 0.02 to 1 in this figure. As the spectral emissivity decreases the actual temperature increases as shown.

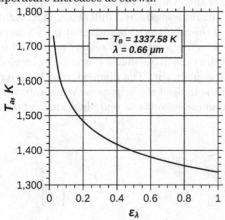

Figure 4.45: *Temperature a surface whose brightness temperature equals gold point temperature of 1337.58 K ($\lambda = 0.66\ \mu m$)*

Vanishing filament pyrometer

This is fairly standard equipment that is used routinely in industrial practice. The principle of operation of the pyrometer is explained by referring to the schematic of

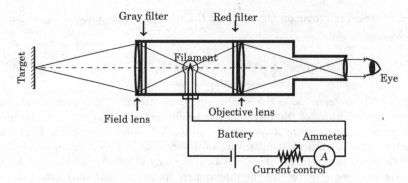

Figure 4.46: *Schematic of a vanishing filament pyrometer*

the instrument shown in Figure 4.46. The pyrometer consists of collection optics (basically a telescope) to gather radiation coming from the target whose temperature is to be estimated. The radiation then passes through an aperture (to reduce the effect of stray radiation), a neutral density or gray filter (to adjust the range of temperature) and is brought to focus in a plane that also contains a source (tungsten filament standard) whose temperature may be varied by varying the current through it. The radiation from the target and the reference then passes through a red filter and is viewed by an observer as indicated in Figure 4.46. The observer adjusts the current through the reference lamp such that the filament brightness and the target brightness are the same.

If the adjustment is such that the filament temperature is greater than the target brightness temperature the setting is referred to as "high". The filament appears as a bright object in a dull background (Figure 4.47 labeled "High"). If the adjustment is such that the filament temperature is lower than the target brightness temperature the setting is referred to as "Low". The filament appears as a dull object in a bright background (Figure 4.47 labeled "Low"). If the adjustment is such that the filament temperature is equal to the target brightness temperature the setting is referred to as "Correct". The filament and the target are indistinguishable (Figure 4.47 labeled "Correct"). Thus the adjustment is a null adjustment. The filament vanishes from the view! The temperature of the filament is in deed the brightness temperature of the object. The wavelength of observation corresponds to roughly a value of 0.66

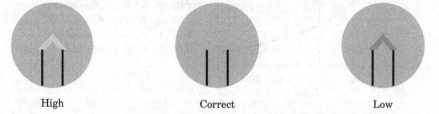

Figure 4.47: *Pyrometer adjustments*

μm. This is achieved by a combination of the response of the red filter and the eye of the observer (Figure 4.48). The average eye response peaks around 0.6 μm and responds very poorly beyond 0.67 μm. The red filter transmits radiation beyond about 0.62 μm. On the whole the red filter - eye combination uses the radiation inside the hatched triangular region formed by the intersections of the two curves. This region corresponds to a mean of 0.66 μm with a spectral width of roughly 0.03 μm. Since the chosen wavelength is in the visible radiation range, only targets at

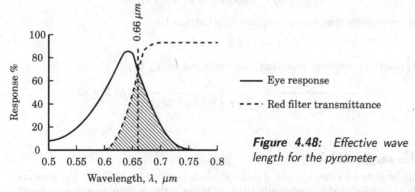

Figure 4.48: *Effective wave length for the pyrometer*

high enough temperature such that there is significant radiation emission in the visible part of the spectrum, are suitable for determination of temperature by the vanishing filament pyrometer. The lower limit of temperature of a target is about 800°C. Such temperatures are usually encountered, for example, in ovens, boiler combustion chambers and in metallurgical applications. Secondly, the emissivity of the target must be known at the wavelength used in the pyrometer. Any uncertainty in the emissivity will lead to a systematic error in the determination of temperature by a pyrometer.

Filament operating at a fixed temperature

Figure 4.45 has indicated that the actual temperature of an object that matches the brightness of a black body at the gold point is a function of the surface emissivity transmittance product (refer Equation 4.51). This at once gives an alternate method of temperature measurement using a vanishing filament pyrometer. If the target is of known emissivity, and is at a temperature above the reference black body temperature (eg. the gold point) which is also the fixed temperature of the filament, then brightness match may be obtained by varying the transmittance instead of the filament temperature. The advantage of this method is that the filament runs at a fixed temperature and does not deteriorate due to thermal cycling. The filament life is enhanced and hence cuts down on the expense incurred in changing the reference lamp very often. The second advantage is that the pyrometer is based on a single reference temperature! The transmittance may be varied by using a variable neutral density filter (gray filter) in the path of the light entering the pyrometer. The variable neutral density filter is in the form of a circular wheel with transmittance marked on the periphery.

Example 4.13

A pyrometer gives the brightness temperature of an object to be 800°C. The optical transmittance for the radiation collected by the pyrometer is known to be 0.965 and the target emissivity is 0.260. Estimate the temperature of the object. Take $\lambda = 0.655\,\mu m$ as the effective wavelength for the pyrometer.

Solution :

The brightness temperature of the target is specified as

$$T_B = 800°C = 800 + 273.15\,K \approx 1073\,K$$

The emissivity transmittance product is obtained as

$$\varepsilon_\lambda \tau_\lambda = 0.260 \times 0.965 = 0.251$$

We use the pyrometer equation 4.51 and solve for T_a to get

$$T_a = \frac{1}{\frac{1}{T_B} + \frac{\lambda}{C_2}\ln(\varepsilon_\lambda \tau_\lambda)} = \frac{1}{\frac{1}{1073} + \frac{0.655}{14390}\ln(0.251)} = 1151\,K = 878°C$$

Example 4.14

The brightness temperature of a metal block is given as $900°C$. A thermocouple embedded in the block reads $1015°C$. What is the emissivity of the surface? The pyrometer used in the above measurement is a vanishing filament type with an effective λ of $0.65\mu m$. Assuming that the thermocouple reading is susceptible to an error of $\pm 5°C$ while the brightness temperature is error free, determine an error bar on the emissivity determined above.

Solution :

The first and second radiation constants, in SI units are:

$$\begin{aligned} C_1 &= 3.743 \times 10^{-16}\,W\mu m^4/m^2 \\ C_2 &= 14387\,\mu m\,K \end{aligned}$$

The data specifies the actual temperature (T_a) as well as the brightness temperature (T_B) and it is desired to determine the emissivity of the metal block at the stated wavelength.

$$\lambda = 0.65\,\mu m; T_a = 1015°C = 1288\,K; T_B = 900°C = 1173\,K$$

We assume the transmittance of the optics as unity since the problem does not explicitly give it. We use the ideal pyrometer equation 4.50 and solve for the target emissivity to get

$$\varepsilon_\lambda = e^{\left\{\frac{C_2}{\lambda}\left(\frac{1}{T_a} - \frac{1}{T_B}\right)\right\}} = e^{\left\{\frac{14387}{0.65}\left(\frac{1}{1288} - \frac{1}{1173}\right)\right\}} = 0.185$$

This is the nominal value of the emissivity. The error in ε_λ is due to error in T_a. This is calculated using the error propagation formula. We need to calculate the appropriate influence coefficient given by the derivative of ε_λ with respect to T_a.

$$I_\varepsilon = \frac{\partial \varepsilon_\lambda}{\partial T_a} = \frac{C_2}{\lambda}\left(\frac{1}{T_a^2}\right)\varepsilon_\lambda = \frac{14387}{0.65}\left(\frac{1}{1288^2}\right)0.185 = -0.00247\,K^{-1}$$

Since error in temperature is $\Delta T_a = \pm 5K$, the error in emissivity is given by

$$\Delta\varepsilon_\lambda = \frac{\partial \varepsilon_\lambda}{\partial T_a}\Delta T_a = -0.00247 \times (\pm 5) = \mp 0.0124 \approx \mp 0.012$$

Emissivity values

Example 4.14 has shown how the spectral emissivity of a target may be estimated if both its actual temperature and the brightness temperature are known. In fact this is one way of obtaining the spectral emissivity (at $\lambda = 0.66\ \mu m$) of a target material. This is possible if an experiment can be set up with an embedded thermocouple to measure the actual temperature at the same instant a pyrometer reading is taken. In fact one may vary the surface temperature over a range of values that is expected to occur in a particular application and measure the emissivity values over this range. An emissivity vs. temperature table may then be made for later use.

Useful emissivity data is given in Table 4.11. These are representative values since the nature of the surface is application specific and much variation is possible. Uncertainty in emissivity affects the measurement of temperature using

Table 4.11: *Useful emissivity values*

Surface	Temperature ° C			
	600	1200	1600	1800
Iron (Un-oxidized)	0.2	0.37	...	
Iron (Oxidized)	0.85	0.89	...	
Molten Cast Iron	...	...	0.29	...
Molten Steel	...	...	0.28	0.28
Nickel (Oxidized)	...	0.75	0.75	...
Fire Clay	...	0.52	0.45	...
Silica Brick	...	0.54	0.46	...
Alumina Brick	...	0.23	0.19	...

a pyrometer. For small errors the relationship is almost linear with $\Delta T_a = \Delta \varepsilon_\lambda$ for a vanishing filament pyrometer operating at 0.65 μm.

4.5.3 Total radiation pyrometer

We have seen in section 4.5.2 that a vanishing filament pyrometer uses a narrow band of radiation in the vicinity of 0.66 μm for its operation. The detector needs to be very sensitive since the amount of radiation in this narrow spectral band is a small fraction of the total radiation emitted by the target as well as the reference. An alternate to this would be to use *all* the radiation emitted by a target at its own temperature T_a. The emitted radiation is much larger and hence the required detector sensitivity is much lower than in the spectral measurement case. But the detector should have a *flat* response over the entire spectral range. This is not such a limitation since a thermal detector like the thermopile that was discussed in section 4.3.3 may be used as the detector (see Figure 4.24).

In this case the comparison will be with the total radiation from a black body at the brightness temperature T_B. The target emissive power (total, in W/m^2 instead of spectral that was in $W/m^2 \mu m$) is given by Stefan Boltzmann law as

$$E_a(T_a) = \varepsilon \sigma T_a^4 \tag{4.52}$$

where ε is the total emissivity, also referred to as the equivalent gray body emissivity of the target and σ is the Stefan Boltzmann constant. The emissive power of the black body that is used for comparison is given by

$$E_b(T_B) = \sigma T_B^4 \tag{4.53}$$

The brightness temperature is again defined such that these two total emissive powers are identical. Thus, we have

$$T_a = \frac{T_B}{\varepsilon^{1/4}} \tag{4.54}$$

It is clear from the above that the target emissivity, in this case the total emissivity, has to be known for successful use of the total radiation pyrometer. The total radiation pyrometer will be inaccurate if the target total emissivity is a strong function of temperature.

An important advantage of the total radiation pyrometry is that the measurement need not be limited to very high temperatures where a significant amount of radiation is emitted in the visible part of the spectrum. Thermal detectors respond to radiation from the target even when no visible radiation is emitted by the target. In principle the entire temperature range of interest to us is accessible to the total radiation pyrometer.

Example 4.15

A black body operating at $T_B = 850\,K$ matches the intensity of a target at an unknown temperature. Both measurements are made using a thermopile that responds to all wavelengths. The target total emissivity is known to be $\varepsilon = 0.82$. What is the temperature of the target? If the target emissivity has a specified uncertainty of ±0.02, what is the uncertainty in the actual temperature?

Solution :

The brightness temperature of the target is the same as the temperature of the black body. data specifies $T_B = 850\,K$, $\varepsilon = 0.82$. Hence the actual temperature of the target, T_a, using Equation 4.54 is

$$T_a = \frac{T_B}{\varepsilon^{1/4}} = \frac{850}{0.82^{1/4}} = 893.2\,K$$

Uncertainty calculation assumes that T_B is error free. The influence coefficient with respect to the emissivity is given by

$$I_\varepsilon = \frac{\partial T_a}{\partial \varepsilon} = -0.25 \times \frac{T_B}{\varepsilon^{1.25}} = -0.25 \times \frac{850}{0.82^{1.25}} = -272.33\,K$$

With the uncertainty in the total emissivity of $\Delta\varepsilon = \pm0.02$, we have

$$\Delta T_a = I_\varepsilon \times \Delta\varepsilon = -272.33 \times (\pm0.02) = \mp5.45\,K$$

4.5.4 Ratio Pyrometry and the two color pyrometer

Uncertainty in target emissivity in the case of vanishing filament pyrometer is a major problem. This makes one look for an alternate way of performing pyrometer measurement, based on the concept of color or ratio temperature. Consider two wavelengths λ_1, λ_2 close to each other. Let the corresponding emissivity values of the target be $\varepsilon_1, \varepsilon_2$. The color temperature T_c of the target is defined by the equality of the ratios of emissive powers given by

$$\frac{E_{a\lambda_1}(T_a)}{E_{a\lambda_2}(T_a)} = \frac{E_{b\lambda_1}(T_c)}{E_{b\lambda_2}(T_c)} \tag{4.55}$$

Thus the color temperature is the temperature of a black body for which the ratio of spectral emissive powers at two chosen wavelengths λ_1 and λ_2 is the same as the corresponding ratio for the actual body. Making use of the Wein's approximation we have

$$\frac{\varepsilon_1 e^{\left(-\frac{C_2}{\lambda_1 T_a}\right)}}{\varepsilon_2 e^{\left(-\frac{C_2}{\lambda_2 T_a}\right)}} = \frac{e^{\left(-\frac{C_2}{\lambda_1 T_a}\right)}}{e^{\left(-\frac{C_2}{\lambda_2 T_a}\right)}}$$

This equation may be rearranged as

$$\frac{\varepsilon_1}{\varepsilon_2} \cdot e^{\left[\frac{C_2}{T_a}\left(\frac{1}{\lambda_1} - \frac{1}{\lambda_2}\right)\right]} = e^{\left[\frac{C_2}{T_c}\left(\frac{1}{\lambda_1} - \frac{1}{\lambda_2}\right)\right]}$$

We may take natural logarithms and rearrange to finally get

$$\frac{1}{T_c} - \frac{1}{T_a} = \frac{1}{C_2} \frac{\ln\frac{\varepsilon_1}{\varepsilon_2}}{\left[\frac{1}{\lambda_1} - \frac{1}{\lambda_2}\right]} \tag{4.56}$$

This equation that links the color and actual temperatures of the target is the counterpart of Equation 4.50 that linked the brightness and actual temperatures of the target. In case the emissivity does not vary with the wavelength, we see that the color and actual temperatures are equal to each other. We also see that the actual temperature may be either *less than* or *greater than* the color temperature. This depends solely on the emissivity ratio for the two chosen wavelengths. Equation 4.56 by itself is not directly useful. We measure the ratio of emissive powers of the target, at its actual temperature, at the two chosen wavelengths. Thus what is measured is the ratio occurring on the left hand side of Equation 4.55. Thus we have

$$\frac{E_{a\lambda_1}}{E_{a\lambda_2}} = \frac{\varepsilon_1}{\varepsilon_2} \frac{\lambda_2^5}{\lambda_1^5} \frac{e^{\left(-\frac{C_2}{\lambda_1 T_a}\right)}}{e^{\left(-\frac{C_2}{\lambda_2 T_a}\right)}}$$

Again we may take natural logarithms and solve for T_a to get

$$T_a = \frac{C_2\left[\frac{1}{\lambda_2} - \frac{1}{\lambda_1}\right]}{\ln\left[\frac{E_{a\lambda_1}\lambda_1^5\varepsilon_2}{E_{a\lambda_2}\lambda_2^5\varepsilon_1}\right]} \tag{4.57}$$

Similarly we may show that the color temperature is given by

$$T_c = \frac{C_2\left[\frac{1}{\lambda_2} - \frac{1}{\lambda_1}\right]}{\ln\left[\frac{E_{a\lambda_1}\lambda_1^5}{E_{a\lambda_2}\lambda_2^5}\right]} \tag{4.58}$$

Equation 4.57 is useful as a means of estimating the actual temperature for a target, whose emissivities at the two chosen wavelengths are known, and the measured values of emissive powers at the two wavelengths are available. The instrument that may be used for this purpose is the two color pyrometer whose schematic is shown in Figure 4.49. The radiation from the target is split up in to two beams by the use

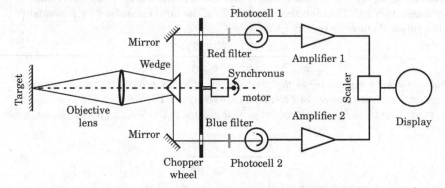

Figure 4.49: *Schematic of a two color pyrometer*

of a wedge, as indicated. Two mirrors redirect the two beams as shown. These pass through a rotating wheel with openings, as indicated. The beams then pass through two filters that allow only a narrow band around a well defined wavelength (indicated as red and blue). The rotating wheel "chops" the two beams and creates ac signals to be detected by the photocells. The ratio of the two signals is equal to the ratio of emissive powers in Equation 4.57 or 4.58. The scaler introduces the other ratios that appear in the same equations. The display then indicates the actual temperature.

Example 4.16

A certain target has a brightness temperature of $1000\,K$ when viewed by a vanishing filament pyrometer. The target emissivity at $0.66\mu m$ is known to be $\varepsilon_1 = 0.8$. What is the true temperature of the target? What is the color temperature of the same target if $\lambda_1 = 0.66\,\mu m$, $\lambda_2 = 0.63\,\mu m$ and $\varepsilon_2 = 0.75$?

Solution :

Given Data: $T_B = 1000\,K$, $\lambda_1 = 0.66\,\mu m$ and $\varepsilon_1 = 0.8$

The ideal pyrometer equation 4.50 may be rearranged as

$$T_a = \left[\frac{1}{T_B} + \frac{\lambda_1}{C_2}\ln\varepsilon_1\right]^{-1}$$

With the numerics in we get

$$T_a = \left[\frac{1}{1000} + \frac{066}{14390}\ln 0.8\right]^{-1} = 1010.3\,K$$

For determining the color temperature the required data is:

$$\lambda_1 = 0.66\,\mu m, \varepsilon_1 = 0.8; \quad \lambda_2 = 0.63\,\mu m, \varepsilon_2 = 0.75$$

We determine the color temperature by using Equation 4.56 and solving for T_c.

$$T_c = \left[\frac{1}{T_a} - \frac{\ln\left(\frac{\varepsilon_2}{\varepsilon_1}\right)}{14390\left(\frac{1}{\lambda_2} - \frac{1}{\lambda_1}\right)} \right]^{-1} = \left[\frac{1}{1010.3} - \frac{\ln\left(\frac{0.75}{0.8}\right)}{14390\left(\frac{1}{0.63} - \frac{1}{0.66}\right)} \right]^{-1} = 950.6 \, K$$

In this case the color temperature is less than the actual as well as the brightness temperatures.

We may also calculate the ratio that would be measured by a two color pyrometer as

$$\frac{E_{a\lambda_1}}{E_{a\lambda_2}} = \frac{\varepsilon_1 e^{\left(-\frac{C_2}{\lambda_1 T_a}\right)}}{\varepsilon_2 e^{\left(-\frac{C_2}{\lambda_2 T_a}\right)}} = \frac{0.8 e^{\left(-\frac{14390}{0.66 \times 1010.3}\right)}}{0.75 e^{\left(-\frac{14390}{0.63 \times 1010.3}\right)}} = 2.981$$

4.5.5   Gas temperature measurement

It is possible to measure temperature of gases using pyrometry. Of special interest is the measurement of high temperature gases such as those that occur in a flame. To make use of radiation in the visible part of the spectrum the gas is made luminous by addition of small quantity of sodium. This may be done by adding small quantities of sodium salts in to the flame. Once sodium is present in the flame it emits a strong doublet at 589.0 nm and 589.6 nm, known as sodium D - lines.[11]

Line reversal technique

Consider the arrangement shown in Figure 4.50. Typical application involving the measurement of temperature of a flame is illustrated here. As mentioned above the flame has a small concentration of sodium atoms in it which are excited thermally so that these atoms emit radiation at wavelengths corresponding to the two sodium D lines.

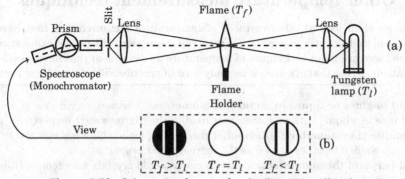

Figure 4.50: Schematic of set up for the 'line reversal' technique

---

[11]First observed by German physicist Joseph von Fraunhofer (1787 - 1826) as dark absorption lines in the solar spectrum.

As indicated the spectroscope slit is illuminated by the radiation from the flame as well as radiation from a tungsten filament placed behind the flame. The two lenses are arranged so as to bring the light gathered from the lamp as well as the flame to come to focus at the slit. Let the flame be at temperature T_f while the filament is at a brightness temperature of T_l.

Let the spectral intensity of radiation leaving the lamp and that is incident on the flame be $I_{\lambda,0} = I_{b,\lambda}(T_l)$. This radiation passes through a distance L in negotiating the flame. If the flame has a constituent that absorbs this radiation the incident intensity will undergo an attenuation because of absorption. At the same time it gets enhanced because of emission by the flame. The process of change in the intensity is governed by the equation of radiation transfer.[12] As derived in the cited reference the intensity leaving the flame would be given by

$$I_{L,\lambda} = I_{b,\lambda}(T_l)\left(1 - e^{\kappa_\lambda L}\right) + I_{b,\lambda}(T_f)e^{\kappa_\lambda L} \qquad (4.59)$$

where κ_λ is the spectral absorption coefficient of the flame.

Equation 4.59 shows that $I_{L,\lambda} > I_{b,\lambda}(T_l)$ if $T_f > T_l$. The sodium D lines appear brighter as compared to the background, as indicated in the first illustration in Figure 4.50(b). Thus the sodium lines appear as bright lines in a dark background. Equation 4.59 indicates that $I_{L,\lambda} < I_{b,\lambda}(T_l)$ if $T_f < T_l$. The sodium D line appears duller as compared to the background, as indicated in the last illustration in Figure 4.50(b). Thus the sodium lines appear as dark lines in a bright background. However when $T_f = T_l$, we have $I_{L,\lambda} = I_{b,\lambda}(T_l)$ and the sodium lines disappear in the background, as shown in the middle illustration in Figure 4.50(b) . The lamp brightness temperature is varied systematically to achieve the disappearance of the lines in the background produced by the radiation from the lamp. As we vary the lamp temperature around the flame temperature the sodium lines change from absorption lines to emission lines and the line radiation is said to reverse and hence the name 'line reversal'. The brightness temperature of the lamp under this condition gives the temperature of the flame.

4.6   Other temperature measurement techniques

Now we shall look at other methods of temperature measurements that have not been placed directly under the ITS90. The discussion will not be exhaustive because there are many techniques of temperature measurement that have specific applications. Those that are more commonly used in practice will be considered here.

Liquid in glass or liquid in metal thermometers: Thermal expansion of volume of a liquid with temperature is used as the thermometric property.

Bimetallic thermometers: Differential thermal expansion between two materials bonded together is used as the thermometric property.

Liquid crystal thermometers: Color change of liquid crystals with temperature is used as the thermometric property.

IC temperature sensor: Junction semiconductor devices for temperature measurement.

[12]Chapter 12 - S.P. Venkateshan, Heat Transfer (Second Edition), Ane Books, 2011

4.6.1 Liquid in glass or liquid in metal thermometers

Liquids contained in a rigid vessel expand much more than the vessel itself and hence there will be a net change in the volume of the liquid contained within the vessel. The volume change is converted to a length change by a suitable arrangement to construct a thermometer. Liquid in glass thermometers cover a wide temperature range using different liquids and are capable of good accuracy. Changes in temperature that are of the order of hundredth of a degree are accurately measurable. These thermometers are used in measuring temperatures of baths and in clinical practice for taking the temperature of patients. These are also used in precision thermometry.

The constructional details of a typical liquid in glass thermometer are given in Figure 4.51. It consists of a glass bulb that holds the liquid and is the part that comes in to contact with the process environment. The bulb is connected to a capillary of uniform bore. When the liquid expands it rises in the capillary. A scale on the stem of the thermometer gives the temperature reading. An immersion ring on the stem indicates the depth to which the stem must be immersed in the process space. This corresponds to the depth of immersion used during the calibration of the thermometer. Liquid in glass thermometer may use either partial immersion as in Figure 4.51 or total immersion in which case the entire thermometer should come in contact with the process space. A total immersion thermometer will give erroneous results if partially immersed. The capillary also has a contraction chamber on the low side and an expansion chamber on the high side. The expansion chamber prevents damage to the thermometer in case the thermometer is exposed to a temperature above the maximum of the useful range.

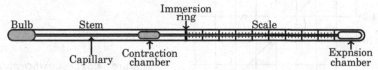

Figure 4.51: *Construction details of a typical liquid in glass thermometer*

The best accuracy for a total immersion thermometer with mercury in glass is from 0.01 °C for low temperature range up to 150°C. For higher temperature range up to 500°C the accuracy is limited to 1°C. The corresponding numbers for partial immersion would, however, be 0.1°C and 2°C respectively. The properties of the liquid such as the freezing point and the boiling point limit the range for each liquid. Table 4.12 lists a few of the liquids along with the allowable temperature ranges. In the case of mercury the upper limit shown is possible if it is under a pressure greater than atmospheric pressure. The coefficient of cubical expansion of liquids that are used in thermometry are temperature dependent and hence this will have to be taken in to consideration while designing a liquid in glass thermometer. For example, the volume of mercury follows the quadratic relation

$$V_t = V_0[1 + 18.18 \times 10^{-5}t + 15.6 \times 10^{-9}t^2] \qquad (4.60)$$

Table 4.12: *Liquids useful in thermometry*

Liquid	Temperature Range in $°C$
Pentane	-200 to 30
Alcohol	-75 to 70
Toluene	-75 to 100
Creosote	-40 to 200
Xylene	-40 to 400
Mercury	-30 to 520

Here V_t is the volume of mercury at temperature $t°C$ and V_0 is the temperature at $0°C$. It is thus seen that the coefficient of cubical expansion defined as $\beta = \dfrac{1}{V}\dfrac{dV}{dt}$ is given, for mercury, by the relation

$$V_t = \frac{18.18 \times 10^{-5} + 15.6 \times 10^{-9}t}{1 + 18.18 \times 10^{-5}t + 15.6 \times 10^{-9}t^2} \tag{4.61}$$

A plot of cubical expansion of mercury over its useful range is shown in Figure 4.52. There is roughly a 5% decrease in the cubical expansion of mercury over the range. Useful data regarding the cubical expansion and bulk modulus of liquids and solids (used as bulb-capillary material) is given in Table 4.13. Liquid expansion

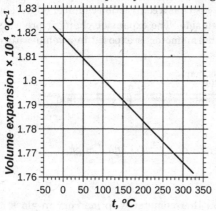

Figure 4.52: *Coefficient of cubical expansion of mercury*

Table 4.13: *Coefficient of cubical expansion and bulk modulus of some useful materials*

Solid	β	Bulk Modulus GPa	Liquid	β	Bulk Modulus GPa
Aluminum	2.3×10^{-5}	75.5	Alcohol	1.12×10^{-3}	1.47
Platinum	9×10^{-6}	228	Benzene	1.24×10^{-3}	1.03
Glass	2.76×10^{-5}	43	Mercury	1.81×10^{-3}	28.6
Steel	2×10^{-5}	160	Water	2.1×10^{-3}	2.24

thermometers may also be used by containing the liquid in a rigid metal bulb connected to a metal tube as shown in Figure 4.53(a). The expansion of the liquid is curtailed by the rigid metal design and hence the pressure increases. The pressure is measured by a pressure sensor such as a Bourdon gage (see section 7.3). The metal bulb is exposed to the process environment, as shown. It is possible to compensate for the effect of ambient temperature on the measurement by using a second capillary that is exposed to the ambient as indicated in Figure 4.53(b). In this case the output is communicated to a differential pressure transducer.

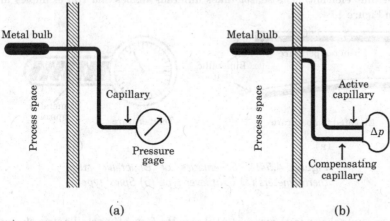

(a)   (b)

Figure 4.53: *Construction details of liquid in metal thermometers*

Example 4.17

Mercury has a cubical expansion of $18.1 \times 10^{-5}\,°C^{-1}$ while glass has a coefficient of cubical expansion of $2.76 \times 10^{-5}\,°C^{-1}$. A thermometer is constructed with the bulb having a volume of $200\ mm^3$. The capillary has a bore of $0.2\ mm$. What is the sensitivity of this thermometer?

Solution :

We are given the following data:

Cubical expansion of mercury: $\beta_{Hg} = 18.1 \times 10^{-5}\,°C^{-1}$

Cubical expansion of Glass: $\beta_{Glass} = 2.76 \times 10^{-5}\,°C^{-1}$

Relative cubic expansion is thus given by

$$\beta_{Rel} = \beta_{Hg} - \beta_{Glass} = (18.1 - 2.76) \times 10^{-5} = 15.34 \times 10^{-5}\,°C^{-1}$$

This is also the fractional change in the volume of mercury for a change of temperature of $1°C$. The change in volume of mercury is then given by

$$\Delta V = \beta_{Rel} V = 15.34 \times 10^{-5} \times 200 = 30.68 \times 10^{-3}\ mm^3\,°C^{-1}$$

This change in volume gives rise to a change in the length of mercury column in the capillary given by $A_c \Delta L$ where A_c is the area of cross section of the capillary, given by $A_c = \pi \dfrac{D^2}{4} = \pi \dfrac{0.2^2}{4} = 0.01\pi\ mm^2$. Noting that ΔL is in fact the sensitivity S of the thermometer, we have

$$S = \Delta L = \frac{\Delta V}{A_c} = \frac{30.68 \times 10^{-5}}{0.01 \times \pi} = 0.977 \approx 1\ mm\,°C^{-1}$$

4.6.2　Bimetallic thermometer

Different materials expand by different extents when heated. The coefficient of thermal expansion defined as $\alpha = \dfrac{1}{L}\dfrac{dL}{dt}$ varies from material to material. This property can be gainfully used by constructing a temperature sensor in the form of a bimetallic element. The sensor takes different shapes and two examples are shown in Figure 4.54.

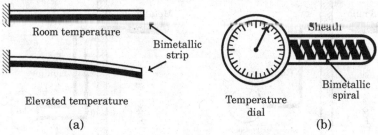

(a)　　　　　　　　　　　　　　　　　　　　　　(b)

Figure 4.54: *Schematics of Bimetallic strip thermometers (a) Cantilever type (b) Spiral type*

In the example shown in Figure 4.54(a), it is in the form of bimetallic strip element that is fixed at one end. The element will be straight at a particular temperature that will be referred to as the reference temperature t_0. When the temperature of the element changes it will bend as indicated, the extent of the bend being a function of temperature. In the second example shown in Figure 4.54(b), the sensing bimetallic element is in the form of a helical coil that winds or unwinds depending on the temperature. In both cases a suitable mechanical arrangement may be used to transmit the movement to a dial that may be marked in temperature units. Thus the bimetallic thermometer does not require any external power for its operation.

Bimetallic element analysis

Consider a bimetallic element fabricated by joining together two strips of materials A and B as shown in Figure 4.55(a). The two strips that make up the bimetallic element have identical lengths at a certain temperature t_0 (this may be the room temperature or some other temperature such as the temperature at which the bimetallic element is fabricated by joining the two strips together). The bimetallic element is flat at this temperature. When the temperature is different from t_0 the bimetallic element takes on a bent or curved shape because of the different coefficients of thermal expansions of the two materials. For example, if the coefficient of thermal expansion α_A of material A is more than α_B, the coefficient of thermal expansion of material B, the bimetallic element will take the bent form as shown in Figure 4.55(a) for $t > t_0$. The bent shape is characterized by the radius of curvature R. Strip A tends to expand more than strip B and hence internal stresses are developed as indicated Figure 4.55(b). For a simplified analysis we assume that both materials have the same Young's modulus and are of the same thickness δ. The outer surface of strip A is in tension while at the interface it is in compression. Strip B experiences just the opposite as shown in Figure 4.55(b). At the interface plane between the two strips the compressive stress in strip A must be equal to the tensile stress in B. Hence the internal stresses developed in the two materials are equal and opposite.

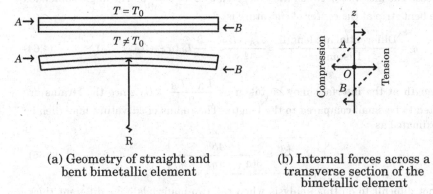

(a) Geometry of straight and bent bimetallic element	(b) Internal forces across a transverse section of the bimetallic element

Figure 4.55: *Bimetallic element analysis*

Consider the forces across the thickness of the strip A. These give rise to a net bending moment leading to the bent shape. The moment is clockwise as indicated in the figure. The forces across strip B also give rise a bending moment in the same sense. We notice that there can be no net moment about O on the interface since there is no external applied moment. Hence each of the moments taken with respect to O (either in material A or material B) should vanish.

Now consider the strain as function of location across the section of material A. Let z represent the coordinate measured with O as the reference. The strain is a linear function of z and may be taken as $s = a - bz$. This form is chosen since the stress will undergo a change in sign somewhere within the thickness of the strip. The corresponding stress is $\sigma = E(a - bz)$ where E is the Young's modulus. The corresponding force is obtained by multiplying this by an elemental thickness dz. The elemental moment about O due to this force is obtained as $dM_O = E(a-bz)zdz = E(az - bz^2)dz$. We integrate this with respect to z from $z = 0$ to $z = t$ (the thickness of strip A or B) and set it equal to zero to get a relation between the coefficients a, b and the thickness δ.

$$M_O = \int_0^\delta E(az - bz^2)dz = E\left(\frac{a\delta^2}{2} - \frac{b\delta^3}{3}\right) = 0 \qquad (4.62)$$

Thus we have

$$\frac{a}{b} = \frac{2}{3}\delta \qquad (4.63)$$

Thus the strain at any location z is given by $s(z) = A\left(1 - \frac{3z}{2\delta}\right)$. The strain is zero at $z = \frac{2}{3}\delta$. Thus the strip A has its original length at this location! Similar analysis would indicate that at $z = -\frac{2}{3}\delta$ in material B it has its original length. Let the bimetallic strip in its flat state have a length equal to L_0. At temperature $t > t_0$ the length of element A would be $L_A = L_0[1 + \alpha_A(t - t_0)]$. Similarly, for the material B,

$L_B = L_0[1 + \alpha_b(t - t_0)]$. These *in fact* are the lengths at the location where there is no stress! The distance between these is given by $2 \times \frac{2}{3}\delta = \frac{4}{3}\delta$. The angle subtended by the bent strip at the center of curvature is

$$\phi = \frac{\text{Differential arc length}}{\text{Distance}} = \frac{L_A - L_B}{\frac{4}{3}\delta} = \frac{3}{4\delta}L_0(\alpha_A - \alpha_B)(t - t_0) \qquad (4.64)$$

The length at the interface may be taken as $\frac{L_A + L_B}{2} \approx L_0$ since the strains are expected to be small compared to the length. The radius of curvature may then be approximated as

$$\boxed{R \approx \frac{L_0}{\phi} = \frac{4\delta}{3(\alpha_A - \alpha_B)(t - t_0)}} \qquad (4.65)$$

It is not difficult to do this analysis when the two materials have different thicknesses and different elastic constants. The ratio of the elastic moduli $r_E = \frac{E_B}{E_A}$ and the ratio of thicknesses $r_\delta = \frac{\delta_B}{\delta_A}$ also appear in the formula for the radius of curvature. Note that the subscript B refers to the material of lower coefficient of thermal expansion. The corresponding expression for the radius of curvature is given by

$$\boxed{R = \frac{(\delta_A + \delta_B)\{3(1 + r_\delta)^2 + (1 + r_\delta r_E)[r_\delta^2 + \frac{1}{r_\delta r_E}]\}}{6(\alpha_A - \alpha_B)(t - t_0)(1 + r_\delta)^2}} \qquad (4.66)$$

Note that Equation 4.66 reduces to Equation 4.65 when $r_E = r_\delta = 1$. Data useful in analyzing bimetallic elements is presented in Table 4.14.

Example 4.18

A bimetallic strip has been constructed of strips of yellow brass and Invar bonded together at room temperature of $t_0 = 25°C$. Each strip is 0.2 *mm* thick. What is the radius of curvature of the strip and the angle subtended at the center of curvature when a length of 25 *mm* is subjected to steam point at 1 atmosphere pressure? If it is used in the cantilever arrangement shown in Figure 4.56 what will be the deflection at the free end?

Solution :

In solving this problem all lengths are used in millimeters. The properties of the two materials are given below. *A* refers to yellow brass and *B* to invar. The properties of the two materials are recalled here:

$$\alpha_A = 2.02 \times 10^{-5°}C^{-1}; \alpha_B = 1.7 \times 10^{-6°}C^{-1}$$

$$E_A = 96.5\,GPa,\ \delta_A = 0.2\,mm;\quad E_B = 147\,GPa,\ \delta_B = 0.2\,mm$$

The reference temperature is $t_0 = 25°C$. The required ratios are calculated as:

$$r_\delta = \frac{t_B}{t_A} = \frac{0.2}{0.2} = 1 \text{ and } r_E = \frac{E_B}{E_A} = \frac{147}{96.5} = 1.523$$

Table 4.14: Coefficient of thermal expansion and Young's modulus of some materials

Substance	Coefficient of Thermal Expansion $\alpha\,°C^{-1}$	Young's Modulus Gpa
Aluminum	24×10^{-6}	69
Brasses	19×10^{-6}	100 - 125
Yellow Brass	20.2×10^{-6}	96.5
Bronzes	18×10^{-6}	100 - 125
Copper	17×10^{-6}	124
Glass (ordinary)	9×10^{-6}	50 - 90
Glass (Pyrex)	3×10^{-6}	67
Inconel 702	12.5×10^{-6}	217
Invar*	1.7×10^{-6}	147
Iron	12×10^{-6}	196
Lead	29×10^{-6}	16
Monel 400**	12.5×10^{-6}	179
Stainless Steel	16×10^{-6}	193
Nickel	13.6×10^{-6}	214

*Invar is an alloy steel containing 36%Nickel

**Monel is a class of stainless metal alloys containing up to 67% Nickel, Copper and some Iron and other trace elements

The temperature at which the radius of curvature of the bimetallic element is desired is $t = 100°C$. We make use of Equation 4.66 to calculate R.

$$R = \frac{(0.2+0.2)\{3(1+1)^2+(1+1\times1.523)\left[1^2+\frac{1}{1\times1.523}\right]\}}{6(2.02\times10^{-5}-1.7\times10^{-6})(100-25)(1+1)^2} = 194.35\,mm$$

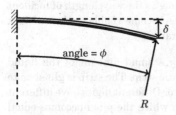

angle = ϕ

Figure 4.56: Bimetallic element geometry in Example 4.18

Length of bimetallic strip is given to be $L = 25mm$. The angle subtended at the center of curvature is then given by

$$\phi = \frac{L}{R} = \frac{25}{194.35} = 0.129\,rad$$

Referring to Figure 4.56 we see that the deflection of the free end is given by

$$\delta = R(1-\cos\phi) = 194.35(1-\cos0.129) = 1.606\,mm$$

The arrangement shown in Example 4.18 is typical of a bimetallic element used as an on-off controller. The current to a heater passes through the bimetallic element that is exposed to the process space. If the temperature of the process space exceeds a set point (set by a mechanical arrangement) the bimetallic strip will open out thus switching off the current. When the process space dips below the set point the bimetallic strip straightens out and turns on the current to the heater. Another arrangement that is possible is shown in Figure 4.57.

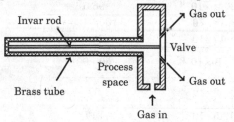

Figure 4.57: *Thermostat for maintaining constant temperature in a process space*

The "thermostat" consists of an invar valve push rod surrounded by a brass tube. The two are intimately connected at one end. The brass tube is fixed at the other end to a rigid support. The Invar push rod is connected to a valve disk at the other end. The differential expansion between the brass tube and the Invar rod controls the valve position and hence the flow rate of the gas that is supplied to the burner that provides the heat to the process space.

4.6.3   Liquid crystal thermometers

The liquid crystal used in a thermometer is a *chiral nematic liquid crystal* that has a natural twisted structure such that the preferred molecular orientation is such as to give it a corkscrew shape. The pitch of the screw structure may be a few nanometers to many microns. When visible light is incident on the liquid crystal it gets reflected if the pitch is equal to the wavelength of the incident light. The pitch of the liquid crystal varies with temperature. The liquid crystal has a narrow temperature range over which its pitch will be the same as the wavelength of incident light. Hence a liquid crystal will reflect light and thus become visible against a black background when the temperature is in the correct range.

A liquid crystal thermometer is made by having a small amount of the liquid crystal behind a digit shaped opening on a black plastic strip. The strip is glued on to the surface whose temperature one wishes to measure. Different digits have different liquid crystals with different temperature ranges in which the pitch becomes equal to the wavelength of the incident light. Hence the appropriate digit will "light up" when the temperature of the object has the corresponding value. Liquid crystal thermometers have a narrow temperature range and are used in applications such as the measurement of a patient's temperature for clinical purpose. It may also be used while performing experiments with temperatures close to the ambient temperature.

Liquid crystal sheets with adhesive on the back are available that may be attached to a surface whose temperature needs to be measured. If the surface has non uniform temperature the reflected light will show variation of hue that may be easily related to the temperature field.

4.6.4   IC temperature sensor

Integrated chip temperature sensors are available to cover a range of temperatures from -55 to +150 $^{\circ}C$. They are fairly linear with a response of 1 $\mu A/K$, for a typical IC AD 590 series available from Analog Devices. Alternately LM 135 series of IC temperature sensor from National Semiconductors may be used. This IC operates as a two terminal Zener (Figure 4.58) with breakdown voltage directly proportional to absolute temperature of +10 mV/K. When calibrated at 25 $^{\circ}C$ it has less than 1$^{\circ}C$ error over a 100$^{\circ}C$ range. The output voltage of the sensor may be represented by

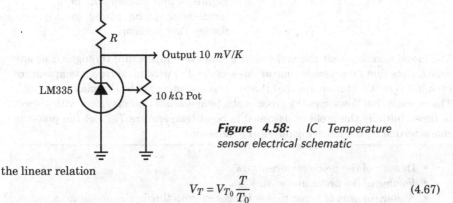

Figure 4.58: IC Temperature sensor electrical schematic

the linear relation

$$V_T = V_{T_0} \frac{T}{T_0} \tag{4.67}$$

Note that the temperatures are in K. The output of the IC sensor is represented as V and the subscript represents the temperature. The reference temperature is usually taken as 25$^{\circ}C$. Calibration of the sensor is done by connecting a 10 $k\Omega$ potentiometer as shown in Figure 4.58 and with the arm tied to the adjustment terminal of the IC. The calibration is done at the single temperature of 25$^{\circ}C$ at which the output should be 2.982 V (= 298.2 K × 10mV/K).

4.7   Measurement of transient temperature

Many processes of engineering relevance involve variations of process parameters with respect to time. The system variables like temperature, pressure and flow rate may vary with time. These are referred to as transients and the measurement of these transients is an important issue while designing or choosing the proper measurement technique and the probe. Here we look at the measurement of temperature transients.

4.7.1   Temperature sensor as a first order system - Electrical analogy

Let us look at a typical temperature measurement situation. We visualize the temperature probe as a system that is subject to the temperature transient. The probe is exposed to the environment *whose* temperature changes with time and it is desired to measure this temperature change by following it as closely as possible. In Figure 4.59 we show the schematic of the thermal model appropriate for this study.

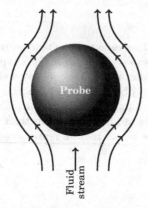

Probe temperature:	$T(t)$
Probe characteristic length:	L_{ch}
Fluid temperature:	T_∞
Heat transfer coefficient:	h

Fluid stream

Figure 4.59: *Schematic of a temperature probe placed in a flowing fluid medium*

The model assumes that the probe is at a uniform temperature throughout at any time t. Note that t represents time in this section. T is used to represent temperature even if it is in $^\circ C$. This means that the probe is considered to be thermally "lumped". The medium that flows over the probe is at a temperature that may vary with respect to time. Initially the probe is assumed to be at temperature T_0. Let the probe be characterized by the following physical parameters:

- Density of the probe material: ρ, kg/m^3
- Volume of the probe material: V, m^3
- Surface of area of probe that is in contact with the flowing medium: S, m^2
- Specific heat of probe material: C, $J/kg\,^\circ C$
- Convection heat transfer coefficient for heat transfer between the probe and the process fluid: h, $W/m^{2\,\circ}C$

By conservation of energy, we have

$$\left[\begin{array}{c}\text{Rate of change of energy}\\\text{stored in the probe}\end{array}\right] = \left[\begin{array}{c}\text{Rate of heat transfer between}\\\text{the probe and the flowing fluid}\end{array}\right] \qquad (4.68)$$

If we assume that the probe is at a higher temperature as compared to the fluid heat transfer will be from the probe to the fluid and the internal energy of the probe will reduce with time. Using the properties of the probe introduced above, the left hand side of Equation 4.68 is given by $-\rho V C \dfrac{dT}{dt}$. The right hand side of Equation 4.68 is given by $hS(T - T_\infty)$. Conservation of energy in equation form is then given by

$$-\rho V C \frac{dT}{dt} = hS(T - T_\infty) \qquad (4.69)$$

After some rearrangement, Equation 4.69 takes the form

$$\frac{dT}{dt} + \frac{hS}{\rho CV}T = \frac{hS}{\rho CV}T_\infty \qquad (4.70)$$

Note that this equation *holds* even when the probe temperature is lower than the fluid temperature. The quantity $\dfrac{\rho CV}{hS}$ has the unit of time and is referred to as the time constant τ of the first order system (first order since the governing differential equation is a first order ordinary differential equation). The first order time constant involves thermal and geometric properties. The volume to surface area ratio is a characteristic length dimension and is indicated as L_{ch} in Figure 4.59. Noting that the product of density and volume is the mass m of the probe, the first order time constant may also be written as $\dfrac{mC}{hS}$. The time constant may be interpreted in a different way also, using electrical analogy. The quantity mC represents the thermal capacity and the quantity $\dfrac{1}{hS}$ represents the thermal resistance. Based on this interpretation an electric analog may be made as shown in Figure 4.60. In the

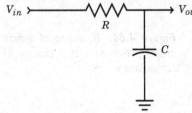

Figure 4.60: *Electrical analog of a first order thermal system*

electric circuit shown in Figure 4.60 the input voltage represents the temperature of the fluid, the output voltage represents the temperature of the probe, the resistance R represents the thermal resistance and the capacitance C represents the thermal mass (mass specific heat product) of the probe. Equation 4.70 may be rewritten as

$$\frac{dT}{dt} + \frac{T}{\tau} = \frac{T_\infty}{\tau} \tag{4.71}$$

Equation 4.71 may be simplified using the integrating factor $e^{t/\tau}$ to write it as

$$\frac{de^{t/\tau}T}{dt} = T_\infty e^{t/\tau} \tag{4.72}$$

This may be integrated to get $T = e^{-t/\tau}\left[\int_0^t T_\infty e^{t/\tau}dt + A\right]$ where A is a constant of integration. Using the initial condition $T(t=0) = T_0$ we get

$$T(t) = e^{-t/\tau}\left[T_0 + \int_0^t T_\infty(t)e^{t/\tau}dt\right] \tag{4.73}$$

This is the general solution to the problem. If the variation of fluid temperature with time is specified, we may perform the indicated integration to obtain the response of the probe as a function of time.

4.7.2 Response to step input

If the fluid temperature T_∞ is constant but different from the initial temperature T_0 of the probe, the solution is easily shown to be represented by

$$\varphi = \frac{T - T_\infty}{T_0 - T_\infty} = e^{-t/\tau} \tag{4.74}$$

Note that the temperature ratio φ is non-dimensional and hence the solution represents a universal solution to a first order system subject to a step input. The temperature difference between the probe and the fluid exponentially decreases with time. The variation is indicated in Figure 4.61. At the end of one time constant the temperature difference is some 37% of the initial temperature difference. After about 5 time constants the temperature difference is quite negligible. This means that the system has come to equilibrium with the medium. A step input may be

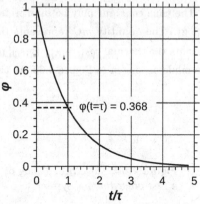

Figure 4.61: *Response of a first order system to a step change in temperature*

experimentally realized by heating the probe to an initial temperature exceeding the fluid temperature and then exposing it quickly to the fluid environment. The probe temperature is recorded as a function of time. If it is plotted in the form $\ln(\varphi)$ as a function of t, the slope of the line is negative reciprocal of the time constant. In fact, this is one method of measuring the time constant. Example 4.19 shows how this is done.

Example 4.19

A temperature probe was heated by immersing it in boiling water and then it was quickly transferred in to a fluid medium at a temperature of $T_{amb} = 25°C$. The temperature difference between the probe and the medium is recorded below at various time instances:

t, s	0.35	0.6	0.937	1.438	2.175	3.25
$T - T_{amb}, °C$	60	50	40	30	20	10

What is the time constant of the probe in this situation?

Solution :

The data is plotted on a semi-log graph as shown in Figure 4.62. It is seen that it is well represented by a straight line whose equation is given as an inset in the plot. EXCEL was used to obtain the best fit. The slope of the trend line is -0.6065 and hence the time constant is obtained as

$$\tau = -\frac{1}{slope} = \frac{1}{0.6065} = 1.6487 \approx 1.65 \; s$$

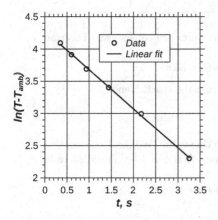

Fit equation:

$$\ln(T - T_{amb}) = 4.2837 - 0.6065t$$
$$R^2 = 0.9989$$

Figure 4.62: *Semi-log plot of data in Example 4.18*

A note on time constant It is clear from our discussion above that the time constant of a system (in this case the temperature probe) is not a property of the system. It depends on parameters that relate to the system as well as the parameters that define the interaction between the system and the surrounding medium (whose temperature we are trying to measure, as it changes with time). The time constant is the ratio of thermal mass of the system to the conductance (reciprocal of the thermal resistance) between the system and the medium. It is also clear now how we can manipulate the time constant. Thermal mass reduction is one possibility. The other possibility is the reduction of the thermal resistance. This may be achieved by increasing the interface area between the system and the medium. In general this means a reduction in the characteristic dimension L_{ch} of the system. A thermocouple attached to a thin foil will accomplish this. The characteristic dimension is equal to half the foil thickness, if heat transfer takes place from both sides of the foil. Another way of accomplishing this is to use very thin thermocouple wires so that the bead at the junction has very small volume and hence the thermal mass. Indeed these are the methods used in practice and thin film sensors are commercially available. Lastly, if time constant data is to be specified for a particular installation it is necessary also to specify the thermal environment in which the time constant has been determined. For example, time constant is specified, for a particular thermistor as follows:

- Time constant is $0.14\,s$ (in water at room temperature)
- Time constant is $2\,2.2\,s$ (in still air at $5-25°C$)

Example 4.20

The time constant of a first order thermal system is given as $0.55\,min$. The uncertainty in the value of the time constant is given to be $\pm0.01\,min$. The initial temperature excess of the system over and above the ambient temperature is $45°C$. Determine the system temperature excess and its uncertainty at the end of $50\,s$ from the start.

Hint: It is known that the temperature excess follows the formula $\dfrac{T(t)}{T(0)} = e^{-t/\tau}$ where $T(t)$ is the temperature excess at any time t, $T(0)$ is the temperature excess at $t = 0$ and τ is the time constant.

Solution :

We shall convert all times given to s so that units are consistent. The time constant is

$$\tau = 0.55\,min = 0.55 \times 60 = 33\,s$$

We need the temperature excess at $t = 50\,s$ from the start. Hence

$$T(50) = T(0)e^{-t/\tau} = 45 \times e^{-\frac{50}{33}} = 9.89°C$$

We would like to calculate the uncertainty in this value. We shall assume that this is due to the error in time constant alone that has been specified to be $\pm 0.1\,min$.

$$\Delta\tau = \pm 0.1\,min = \pm 0.1 \times 60 = \pm 0.6\,s$$

The influence coefficient I_τ is given by

$$I_\tau = \left.\frac{\partial T}{\partial \tau}\right|_{t=50\,s} = \left.T(0)e^{-t/\tau}\frac{t}{\tau^2}\right|_{t=50\,s} = 45 \times e^{-\frac{50}{33}} \times \left[\frac{50}{33^2}\right] = 0.454°C/s$$

Hence the uncertainty in the estimated temperature excess is:

$$\Delta T(50) = \pm I_\tau \Delta\tau = \pm 0.454 \times 0.6 = \pm 0.272°C$$

Example 4.21

A certain first order system has the following specifications:

- Material in the form of a shell: copper
- Shell wall thickness: $\delta = 1\,mm$
- Shell outer radius: $R = 6\,mm$
- Initial temperature of shell: $T_0 = 50°C$
- Fluid; Air at $T_{amb} = 30°C$

How long should one wait for the temperature of the shell to reach $40°C$? Assume that heat transfer is by free convection. Use suitable correlation (from a heat transfer text) to solve the problem.

Solution :

Step 1 Heat transfer coefficient calculation

Heat transfer from shell to air is by natural convection. The appropriate correlation for the Nusselt number is given by $Nu = 2 + 0.43Ra^{0.25}$ where Ra is the Rayleigh number. The characteristic length scale is the sphere (shell) diameter $D = 2R$. The air properties are calculated at the mean temperature $T_m(0) = \frac{T_0 + T_{amb}}{2}$ at $t = 0$. From the given data, we have

$$
\begin{aligned}
OD &= 2R = 2 \times 6 = 12\,mm = 0.012\,m \\
ID &= OD - 2\delta = 0.012 - 2 \times 0.001 = 0.01\,m \\
T_m(0) &= \frac{50 + 30}{2} = 40°C
\end{aligned}
$$

The air properties required are read off Table B.7 in Appendix B at $40°C$:

$$v = 17.07 \times 10^{-6} \, m^2/s; \ Pr = 0.0.699 \ \text{ and } \ k = 0.0274 \, W/m°C$$

In the above Pr is the Prandtl number of air. The isobaric compressibility β of air is calculated based on the ideal gas assumption. Thus

$$\beta = \frac{1}{T_{amb}} = \frac{1}{273 + 30} = 3.3 \times 10^{-3} \, K^{-1}$$

The temperature difference for calculating the Rayleigh number is taken as the mean shell temperature during the cooling process minus the ambient temperature. We are interested in determining the time to cool from 50 to $40°C$. Hence the mean shell temperature is $\overline{T}_{shell} = \dfrac{50 + 40}{2} = 45°C$. The temperature difference is $\Delta T = \overline{T}_{shell} - T_{amb} = 45 - 30 = 15°C$. The value of the acceleration due to gravity is taken as $g = 9.8 \, m/s^2$. The Rayleigh number is then calculated as

$$Ra = \frac{g\beta\Delta T D^3}{v^2} Pr = \frac{9.8 \times 3.3 \times 10^{-3} \times 15 \times 0.012^3}{(17.07 \times 10^{-6})^2} \times 0.699 = 2011$$

The Nusselt number is then calculated as

$$Nu = 2 + 0.43 Ra^{0.25} = 2 + 0.43 \times 2011^{0.25} = 4.88$$

Thus the heat transfer coefficient is

$$h = \frac{Nu \cdot k}{D} = \frac{4.88 \times 0.0274}{0.012} = 11.14 W/m^2°C$$

Step 2 Time constant calculation
Copper shell properties are

$$\rho = 8954 \, kg/m^3; \ C = 383.1 \, J/kg°C$$

With copper shell thickness of $\delta = 0.001 \, m$ mass of the copper shell is calculated as

$$m = \rho\pi \frac{OD^3 - ID^3}{6} = 8954 \times \pi \times \frac{0.012^2 - 0.01^2}{6} = 3.413 \times 10^{-3} \, kg$$

Surface area of shell exposed to the fluid is calculated as

$$S = \pi OD^2 = \pi \times 0.012^2 = 4.524 \times 10^{-4} \, m^2$$

The time constant is then estimated as

$$\tau = \frac{mC}{hS} = \frac{3.413 \times 10^{-3} \times 383.1}{11.14 \times 4.524 \times 10^{-4}} = 259.5 \, s$$

Step 3 Time required for cooling
Cooling follows an exponential process. We take natural logarithms of Equation 4.74 (after replacing T_∞ by T_{amb}) and solve for t to get

$$t = -259.5 \times \ln\left[\frac{40 - 30}{50 - 30}\right] = 179.8 \, s \approx 180 \, s$$

/

4.7.3   Response to a ramp input

In applications involving material characterization heating rate is controlled to follow a predetermined heating program. The measurement of the corresponding temperature is to be made so that the temperature sensor follows the temperature very closely. Consider the case of linear heating and *possibly* linear temperature rise of a medium. Imagine an oven being turned on with a constant amount of electrical heat input. We would like to measure the temperature of the oven given by

$$T_\infty(t) = T_0 + Rt \tag{4.75}$$

The general solution to the problem is given by (using Equation 4.73)

$$e^{t/\tau}T = A + \frac{T_0}{\tau}\int_0^t e^{t/\tau}dt + \frac{R}{\tau}\int_0^t te^{t/\tau}dt \tag{4.76}$$

where A is a constant of integration. The first integral on the right hand side of Equation 4.76 is easily obtained as $\tau(e^{t/\tau} - 1)$. Second integral on the right hand side is obtained by integration by parts, as follows.

$$\begin{aligned}\int_0^t te^{t/\tau}dt &= t\tau e^{t/\tau}\Big|_0^t - \tau\int_0^t e^{t/\tau}dt \\ &= t\tau e^{t/\tau} - \tau^2(e^{t/\tau} - 1) \end{aligned} \tag{4.77}$$

If the initial temperature of the first order system is T_i, then $A = T_i$, since both the integrals vanish for $t = 0$ (the lower and upper limit will be the same). On rearrangement, the solution is

$$\boxed{T(t) = (T_i - T_0 + R\tau)e^{-t/\tau} + (T_0 + Rt - R\tau)} \tag{4.78}$$

We notice that as $t \to \infty$ the transient exponentially decaying part tends to zero and the steady part (this part survives for $t \gg \tau$) yields

$$T_0 + Rt - T(t) = T_\infty(t) - T(t) \approx R\tau \tag{4.79}$$

The steady state response has a lag equal to $R\tau$ with respect to the input. This may be treated as a "systematic error"! Figure 4.63 shows the response of first order system to a ramp input. The case shown corresponds to $T_i = 20°C, T_0 = 35°C, R = 0.15°C/s$ and $\tau = 10\ s$. For $t \to \infty$ (say, $t > 5\tau = 50\ s$) the probe follows the linear temperature rise with a lag of $R\tau = 0.15 \times 10 = 1.5°C$. In this case it is advisable to treat this as a systematic error and add it to the indicated temperature to get the correct oven temperature, after the oven is on for more than a minute. Note that the large time behavior of the temperature sensor is independent of its initial temperature. If $T_i > T_0$ the sensor response will cross over, after a short time such that the input is *higher* than the output! The reader should make calculations for the case shown in Figure 4.63 but with $T_i = 45°C$ to appreciate this.

Example 4.22

A thermistor has a dissipation constant of $P_D = 2\ mW/°C$ and a time constant of $\tau = 1\ s$. This thermistor is attached to an electric heater. Initially both the heater and the thermistor are at room temperature of $T_0 = T_i = 30°C$. The

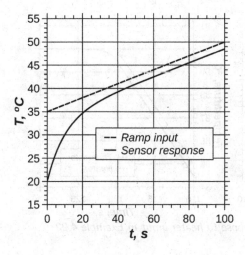

Figure 4.63: *Typical response of a first order system to ramp input*

heater is turned on at a certain instant. The thermistor starts recording the temperature at the same time. The heater temperature is expected to increase linearly with time at a constant rate of $R = 0.5°C/s$. Heating continues till the temperature of heater rises to $90°C$ after which it remains constant. Make a plot of thermistor response with time.

Solution :

Since the heater temperature increases at a uniform rate till it reaches a temperature of $90°C$, the response of the thermistor during this time follows that of a first order system subject to ramp input. Since $T_i = T_0 = 30°C$ the solution is given by Equation 4.78 which modifies to

$$T(t) = (T_i - T_0 + R\tau)e^{-t/\tau} + (T_0 + Rt - R\tau) = 0.5e^{-t} + [30 + 0.5(t - 1)]$$

The heater temperature reaches $90°C$ at $t = 120\ s$ as may easily verified. At this time, the temperature of the thermistor is

$$T(t) = 0.5e^{-120} + [30 + 0.5(120 - 1)] = 89.5°C$$

The temperature of thermistor follows the step input response for $t > 120\ s$. The solution is given by (based on exponential response)

$$T(t \geq 120\ s) = 90 - 0.5e^{(t-120)}$$

The response of the thermistor is interesting initially up to $t = 10\ s$ and again for $t > 120\ s$. These two responses are shown in Figure 4.64. The temperature difference between the thermistor and the heater shows the details of the transients near $t = 0\ s$ and near $t = 120\ s$ more clearly. The transients are exponentials that are similar.

4.7.4   Response to a periodic input

There are many applications that involve periodic variations in temperature. For example, the walls of an internal combustion engine cylinder are exposed to periodic

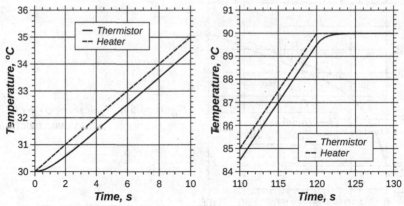

Figure 4.64: *Thermistor response to heater input in Example 4.22*

h]

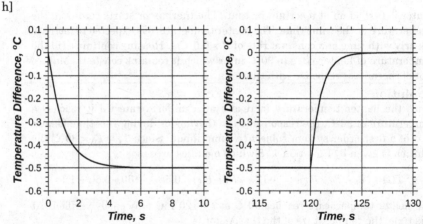

Figure 4.65: *Temperature difference between thermistor and heater vs time in Example 4.22*

heating and hence will show periodic temperature variation. Of course, the waveform representing the periodic temperature variation may be of a complex shape (non sinusoidal). In that case the waveform may be split up in to its Fourier components. The response of the probe can also be studied as that due to a typical Fourier component and combine such responses to get the actual response. Hence we look at a periodic sinusoidal input given by

$$T_\infty(t) = T_a \cos(\omega t) \tag{4.80}$$

In the above expression T_a is the amplitude of the input wave and ω is the circular frequency. We may use the general solution given by Equation 4.73 and perform the indicated integration to get the response of the probe. The steps are left as exercise to the reader. Finally the response is given by

$$T(t) = \left\{ T_0 - \frac{T_a}{1+\omega^2\tau^2} \right\} e^{-t/\tau} + \frac{T_a}{\sqrt{1+\omega^2\tau^2}} \cos\left[\omega\tau - \tan^{-1}\omega\tau \right] \tag{4.81}$$

The term with single underline in Equation 4.81 is the transient part that drops off as $t \to \infty$ and the term with the double underline is the steady state response of the temperature sensor. There is a reduction in the amplitude of the response and also a time lag with respect to the input wave. Amplitude reduction and the time lag (or phase lag) depend on the product of the circular frequency and the time constant. The variations of these with $\omega\tau$ product are as shown in Figure 4.66.

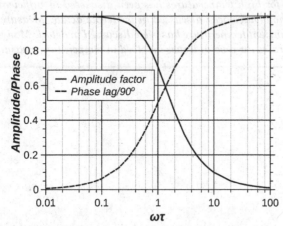

Figure 4.66: Amplitude and phase of a first order system subject to periodic input

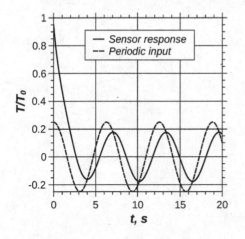

Figure 4.67: Response of a first order system to periodic input

In order to bring out the features of the response of the probe, we make a plot (Figure 4.67) that shows both the input and output, for a typical case. The case shown corresponds to $\dfrac{T_a}{T_0} = 0.25$, $\omega = 1 \, rad/s$ and $\tau = 1 \, s$. The output response has an initial transient that adjusts the initial mismatch between the probe temperature and the imposed temperature. By about 5 time constants (5 s since the time constant has been taken as 1 s) the probe response has settled down to a response that follows the input but with a time lag and an amplitude reduction, as is clear from Figure 4.67.

Concluding remarks

. *Temperature measurement is a well developed experimental activity with many different "thermometers" available for measuring temperature over a wide range. Two important techniques viz. thermoelectric thermometry and resistance thermometry have been discussed in great detail. Pyrometry which is a technique normally reserved for high temperatures has been discussed in sufficient detail. Other special methods have been discussed in moderate detail. Measurement of time varying or transient temperature has been discussed in detail. Many worked examples demonstrates the several issues involved in temperature measurement.*

Chapter 5

Systematic errors in temperature measurement

5.1 Introduction

Systematic errors in temperature measurement are inevitable because of finite heat transfer rate between the system whose temperature is being measured and the sensor that is interrogating it. Earlier, in sections 4.7.3 and 4.7.4 we have seen how there is a systematic temperature difference when a transient temperature is measured using a temperature probe.

Systematic errors are situation dependent. We look at typical temperature measurement situations and discuss qualitatively the errors before we look at the quantitative estimation of these errors. The situations of interest are the measurement of temperature:

- at a surface
- inside a solid
- of a flowing fluid

5.2 Examples of temperature measurement

5.2.1 Surface temperature measurement using a compensated probe

Consider the measurement of the temperature of a surface by attaching a thermocouple sensor normal to it, as shown in Figure 5.1. Of course, this arrangement may not be the best method of measuring the surface temperature. In some situation it may not be possible to use other better methods and hence this is a possible arrangement.

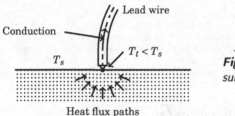

Figure 5.1: *Temperature measurement of a surface*

Lead wires conduct heat away from the surface and this is compensated by heat transfer to the surface from within, as shown. This sets up a temperature field within the solid such that the temperature of the surface where the thermocouple is attached is depressed and hence is less than the surface temperature elsewhere on the surface. This introduces an error in the surface temperature measurement. One way of reducing or altogether eliminating the conduction error is by the use of a compensated sensor as indicated schematically in Figure 5.2.

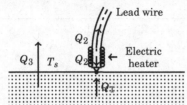

Figure 5.2: *Schematic of a compensated probe*

This figure is based on "Industrial measurements with very short immersion & surface temperature measurements" by J.P. Tavener, D. Southworth, D. Ayres, N. Davies, Isothermal Technology Limited. The surface temperature, in the absence of the probe is at an equilibrium temperature under the influence of steady heat loss Q_3 to an environment. The probe would involve an additional heat loss due to conduction. If we supply heat Q_2 by heating the probe such that there is no temperature gradient along the thermocouple probe then $Q_1 - Q_2 = Q_3$ and the probe temperature is the same as the surface temperature. The above reference also demonstrates that the compensated probe indicates the actual surface temperature with negligible error. Compensated probes as described above are commercially available from ISOTECH (Isothermal Technology Limited, Pine Grove, Southport, Merseyside, England) and described as "944 True Surface temperature measurement systems".

5.2.2   Measurement of temperature inside a solid

Figure 5.3 shows how one can arrange a thermocouple to measure the temperature inside a solid. The thermocouple junction is placed at the bottom of a blind hole drilled into the solid. The gap between the thermocouple lead wires and the hole is filled with a heat conducting cement. The lead wire is exposed to the ambient as it emerges from the hole. It is easy to visualize that the lead wire conduction must be compensated by heat conduction into the junction from within the solid. Let the solid be at a temperature higher than the ambient temperature. Then we expect the solid temperature to be greater than the junction temperature (this is the temperature that is indicated) which is greater than the ambient temperature.

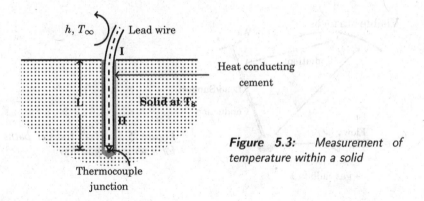

Figure 5.3: *Measurement of temperature within a solid*

5.2.3 Measurement of temperature of a moving fluid

Often it is necessary to measure the temperature of a fluid flowing through a duct. In order to prevent leakage of the fluid or prevent direct contact between the fluid and the temperature sensor, a thermometer well is provided as shown in Figure 5.4. The sensor is attached to the bottom of the well as indicated. The measured temperature is the temperature of the bottom of the well and what is desired to be measured is the fluid temperature.

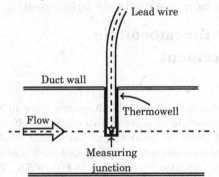

Figure 5.4: *Measurement of temperature of a moving fluid*

If the duct wall temperature is different from the fluid temperature, heat transfer takes place by conduction between the fluid and the duct wall, via the well and hence the well bottom temperature will be at a value in between that of the fluid and the wall. There may also be radiation heat transfer between surfaces, further introducing errors. If the fluid flows at high speed (typical of supersonic flow of air) viscous dissipation - conversion of kinetic to internal energy - may also be important. With this background we generalize the thermometer error problem in the case of measurement of temperature of a gas flow as indicated in Figure 5.5.

The temperature of the sensor is determined, under the steady state, by a balance of the different heat transfer processes that take place, as indicated in Figure 5.5. Not all the heat transfer processes may be active in a particular case. The thermometer error is simply the difference between the gas temperature and the sensor temperature. Estimation of the error will be made later on.

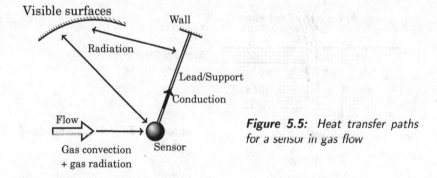

Figure 5.5: *Heat transfer paths for a sensor in gas flow*

5.2.4   Summary of sources of error in temperature measurement

1. Sensor interferes with the process - Conduction error in surface temperature measurement
2. Sensor interferes with the process as well as other environments - Radiation error
3. While measuring temperature of moving fluids convection and conduction processes interact and lead to error
4. In case of high speed flow, viscous dissipation effects may be important

5.3   Conduction error in thermocouple temperature measurement

5.3.1   Lead wire model

Heat transfer through the lead wires of a thermocouple introduces error in the measured temperature. Since a thermocouple consists of two wires of different materials covered with insulation, and since the error estimation should involve a simple procedure, we replace the actual thermocouple by a single wire thermal equivalent. How this is accomplished is indicated by referring to Figure 5.6. The cross section of an actual thermocouple is shown at the left in Figure 5.6. It consists of two wires of different materials with the indicated radii and thermal conductivity values. The insulation layer encloses the two wires as indicated. We replace the two wires and the insulation by a single wire of radius r_1 and a coaxial insulation layer of outer radius r_2.

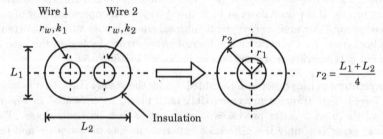

Figure 5.6: *Single wire equivalent of a thermocouple*

5.3.2 The single wire model

The single wire model simplifies the lead wire conduction analysis. This model consists in replacing the two wire - insulation combination (twc) by a single wire - insulation combination (swm) according to the following scheme.

1. The area thermal conductivity product is the same for the actual and the single wire model. Thus

$$\underbrace{\widetilde{kA}}_{\text{swm}} = \underbrace{k_1 A_{w1} + k_2 A_{w2}}_{\text{twc}}$$

   If the two wires have the same diameter (this is normally the case) we may replace the above by

$$\underbrace{\widetilde{kA}}_{\text{swm}} = \underbrace{A_w(k_1 + k_2)}_{\text{twc}} = \pi r_w^2 (k_1 + k_2) = \underbrace{2\pi r_w^2}_{\substack{\text{Area of} \\ \text{two wires}}} \times \underbrace{\frac{k_1 + k_2}{2}}_{\substack{\text{Mean thermal} \\ \text{conductivity}}}$$

   where A_w is the cross section area of each wire of radius r_w and hence equal to πr_w^2. Thus the thermal conductivity of the single wire equivalent is equal to the mean of the thermal conductivities of the two wires and the area of cross section of the single wire equivalent is twice the area of cross section of either wire. Hence the radius of the single wire equivalent r_1 indicated in Figure 5.6 is given by $r_1 = \sqrt{2} r_w$.
2. The insulation layer is to be replaced by a coaxial cylinder of inner radius r_1 and the outer radius r_2. The outer radius is taken as

$$r_2 = \frac{L_1 + L_2}{4}$$

   Note that if $L_1 = L_2 = 2r$ (true for a cylinder of radius r), this formula gives $r_2 = r$, as it should.
3. Since the insulation layer is of low thermal conductivity while the wire materials have high thermal conductivities, it is adequate to consider heat conduction to take place along the single wire axially and across the insulation radially, as indicated in Figure 5.7(a).
4. Finally the outer surface of insulation transfers heat to an ambient by convection as indicated in Figure 5.7(a).

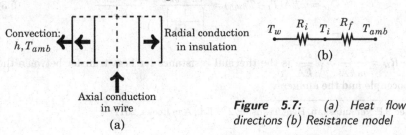

Convection:
h, T_{amb} Radial conduction in insulation

T_w R_i T_i R_f T_{amb}

(b)

Axial conduction in wire

(a)

Figure 5.7: (a) Heat flow directions (b) Resistance model

5.3.3   Heat loss through lead wire

Heat loss through the lead wire is modeled using fin analysis, familiar to us from the study of heat transfer.[1] Since the wire is usually very long (essentially extremely long compared to its diameter), it may be assumed to be infinitely long.

Referring to Figure 5.7(b) we see that heat transfer from the thermocouple wire (temperature T_w) to the ambient (temperature T_{amb}) takes place through two thermal resistance in series viz. R_i the conduction resistance of the insulation layer and R_f the film resistance at the outer surface of insulation. These resistances are given by

$$R_i = \frac{1}{2\pi k_i} \ln\left(\frac{r_2}{r_1}\right) \quad \text{and} \quad R_f = \frac{1}{2\pi r_2 h}$$

The overall resistance R_o is then given by

$$R_o = R_i + R_f = \frac{1}{2\pi k_i} \ln\left(\frac{r_2}{r_1}\right) + \frac{1}{2\pi r_2 h} \tag{5.1}$$

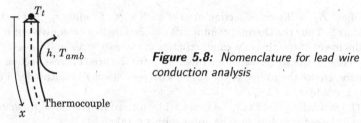

Figure 5.8: *Nomenclature for lead wire conduction analysis*

Now consider the model shown in Figure 5.8. The thermocouple is at temperature T_t i.e. $T = T_t$ at $x = 0$ where x is measured along the thermocouple wire as indicated in the figure. The thermocouple wire temperature varies with x. Heat transfer in the thermocouple may easily be obtained by fin analysis. The appropriate fin parameter m is given by

$$m = \frac{1}{\sqrt{R_o \widetilde{kA}}} \tag{5.2}$$

Assuming the lead wire to be infinitely long, the heat loss through the lead wire is given by

$$Q_{tc} = m\widetilde{kA}(T_t - T_{amb}) = \frac{T_t - T_{amb}}{\left[\dfrac{1}{m\widetilde{kA}}\right]} = \frac{T_t - T_{amb}}{R_{tc}} \tag{5.3}$$

where $R_{tc} = \dfrac{1}{m\widetilde{kA}} = \sqrt{\dfrac{R_o}{\widetilde{kA}}}$ is the thermal resistance for heat transfer between the thermocouple and the ambient.

---

[1] See S.P. Venkateshan, Heat Transfer, 2nd Ed., Ane Books, 2011

5.3.4 Typical application and thermometric error

Consider a typical application such as the measurement of surface temperature by a surface mounted thermocouple as shown in Figure 5.9(a).

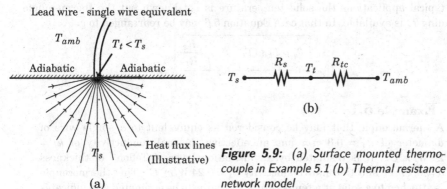

Figure 5.9: *(a) Surface mounted thermo-couple in Example 5.1 (b) Thermal resistance network model*

Heat loss from the thermocouple is modeled as shown above by discussion leading to Equation 5.3. Thermocouple is attached to the surface of a massive solid whose temperature we would like to measure. Excepting where the thermocouple has been attached the rest of the surface of the solid is assumed to be perfectly insulated. Since the thermocouple conducts away some heat from the solid there is a depression of temperature of the solid where the thermocouple has been attached. Thus $T_t < T_s$ if $T_{amb} < T_s$. The temperature field within the solid is axi-symmetric with the small area on the surface at temperature T_t while the solid temperature far away from it is T_s. Heat transfer by conduction is quite closely represented by the relation

$$Q_s = \frac{T_s - T_t}{\left(\frac{1}{4k_s r_1}\right)} = \frac{T_s - T_t}{R_s} \tag{5.4}$$

where R_s is the thermal resistance for heat transfer between the solid and the spot where the thermocouple is attached.

Under steady conditions $Q_s = Q_{tc}$ and the thermal resistance network in Figure 5.9(b) represents the situation. Using this condition (or Ohm's law applied to the network) we calculate the thermocouple indicated temperature as

$$T_t = \frac{\left(\dfrac{T_s}{R_s} + \dfrac{T_{amb}}{R_{tc}}\right)}{\left(\dfrac{1}{R_s} + \dfrac{1}{R_{tc}}\right)} \tag{5.5}$$

Equation 5.5 shows that the sensor temperature is a weighted mean of the solid temperature and the ambient temperature. It is clear that the smaller the weight on the ambient side better it is from the point of view of temperature measurement. This is a general feature, as we shall see later, in all cases involving temperature measurement. The thermometer error is then given by the following expression:

$$T_e = T_t - T_s = -\frac{(T_s - T_{amb})}{\left(1 + \frac{R_{tc}}{R_s}\right)} \qquad (5.6)$$

In typical applications the solid temperature is not known but the thermocouple reading T_t is available. In that case Equation 5.6 may be rearranged to get

$$T_s = T_t + (T_t - T_{amb})\frac{R_s}{R_{tc}} \qquad (5.7)$$

Example 5.1

A thermocouple that may be considered as equivalent to a single wire of diameter of $d_1 = 0.3$ mm has an effective thermal conductivity of $k_t = 20$ $W/m°C$. It is insulated by an insulation layer of Teflon® of thickness $\delta = 0.2$ mm and thermal conductivity $k_i = 0.24$ $W/m°C$. This thermocouple is attached to a solid at a temperature T_s that is to be estimated and indicates a temperature of $149.2°C$. The solid thermal conductivity is $k_s = 14$ $W/m°C$. The thermocouple loses heat from its insulated outer surface to an ambient at $T_{amb} = 30°C$ via a heat transfer coefficient $h = 1$ $W/m^2°C$. Assume that there is no heat loss from the solid excepting via the thermocouple attached to it as shown in Figure 5.9. Estimate the value of T_s.

Solution :

Solution to this problem follows the procedure discussed above.

Step 1 Solid side resistance:

Solid side resistance is calculated as

$$R_s = \frac{1}{4k_s r_1} = \frac{1}{2k_s d_1} = \frac{1}{2 \times 14 \times 0.0003} = 119.05\ °C/W$$

Step 2 Thermocouple resistance R_{tc}:

With $r_1 = \dfrac{d_1}{2} = 0.00015$ m and insulation thickness $\delta = 0.0002$ m the outer radius of insulation is given by

$$r_2 = r_1 + \delta = 0.00015 + 0.0002 = 0.00035\ m$$

Considering the case where $h = 1$ $W/m^2°C$, the overall resistance may be calculated using Equation 5.1 as

$$R_o = \frac{1}{2 \times \pi \times 0.24}\ln\left(\frac{0.00035}{0.00015}\right) + \frac{1}{2\pi \times 0.00035} = 455.29\ m°C/W$$

The fin parameter is then obtained using Equation 5.2 as

$$m = \frac{1}{\sqrt{455.29 \times \pi \times 0.00015^2 \times 20}} = 39.42\ m^{-1}$$

Finally we obtain the thermocouple thermal resistance as

$$R_{tc} = \frac{1}{39.42 \times 20 \times \pi \times 0.00015^2} = 17943.7\ °C/W$$

Step 3 Solid temperature T_s:
Solid temperature is then obtained using Equation 5.7 as

$$T_s = 149.2 + (149.2 - 30) \times \frac{119.05}{17943.7} = 149.99 \approx 150°C$$

5.3.5 Measurement of temperature within a solid

Now we shall look at the situation depicted in Figure 5.3. Temperature error is essentially due to conduction along the lead wires. However, the portion embedded within the solid (II) has a different environment as compared to the part that is outside (I). Both of these may be treated by the single wire model introduced earlier. Assume that the solid is at a temperature higher than the ambient. The thermocouple junction will then be at an intermediate temperature between that of the solid and the ambient. Heat transfer to the embedded thermocouple is basically by conduction while the heat transfer away from the part outside the solid is by conduction and convection. The embedded part is of finite length L while the portion outside may be treated as having an infinite length.

Represent the temperature of the single wire equivalent as T_i in a plane coinciding with the surface of the solid. Let T_t be the temperature of the junction while T_s is the temperature of the solid. Let the ambient temperature be T_{amb}. The fin parameter for the embedded part may be calculated based on the overall thermal resistance given by

$$R_{II} = \frac{1}{2\pi k_c} \ln\left(\frac{r_3}{r_2}\right) + \frac{1}{2\pi k_i} \ln\left(\frac{r_2}{r_1}\right) \tag{5.8}$$

In the above, k_c is the thermal conductivity of the heat conducting cement, r_3 is · the radius of the hole and the other symbols have the earlier meanings. Note that Equation 5.8 is based on two conductive resistances in series. The corresponding fin parameter is

$$m_{II} = \frac{1}{\sqrt{R_{II}\widetilde{k}A}} \tag{5.9}$$

The overall resistance for the exposed part of the thermocouple is given by the expression given earlier in Equation 5.1 which is reproduced here for easy reference:

$$R_I = \frac{1}{2\pi k_i} \ln\left(\frac{r_2}{r_1}\right) + \frac{1}{2\pi r_2 h} \tag{5.10}$$

The fin parameter is again based on Equation 5.2 and is reproduced here for easy reference:

$$m_I = \frac{1}{\sqrt{R_I\widetilde{k}A}} \tag{5.11}$$

Referring now to Figure 5.10 we see that the heat transfer across the surface through

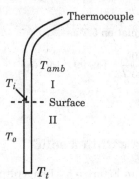

Figure 5.10: *Nomenclature for thermal analysis*

the thermocouple should be the same i.e. $Q_{II} = Q_I$. Using familiar fin analysis, we have

$$Q_{II} = \sqrt{\frac{\widehat{kA}}{R_{II}}}(T_s - T_i)\tanh\left\{\frac{L}{\sqrt{R_{II}\widehat{kA}}}\right\} \qquad (5.12)$$

For the exposed part, we have

$$Q_I = \sqrt{\frac{\widehat{kA}}{R_I}}(T_i - T_{amb}) \qquad (5.13)$$

Equating the above two expressions we solve for the unknown temperature T_i to get

$$T_i = \frac{w_1 T_s + w_2 T_{amb}}{w_1 + w_2} \qquad (5.14)$$

where the w's are weights given by

$$w_1 = \sqrt{\frac{\widehat{kA}}{R_{II}}}\tanh\left\{\frac{L}{\sqrt{R_{II}\widehat{kA}}}\right\}$$

$$w_2 = \sqrt{\frac{\widehat{kA}}{R_I}} \qquad (5.15)$$

Having found the unknown temperature T_i, we make use of fin analysis for the embedded part to get the temperature T_t. In the familiar fin analysis the thermocouple indicated temperature corresponds to the "tip" temperature while the surface temperature corresponds to the "base" temperature. Assuming insulated tip condition, the appropriate expression is

$$T_t = T_s + \frac{T_i - T_s}{\cosh\left\{\frac{L}{\sqrt{R_{II}\widehat{kA}}}\right\}} \qquad (5.16)$$

Following points may be made in summary

- The longer the depth of embedding smaller the thermometric error

- Higher the thermal conductivity of the epoxy filling the gap between the thermocouple and the hole the smaller the thermometric error
- The smaller the diameter of the thermocouple wires smaller is the thermometric error
- Smaller the thermal conductivity of the thermocouple wires smaller the thermometric error
- If it is possible the insulation over the thermocouple wires should be as thin as possible in the embedded portion

Example 5.2

A thermocouple that may be considered as equivalent to a single wire of diameter of $d_1 = 1\ mm$ has an effective thermal conductivity of $k_t = 25\ W/m°C$. It is insulated by an insulation layer of Teflon® of thickness $\delta = 1\ mm$ and thermal conductivity $k_i = 0.24\ W/m°C$. This thermocouple is used to measure the temperature of a solid of thermal conductivity $k_s = 14\ W/m°C$ by embedding it in a $d_h = 4\ mm$ diameter hole that is $L = 10\ mm$ deep. The space between the thermocouple and the hole is filled with a heat conducting epoxy that has a thermal conductivity of $5\ W/m°C$. The lead wires coming out of the hole are exposed to an ambient at $30°C$ with a heat transfer coefficient of $5\ W/m^2°C$. If the temperature of the solid is $80°C$, estimate the temperature indicated by the thermocouple.

Solution :

We use the data provided in the problem and calculate the following using the nomenclature provided in Figures 5.6 and 5.10.

$$r_1 = \frac{d_1}{2} = \frac{0.001}{2} = 0.0005\ m$$

$$r_2 = r_1 + \delta = 0.0005 + 0.001 = 0.0015\ m$$

$$d_h = 4\ mm = 0.004\ m \text{ (Embed hole diameter)}$$

$$r_3 = \frac{d_h}{2} = \frac{0.004}{2} = 0.002\ m$$

$$\widetilde{kA} = k_t \pi r_1^2 = 25 \times \pi \times 0.0005^2 = 1.9635 \times 10^{-5}\ Wm/°C$$

Step 1 Calculation of thermal resistances:
Exposed lead wire I: Thermocouple insulation resistance is given by

$$R_{ins} = \frac{\ln\left(\frac{r_2}{r_1}\right)}{2\pi k_i} = \frac{\ln\left(\frac{0.0015}{0.0005}\right)}{2 \times \pi \times 0.24} = 0.7285\ m°C/W$$

Film resistance is given by

$$R_f = \frac{1}{2\pi r_2 h} = \frac{1}{2 \times \pi \times 0.0015 \times 5} = 21.2207\ m°C/W$$

The overall resistance is then given by

$$R_I = R_{ins} + R_f = 0.7285 + 21.2207 = 21.9492\ m°C/W$$

Step 2 Calculation of thermal resistance:
Embedded lead wire II: Use Equation 5.8 to calculate the thermal resistance for the embedded part of the thermocouple.

$$R_{II} = \frac{1}{2 \times \pi \times 5} \ln\left(\frac{0.002}{0.0015}\right) + \frac{1}{2 \times \pi \times 0.24} \ln\left(\frac{0.0015}{0.0005}\right) = 0.7377 \ m°C/W$$

Step 3 Calculation of weights:
Weights are calculated using Equations 5.15.

$$w_1 = \sqrt{\frac{1.9635 \times 10^{-5}}{0.7377}} \tanh\left\{\frac{0.01}{\sqrt{0.7377 \times 1.9635 \times 10^{-5}}}\right\} = 5.1552 \times 10^{-3}$$

$$w_2 = \sqrt{\frac{1.9635 \times 10^{-5}}{21.9492}} = 9.4581 \times 10^{-4}$$

Step 4 Calculation of temperature T_i:
Temperature T_i is calculated using Equation 5.14 as

$$T_i = \frac{5.1552 \times 10^{-3} \times 80 + 9.4581 \times 10^{-4} \times 30}{5.1552 \times 10^{-3} + 9.4581 \times 10^{-4}} = 72.25°C$$

Step 5 Calculation of thermocouple indicated temperature T_t:
We now make use of Equation 5.16 to get

$$T_t = 80 + \frac{72.25 - 80}{\cosh\left\{\frac{0.01}{\sqrt{0.7377 \times 1.9635 \times 10^{-5}}}\right\}} = 79.70°C$$

Step 6 Thermometer error:
Thus the thermometer error in this installation is $T_e = T_t - T_s = 79.70 - 80 = -0.30°C$.

5.4 Measurement of temperature of a moving fluid

There are many engineering applications where it is necessary to measure the temperature of a moving fluid. The temperature of the moving fluid may span a large range of temperatures and hence may require different temperature measurement techniques. Systematic errors in such measurements are discussed in this section.

5.4.1 Temperature error due to radiation

Errors in temperature measurement may occur due to surface radiation, especially at elevated temperatures. We consider a typical example shown in Figure 5.11. Temperature sensor is considered to be a small spherical structure to which a thermocouple has been attached. The sensor is situated in a duct whose walls are at a temperature T_w which is less than the temperature of a fluid at temperature T_f that is flowing axially along the duct with a velocity V. Outer surface of the sphere is gray and has an emissivity of ε. The sphere loses heat because of lead wire

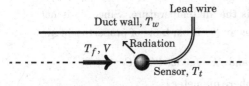

Lead wire

Duct wall, T_w

T_f, V Radiation

Sensor, T_t

Figure 5.11: *Radiation and lead wire conduction errors in temperature measurement with $T_f > T_t > T_w$*

conduction and radiation to the duct wall. Heat loss due to radiation is given by

$$Q_R = \varepsilon \sigma S \left(T_t^4 - T_w^4 \right) \tag{5.17}$$

where S is the surface area of the sphere. Note that the temperatures are to be expressed in K in Equation 5.17 and σ is the Stefan Boltzmann constant. The sphere gains heat by convection from the fluid. With h_f the heat transfer coefficient heat gain from the fluid is

$$Q_f = h_f S (T_f - T_t) \tag{5.18}$$

Heat loss through the lead wires is modeled as explained earlier and is given by

$$Q_{tc} = \frac{(T_t - T_{amb})}{R_{tc}} \tag{5.19}$$

The temperature of the sensor is determined by equating heat gain by convection to heat loss by conduction and radiation. Thus, we have

$$Q_R + Q_{tc} = Q_f \tag{5.20}$$

Thus we get

$$\frac{(T_t - T_{amb})}{R_{tc}} + \varepsilon \sigma S (T_t^4 - T_w^4) = h_f S (T_f - T_t) \tag{5.21}$$

The fluid temperature T_f may then be estimated with the measured values of T_t, T_w and T_{amb} as

$$\boxed{T_f = T_t + \frac{(T_t - T_{amb})}{h_f S R_{tc}} + \frac{\varepsilon \sigma}{h_f} (T_t^4 - T_w^4)} \tag{5.22}$$

Example 5.3

A sensor in the form of a sphere of diameter $D = 3mm$ is exposed to a fluid flowing in a large duct. The fluid is air that flows with a velocity of $V = 1\,m/s$ across the sphere. The walls of the duct are at a temperature of $T_w = 30°C$. The sensor temperature is measured using a thermocouple whose wires are so thin that any conduction heat transfer through them may be ignored. What is the air temperature when the sensor temperature is measured as $T_t = 120°C$ if the surface of the sphere has an emissivity of $\varepsilon = 0.9$?

Solution :

Step 1 Flow Reynolds number:
Convection heat transfer between air and the sphere is dependent on the Reynolds number which is to be calculated using the properties of air at film

temperature viz. $\dfrac{T_f + T_t}{2}$ where T_f is the air temperature. Since T_f is not known we shall take the air properties at T_t. From table of properties of air, we have

$$v = 2.09 \times 10^{-5}\, m^2/s \quad \text{and} \quad k = 0.0328\, W/m^\circ C$$

The Reynolds number based on the sphere diameter is

$$Re_D = \frac{VD}{v} = \frac{1 \times 0.003}{2.09 \times 10^{-5}} = 143.3$$

Step 2 Nusselt number Nu:
McAdams has proposed a correlation for the convection heat transfer between a sphere and a gas as $Nu_D = 0.37 Re_D^{0.6}$. Hence we have

$$Nu_D = 0.37 \times 143.3^{0.6} = 7.28$$

Step 3 Heat transfer coefficient h_f:
The heat transfer coefficient is then calculated as

$$h_f = \frac{Nu_D k}{D} = \frac{7.28 \times 0.0328}{0.003} = 79.6\, W/m^2\,^\circ C$$

Step 4 Estimation of air temperature T_f:
We make use of Equation 5.22, set the lead wire conduction part to zero, to obtain

$$T_f = 120 + \frac{0.9 \times 5.67 \times 10^{-8}}{79.6}\left((120 + 273.15)^4 - (30 + 273.15)^4\right) = 404.9\,K = 131.7^\circ C$$

Step 5 Systematic error due to radiation
It is seen that the systematic error due to radiation is substantial and is given by

$$T_e = T_t - T_f = 120 - 131.7 = -11.7^\circ C$$

Such a large correction is unacceptable in practice. We look for some method by which radiation error may be reduced.

5.4.2 Reduction of radiation error: use of radiation shield

Radiation error may be reduced by increasing the convection heat transfer coefficient or/and reducing the radiation heat transfer. The former may be done by increasing the fluid velocity in the vicinity of the sensor. The radiation loss may be reduced by providing a radiation shield. Both of these are achieved by the arrangement shown in Figure 5.12.

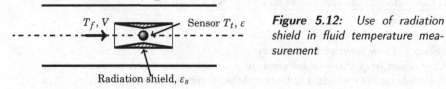

Figure 5.12: Use of radiation shield in fluid temperature measurement

Radiation shield is essentially a thin tube of polished metal so that the surface emissivity is low. Simple buffing operation of a shield made of a metal will give an emissivity value (smaller the better) of around 0.1 . This may be sufficient for most practical purposes. The cross section is chosen to be in the form of a converging diverging nozzle to increase the flow velocity of the fluid as it passes across the sensor. In the illustration shown the connecting lead wire has been left out since it is expected to not participate in the heat transfer process to a significant extent.

Assume that the shield is very thin and the inside and outside surface areas (A_s) are about the same. Let the convective heat transfer coefficient for heat transfer between the fluid and the shield be h_s. The equilibrium temperature of the shield will then be given by requiring that the convective heat transfer balance exactly the radiation heat transfer. Thus

$$2h_s(T_f - T_s) = \varepsilon_s \sigma(T_s^4 - T_w^4) \qquad (5.23)$$

The sensor temperature is determined by the equation

$$h_f(T_f - T_t) = \varepsilon \sigma(T_t^4 - T_s^4) \qquad (5.24)$$

which assumes that heat loss from the sensor is now solely due to radiation interaction with the shield. These two equations are simultaneously solved to get the fluid and shield temperatures.

Example 5.4

In Example 5.3 the sensor is surrounded by a $L_s = 4$ *cm* long and $d_s = 0.75$ *cm* diameter thin cylinder of aluminum. Its surface is buffed and has an emissivity of $\varepsilon_s = 0.1$. Estimate the thermometer error if all other data remains the same.

Solution :

The heat transfer coefficient h_f calculated in Example 5.3 needs no revision.
Step 1 Calculation of h_s:
Heat transfer between air and the shield is modeled as internal tube flow on the sensor side. The Reynolds number is based on the air velocity and the shield inner diameter. We have

$$Re_{d_s} = \frac{Vd_s}{\nu} = \frac{0.0075}{2.09 \times 10^{-5}} = 1910$$

Since $Re_{d_s} < 2300$ flow is laminar. The length to diameter ratio of the shield is

$$\frac{L_s}{d_s} = \frac{4}{0.75} = 5.33$$

Since this is fairly small compared to the development length flow is treated as developing flow. We make use of Nusselt number correlation due to Hausen as

$$Nu_s = 3.66 + \frac{0.0668 Re_s Pr}{1 + 0.04 \left(Re Pr \frac{d_s}{L_s}\right)^{2/3}} = 3.66 + \frac{0.0668 \times 1910 \times 0.703}{1 + 0.04 \left(1910 \times 0.703 \times \frac{0.75}{4}\right)^{2/3}} = 9.315$$

The heat transfer coefficient is then calculated as

$$h_s = \frac{Nu_s k}{d_s} = \frac{9.315 \times 0.0328}{0.0075} = 7.64 \, W/m^2{}^\circ C$$

Step 2 Energy balance on the shield:

The worst case scenario is when the outer surface of the shield is assumed to not receive *any* heat from air by convection. We then recast Equation 5.23 as

$$h_s(T_f - T_s) = \varepsilon_s \sigma (T_s^4 - T_w^4)$$

or, solving for T_s we get

$$T_f = T_s + \left(\frac{\varepsilon_s \sigma}{h_s}\right)(T_s^4 - T_w^4) = T_o + C_s\left(T_s^4 - T_w^4\right)$$

where $C_s = \dfrac{\varepsilon_s \sigma}{h_s}$.

Step 3 Energy balance on the sensor:

Similarly we may recast energy balance equation on the sensor in the form

$$T_f = T_t + \left(\frac{\varepsilon \sigma}{h_f}\right)(T_t^4 - T_s^4) = T_t + C_t\left(T_t^4 - T_s^4\right)$$

where $C_t = \dfrac{\varepsilon \sigma}{h_f}$.

Step 4 Equation for shield temperature T_s:

Equating the two expressions for T_f and on rearrangement the shield temperature is governed by the non-linear equation

$$T_s + (C_s + C_t)T_s^4 - \left(T_t + C_t T_t^4 + C_s T_w^4\right) = 0$$

Using the numerical data, the various quantities occurring in the above equation are

$$C_s = 7.4229 \times 10^{-10}\,K^{-3},\ C_t = 6.4148 \times 10^{-10}\,K^{-3},\ T_t = 393.15\,K,\ T_w = 303.15\,K$$

Non-linear equation for T_s may easily be solved by the Newton Raphson technique. Starting with a guess value of $T_s = 386\,K$ the solution converges to $T_s = 384.50\,K$ after two iterations.

Step 5 Estimated air temperature T_f:

Having calculated the equilibrium temperature of the shield we use any one of the two expressions for T_f to get the air temperature as

$$T_f = T_t + C_t(T_t^4 - T_s^4) = 393.15 + 6.4148 \times 10^{-10}\left(393.15^4 - 384.50^4\right) = 393.82°C$$

Step 6 Radiation error:

The radiation error is now given by

$$T_e = T_t - T_f = 393.15 - 393.82 = -0.67\,K$$

Thus the radiation shield has been successful in reducing the error to a large extent.

5.4.3 Analysis of thermometer well problem

This is a fairly common situation as has been mentioned earlier. The well acts as a protection for the temperature sensor but leads to error due to axial conduction along the well. The external appearance of a thermometer well probe is shown in Figure 5.13. The threaded arrangement is to facilitate the mounting of the probe through a side hole in to the process space. It is easily recognized that the well may be treated as a fin and the analysis made earlier will be adequate to estimate the thermometric error.

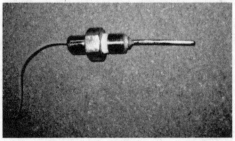

Figure 5.13: *External appearance of a thermometer well probe using a thermocouple*

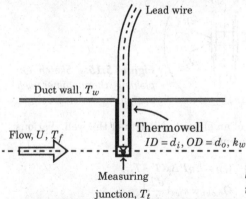

Figure 5.14: *Nomenclature for the thermometer well problem*

Nomenclature used in the analysis that follows is introduced in Figure 5.14. The following assumptions are made:

1. Since the thermometer well has a much larger cross section area than the thermocouple wires conduction along the wires is ignored.
2. The thermometer well is heated by the gas while it cools by radiation to the walls of the duct (based on $T_f > T_t > T_w$).
3. Well is treated as a cylinder in cross flow for determining the convection heat transfer coefficient between the gas and the well surface.

The heat transfer coefficient is calculated based on the Zhukaskas correlation given by

$$Nu = CRe^m Pr^n \tag{5.25}$$

In this relation Re, the Reynolds number is based on the outside diameter of the well and all the properties are evaluated at a suitable mean temperature. The constants C, m and n are given in Table 5.1. They depend on the Reynolds number range. Radiation heat transfer may be based on a linearized model if the gas

Table 5.1: *Constants in the Zhukaskas correlation*

Re	C	m
1-40	0.75	0.4
40-10^3	0.51	0.5
10^3-2×10^5	0.26	0.6
2×10^5-10^6	0.076	0.7

	n
$Pr < 10$	0.37
$Pr > 10$	0.36

and wall temperatures are close to each other. In that case the well temperature variation along its length is also not too large. Thus we approximate the radiant flux $q_R = \varepsilon\sigma(T^4 - T_w^4)$ by the relation $q_R \approx 4\varepsilon\sigma T_w^3(T - T_w)$ where the factor $4\varepsilon\sigma T_w^3$ is referred to as h_R, the radiation heat transfer coefficient. Refer to Figure 5.15 and

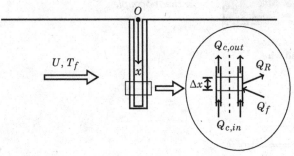

Figure 5.15: *Sketch explaining thermometer well analysis*

the inset that shows an expanded view of an elemental length of the well. Various fluxes crossing the boundaries of the element are:

$$Q_f = h_f P \Delta x (T_f - T) \quad Q_R = h_R P \Delta x (T - T_w)$$
$$Q_{c,in} = k_w A \left.\frac{dT}{dx}\right|_x \quad Q_{c,out} = k_w A \left.\frac{dT}{dx}\right|_{x-\Delta x}$$

In the above the perimeter P is given by πd_o and area of cross section A is given by $\pi \dfrac{d_o^2 - d_i^2}{4}$. Energy balance requires that

$$Q_{c,in} + Q_f = Q_{c,out} + Q_R$$

Substituting the expressions for the fluxes and using Taylor expansion of the derivative, retaining only first order terms in the Taylor expansion, we have

$$k_w A \left.\frac{dT}{dx}\right|_x + h_f P \Delta x (T_f - T) = k_w A \left.\frac{dT}{dx}\right|_x - k_w A \left.\frac{d^2T}{dx^2}\right|_x \Delta x + h_R P \Delta x (T - T_w)$$

This equation may be rearranged as

$$\frac{d^2T}{dx^2} - \frac{h_f + h_R}{k_w A} T + \frac{h_f P}{k_w A} T_f + \frac{h_R P}{k_w A} T_w = 0 \tag{5.26}$$

Let $T_{ref} = \dfrac{h_f T_f + h_R T_w}{h + h_R}$ be a reference temperature. Then Equation 5.26 is rewritten as

$$\frac{d^2\theta}{dx^2} - m_{eff}^2 = 0 \tag{5.27}$$

where $\theta = T - T_{ref}$ and $m_{eff} = \sqrt{\dfrac{(h_f + h_R)P}{k_w A}}$ is the effective fin parameter. Equation 5.27 is the familiar fin equation whose solution is well known. Assuming insulated boundary condition at the sensor location, the indicated sensor temperature is given by

$$\theta_t = T_t - T_{ref} = \frac{T_w - T_{ref}}{\cosh(m_{eff}L)} \tag{5.28}$$

The thermometric error is thus given by

$$T_t - T_f = (T_{ref} - T_f) + \frac{T_w - T_{ref}}{\cosh(m_{eff}L)} \tag{5.29}$$

Following points may be made in summary

- The longer the depth of immersion L smaller the thermometric error
- Lower the thermal conductivity of the well material the smaller the thermometric error
- Smaller the emissivity of the well and hence the h_R smaller the thermometric error
- Larger the fluid velocity and hence the h smaller the thermometric error

Example 5.5

Air at a temperature of $100°C$ is flowing in a tube of diameter 10 cm at an average velocity of 0.5 m/s. The tube walls are at a temperature of $90°C$. A thermometer well of outer diameter 4 mm and wall thickness 0.5 mm made of stainless steel is immersed to a depth of 5 cm, perpendicular to the tube axis. The steel tube is dirty because of usage and has a surface emissivity of 0.2. What will be the temperature indicated by a thermocouple that is attached to the bottom of the thermometer well? What is the consequence of ignoring radiation?

Solution :

Step 1   Well outside convective heat transfer coefficient:
Given data:

$$d_o = 0.004\,m;\ U = 0.5\,m/s;\ T_f = 100 \circ C;\ T_w = 90 \circ C$$

The fluid properties are taken at the mean temperature given by

$$T_m = \frac{T_f + T_w}{2} = \frac{100 + 90}{2} = 95 \circ C$$

From table of air properties (Table B.7, Appendix B) the desired properties at T_m are:

$$\nu = 22 \times 10^{-6}\ m^2/s; k = 0.0307\ W/m°C; Pr = 0.706$$

The Reynolds number based on outside diameter of thermometer well is

$$Re = \frac{Ud_o}{\nu} = \frac{0.5 \times 0.004}{22 \times 10^{-6}} = 90.9$$

Zhukauskas correlation is used now. For the above Reynolds number the appropriate constants in the Zhukauskas correlation are

$$C = 0.51;\ m = 0.5;\ n = 0.37$$

The convection Nusselt number is

$$Nu = CRe^m Pr^n = 0.51 \times 91.9^{0.5} \times 0.7^{0.37} = 4.27$$

Hence the convective heat transfer coefficient is

$$h_f = \frac{Nu\,k}{d_o} = \frac{4.27 \times 0.0307}{0.004} = 32.8\ W/m^2 K$$

Step 2 Radiation heat transfer coefficient:
Linear radiation is used since the fluid and wall temperatures are close to each other. The pertinent data is

$$\sigma = 5.67 \times 10^{-8}\ W/m^2 K^4;\ \varepsilon = 0.2$$

The radiation heat transfer coefficient is

$$h_R = 4\varepsilon\sigma T_w^3 = 4 \times 0.2 \times 5.67 \times 10^{-8} \times (273.15 + 90)^3 = 2.2\ W/m^2 K$$

Step 3 Calculation of reference temperature:

$$T_{ref} = \frac{h_f T_f + h_R T_w}{h_f + h_R} = \frac{32.8 \times 100 + 8.5 \times 90}{32.6 + 8.5} = 99.4°C$$

Step 4 Well treated as a fin:
Well material has a thermal conductivity of $k_w = 14\ W/m\,K$. Internal diameter of well is equal to outside diameter (d_o) minus twice the wall thickness (t) and is given by $d_i = d_o - 2t = 0.004 - 2 \times 0.0005 = 0.003\ m$. The fin parameter is then calculated as

$$m_{eff} = \sqrt{\frac{4d_o(h + h_R)}{k_w(d_o^2 - d_i^2)}} = \sqrt{\frac{4 \times 0.004 \times (32.6 + 2.2)}{45 \times (0.004^2 - 0.003^2)}} = 75.518\ m^{-1}$$

Since the well length is $L = 0.05\ m$ the non-dimensional fin parameter is

$$\mu_{eff} = m_{eff}L = 75.518 \times 0.05 = 3.776$$

Step 5 Well bottom temperature:

The well bottom temperature is given by

$$T_t = T_{ref} + \frac{T_w - T_{ref}}{\cosh(\mu_{eff})} = 99.4 + \frac{90 - 99.4}{\cosh 3.776} = 98.95°C$$

Thus the thermometric error is $T_e = T_t - T_f = 98.95 - 100 = -1.05°C$.

If radiation is ignored the above calculations should be done by taking $h_R = 0$ and $T_{ref} = T_f$. This is left as an exercise to the reader.

Concluding remarks

Systematic errors in temperature measurement are intimately linked with heat loss due to conduction, convection and radiation. These effects depend on the way the sensor is attached to a system whose temperature is to be measured. Typical applications have been considered in this chapter. Estimation of systematic errors is possible with basic knowledge and equations sourced from heat transfer literature.

Chapter 6

Heat flux and Heat Transfer Coefficient

Heat flux and heat transfer coefficient may either be measured by a *single* measurement process or they may require the measurement of several quantities before they may be estimated. Of course, it is best if the required quantity is directly measurable. We will give as much detail as possible, keeping in mind the target reader of the book. Many techniques are beyond the scope of the book and require direct reference to the appropriate research literature.

6.1 Measurement of heat flux

Heat flux is defined as the amount of heat transferred per unit area per unit time from or to a surface. In a basic sense it is a derived quantity since it involves, in principle, two quantities viz. the amount of heat transfer per unit time and the area from/to which this heat transfer takes place. In practice, the heat flux is measured by the temperature field it creates within a sensor of known area. The heat flux to be estimated may set up, either a steady state temperature field or a transient temperature field, within the sensor. The temperature field set up may be either perpendicular to the direction of heat flux or parallel to the direction of heat flux.

6.1.1 Foil type heat flux gauge

The foil type heat flux gauge (also known as the Gardon gauge after its inventor),[1] useful for measuring radiant heat flux, consists of a thin circular foil of constantan stretched tightly over a cooled copper annulus as shown in Figure 6.1. Nomenclature

---

[1]R.Gardon, "An instrument for the direct measurement of intense thermal radiation", Rev. Sci. Instrum., 24, 366-370, 1953
Gardon, R., "A Transducer for the Measurement of Heat-Flow Rate," J Heat Transfer, 396-398,1960.

used in the analysis is also indicated in this figure. One surface of the foil is exposed to the heat flux that is to be measured while the other surface may be *assumed* to be insulated. A copper wire is attached at the geometric center of the foil as indicated in the figure. A second copper wire is attached to the cooled copper annulus. The constantan foil forms two junctions with copper, the first one at its center and the second one at its periphery. Under steady state, the thermoelectric voltage across the two copper leads is a direct measure of the temperature difference set up between the center and the periphery of the constantan disk. The temperature field set up within the sensor is thus *perpendicular* to the direction of the incident heat flux. The temperature difference itself is a measure of the incident heat flux as will be shown by the following analysis.

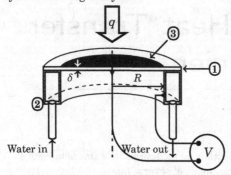

1   Constantan foil, conductivity k
2   Cooled copper annulus
3   Blackened active part of foil

Figure 6.1: *Schematic of a Gardon or foil type heat flux gauge*

Heat transfer analysis of an annular element of the foil shown in Figure 6.2 is made now. Heat is gained by the foil element because of the incident heat flux. It is given by

$$Q_g = q \times (2\pi r \Delta r) \qquad (6.1)$$

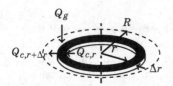

Figure 6.2: *Energy balance over a foil element in the form of an annular ring*

Heat is conducted in the radial direction because of the temperature field set up under the influence of the incident heat flux since the cooled copper annulus is at a lower temperature. The analysis is similar to the fin type analysis presented earlier in section 5.4.3. We have, by Fourier law

$$Q_{c,r} = -2\pi k \delta \left(r \frac{dT}{dr} \right)\bigg|_r \,, \, Q_{c,r+\Delta r} = -2\pi k \left(r \frac{dT}{dr} \right)\bigg|_{r+\Delta r}$$

We Taylor expand the second conduction flux, retain two terms, to get

$$Q_{c,r+\Delta r} = -2\pi k \delta \left[\left(r \frac{dT}{dr} \right)\bigg|_r + \frac{d}{dr}\left(r \frac{dT}{dr} \right)\bigg|_r \Delta r \right]$$

Net heat transfer by conduction $Q_{c,net}$ across the element is then given by $Q_{c,r} - Q_{r+\Delta r}$ which leads to the following expression.

$$Q_{c,net} = 2\pi k\delta \frac{d}{dr}\left(r\frac{dT}{dr}\right)\Delta r \qquad (6.2)$$

For steady state to prevail in the foil $Q_g + Q_{c,net} = 0$. We cancel the common factor $2\pi\Delta r$ to get

$$\frac{d}{dr}\left(r\frac{dT}{dr}\right) + \frac{q}{k\delta}r = 0 \qquad (6.3)$$

The boundary conditions are now specified as

$$T \text{ is finite at } r = 0; \quad T(r = R) = T_R \qquad (6.4)$$

Integrate Equation 6.3 once with respect to r to get

$$r\frac{dT}{dr} + \frac{q}{k\delta}\frac{r^2}{2} = A$$

where A is a constant of integration. This may be rearranged to get

$$\frac{dT}{dr} + \frac{q}{k\delta}\frac{r}{2} = \frac{A}{r} \qquad (6.5)$$

Integrate this once more with respect to r to get

$$T + \frac{qr^2}{4k\delta} = A\ln(r) + B \qquad (6.6)$$

where B is a second constant of integration. The constant A has to be chosen equal to zero in order that the solution does not diverge at $r = 0$. The constant B is obtained from the boundary condition at $r = R$ as $B = T_R + \frac{qR^2}{4k\delta}$. With this the solution for the temperature is obtained as

$$T - T_R = \frac{q}{4k\delta}\left(R^2 - r^2\right) \qquad (6.7)$$

It may be noted that the constant B is nothing but the temperature at the center of the constantan disk. All we do is to put $r = 0$ in Equation 6.7 to verify this. In view of this, Equation 6.7 may be rearranged as

$$q = \frac{T_0 - T_R}{\left(\dfrac{R^2}{4k\delta}\right)} = K\Delta T \qquad (6.8)$$

In Equation 6.8 T_0 is the temperature at the center of the disk, coefficient K is the gauge constant given by $K = \left(\dfrac{4k\delta}{R^2}\right)$ with units of $\dfrac{W}{m^2\,K}$ and ΔT is the temperature difference between the center of the disk and the periphery. ΔT is also the output that appears as a proportional voltage ΔV across the terminals of the differential thermocouple. There is thus a linear relationship between the heat flux and the output of the heat flux gauge. The gauge sensitivity is simply the reciprocal of the gauge constant. Thus the sensitivity is given in thermal units $\left(\dfrac{^\circ C}{\frac{W}{m^2}}\right)$ as

$$S_t = \frac{1}{K} = \frac{R^2}{4k\delta} \qquad (6.9)$$

The sensitivity may also be represented in electrical units $\left(\dfrac{\mu V}{\frac{W}{m^2}}\right)$ as

$$S_e = \frac{\alpha_S}{K} = \frac{\alpha_S R^2}{4k\delta} \qquad (6.10)$$

where α_S is the Seebeck coefficient.

Example 6.1

Gardon gage of diameter 5 mm is 50 μm thick. It is exposed to an incident heat flux of 1 W/cm^2. What is the temperature difference developed by the gauge? If the Seebeck coefficient is assumed to be constant at $\alpha_S = 39\ \mu V/^\circ C$ what is the electrical output of the gauge when exposed to the heat flux indicated above? If the diameter of the gage is uncertain to an extent of 1 % and the thickness is uncertain to an extent of 2 %, what will be the uncertainty in the output when exposed to the heat flux mentioned above? At what rate do we have to remove heat from the copper annulus?

Solution :

Step 1 Gauge output (thermal) for specified heat flux:
The diameter and thickness of the foil are given as $D = 5\ mm = 0.005m$ and $\delta = 50\ \mu m = 5 \times 10^{-5}\ m$. Hence the radius of the gage is $R = \dfrac{D}{2} = \dfrac{0.005}{2} = 0.0025\ m$. The incident heat flux is given to be $q = 1W/cm^2 = 10^4\ W/m^2$. Foil thermal conductivity is taken as $k = 22.5\ W/mK$. Using Equation 6.8 the temperature difference developed by the gauge (i.e. gauge output) is

$$\Delta T = \frac{qR^2}{4k\delta} = \frac{10^4 \times 0.0025^2}{4 \times 22.5 \times 50 \times 10^{-6}} = 13.89^\circ C$$

Step 2 Gauge output (electrical) for specified heat flux:
Seebeck coefficient is specified to be $\alpha_S = 39\mu V/^\circ C$. Hence the electric output of the gauge is given by

$$\Delta V = \alpha_S \Delta T = 39 \times 10^{-6} \times 13.89 = 541.7\ \mu V = 0.542\ mV$$

Step 3 Uncertainty in gauge output for specified heat flux:
Uncertainties have been specified in percentages given by $u_D = u_R = 1\%$; $u_\delta = 2\%$. Logarithmic differentiation is appropriate to calculate the uncertainty in the gauge output because of the uncertainties in the values of the gauge parameters. Uncertainty in the gauge output is given by

$$u_{\Delta T} = \sqrt{(2 \times u_R)^2 + (u_\delta)^2} = \sqrt{(2 \times 1)^2 + (2)^2} = 2.83\%$$

The uncertainty in terms of the temperature difference developed by the gauge may then be calculated as

$$u_{\Delta T} = \frac{2.83 \times 13.89}{100} = 0.39°C$$

Step 4 Heat removal rate from the gauge:
Rate at which heat is to be removed at the periphery is simply the product of heat flux and the frontal area of the gauge.

$$Q_r = q \cdot \pi \cdot R^2 = 10^4 \times \pi \times 0.0025^2 = 0.196 \approx 0.2 \, W$$

The heat flux encountered in Example 6.1 is roughly 8 times the solar flux that arrives at the outer edges of the earth's atmosphere. In typical applications involving rocket nozzles the heat flux 10 times this (i.e.$10^5 \, W/m^2$)is possible. In reentry applications the heat flux on the surface of a reentry vehicle due to kinetic heating may be some 10 times this i.e. $10^6 \, W/m^2$ or $1 \, MW/m^2$. A heat flux gauge suitable for such high heat flux values should be designed with more robust materials and one would have to remove much larger amount of heat to keep the gauge "cool"! If we were to use the gauge described in Example 6.1 the temperature difference is 100 times the value obtained therein, and will equal $1389°C$. This is certainly an impossible situation! The heat removal from the gauge would correspondingly be $200 \, W$ which would be quite impossible to remove from such a small gauge!

We have defined the gauge sensitivity earlier in Equations 6.9 and 6.10. We calculate the sensitivity for the gauge considered in Example 6.1 here as:

$$S_t = \frac{0.0025^2}{4 \times 22.5 \times 50 \times 10^{-6}} = 0.001389 \, \frac{°C}{\frac{W}{m^2}}$$

or

$$S_e = 39 \cdot \frac{0.0025^2}{4 \times 22.5 \times 50 \times 10^{-6}} = 0.0542 \, \frac{\mu V}{\frac{W}{m^2}}$$

One may note that the unit of sensitivity will be $\frac{mV}{W/cm^2}$ if both R and δ are specified in mm. In Example6.1 the sensitivity is also equal to 0.542 $\frac{mV}{W/cm^2}$, as may easily be verified. In fact, it is customary to specify large heat flux values in the units of W/cm^2, such as in aerospace applications. The above sensitivity unit will then be very appropriate. We see that the sensitivity depends on geometrical as well as thermal properties of the foil material. The advantage of using constantan is that it has a high melting point and also forms thermocouple junctions with copper wires. Of course, we may also use other foil materials, if necessary.

Foil size and thickness may be chosen using Figure 6.3 when it is made of constantan foil. The three lines shown are for the indicated sensitivity values in units of $\frac{mV}{W/cm^2}$.

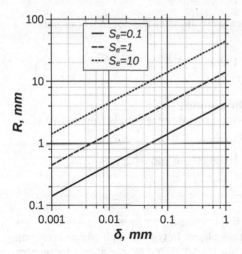

Figure 6.3: *Proportions of foil gauge for desired sensitivity values*

6.1.2 Transient analysis of foil gauge

When the incident heat flux varies with time or when a steady heat flux is "turned on" it would be necessary to find out how the gauge responds to the incident flux. A transient analysis is necessary for this purpose. An approximate analysis is presented below for a thin foil gauge. We shall start from a steady state when the gauge is actually responding to a steady incident flux. We have seen above that the temperature distribution in the foil is a quadratic function of r and is given by Equation 6.7. We divide this by the temperature difference between the center and the periphery of the foil, given by $T - T_R = \dfrac{q}{4k\delta}R^2$ to get

$$\frac{T - T_R}{T_0 - T_R} = \left(1 - \frac{r^2}{R^2}\right) \tag{6.11}$$

In the steady sate the energy stored in the foil is given by

$$E = \int_0^R \rho C\delta(T - T_R)2\pi r dr \tag{6.12}$$

where C is the specific heat of the foil material. We make use of Equation 6.11 to write Equation 6.12 as

$$E = 2\rho C\delta\pi(T_0 - T_R)\int_0^R r\left(1 - \frac{r^2}{R^2}\right)dr \tag{6.13}$$

The indicated integration may easily be performed to show that

$$\int_0^R r\left(1 - \frac{r^2}{R^2}\right)dr = \frac{R^2}{4} \tag{6.14}$$

Thus we finally get

$$E = \rho C\delta\pi(T_0 - T_R)\frac{R^2}{2} \tag{6.15}$$

Consider now the unsteady state heat transfer in the foil as the applied heat flux changes with time as $q(t)$. The input heat flux is partially stored in the foil and partially removed by the coolant at the foil periphery. The stored energy is the change in E with respect to time obtained by differentiating Equation 6.15 with respect to time. Thus,

$$\frac{dE}{dt} = \rho C \delta \pi \frac{R^2}{2} \frac{d(T_0 - T_R)}{dt} \tag{6.16}$$

The heat loss at $r = R$, is in fact given by the instantaneous conductive heat transfer at the periphery , obtained by the use of Fourier law of heat conduction as $Q(r = R) = -2\pi R \delta k \dfrac{dT}{dr}\Big|_{r=R}$. We make use of Equation 6.11 to obtain $\dfrac{dT}{dr}\Big|_{r=R} = \dfrac{-2(T_0 - T_R)}{R}$.
With these we get

$$Q(r = R) = 4\pi k \delta (T_0 - T_R) \tag{6.17}$$

The sum of Equations 6.16 and 6.17 must equal the heat gain by the foil given by $q(t)\pi R^2$. Thus the following equation results after some simplification:

$$\frac{d(T_0 - T_R)}{dt} + \frac{8k}{\rho C R^2}(T_0 - T_R) = \frac{q(t)}{\rho C \delta} \tag{6.18}$$

We notice at once that the governing equation is a first order ordinary differential equation. The coefficient of the second term on the left hand side, in fact, is related to the first order time constant of the heat flux gauge given by

$$\tau = \frac{R^2 \rho C}{8k} = \frac{R^2}{8\alpha} \tag{6.19}$$

where α is the thermal diffusivity of the foil material. Figure 6.4 shows the relationship (Equation 6.19) between the foil radius and the first order time constant of the sensor in graphical form. This figure may be used in tandem with Figure 6.3 for determining the proportions and the time constant for the chosen dimensions of the gauge. Figure 6.5 shows the external appearance of a commercial heat flux gauge which has arrangement for cooling of the annulus by water.

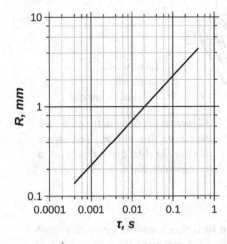

Figure 6.4: *Time constant for a foil type heat flux gauge*

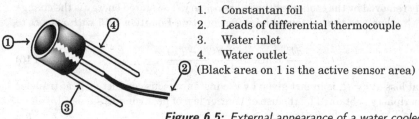

1.   Constantan foil
2.   Leads of differential thermocouple
3.   Water inlet
4.   Water outlet

② (Black area on 1 is the active sensor area)

Figure 6.5: *External appearance of a water cooled Gardon gauge*

6.1.3   Thin film sensors

The operation of a thin film heat flux sensor, shown schematically in Figure 6.6, and as a photograph in Figure 6.7 is very simple. A thin barrier of known thermal conductivity is attached to a surface that is receiving the heat flux to be measured. The barrier imposes a thermal resistance *parallel* to the direction of the heat flux and the heat conduction in the barrier is one-dimensional. The temperature difference across the barrier is measured using a differential thermopile arrangement wherein several hot and cold junctions are connected in opposition. The illustration shows only four hot and four cold junctions. In practice thermocouple materials are deposited as very thin but wide films of very small thickness of a few micrometers thickness. There may be hundreds of hot and cold junctions in a very small area. The output is proportional to the heat flux. Example 6.2 brings out the typical characteristics of such a heat flux gauge.

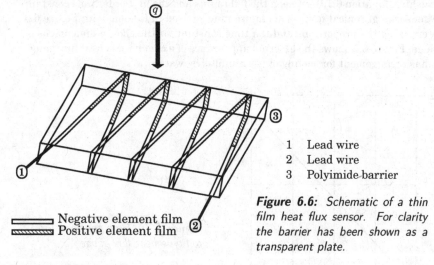

1   Lead wire
2   Lead wire
3   Polyimide barrier

════ Negative element film
▨▨▨▨ Positive element film ②

Figure 6.6: *Schematic of a thin film heat flux sensor. For clarity the barrier has been shown as a transparent plate.*

Example 6.2

The geometrical description of a thin film heat flux sensor is given in Figure 6.6. The thermal conductivity of the barrier material viz. polyimide of thickness $\delta = 0.25\ mm$ is known to be $k = 0.2\ W/m\ K$. Output of the gauge when exposed to a certain heat flux is $V = 0.01\ V$ i.e. $10\ mV$. Determine the

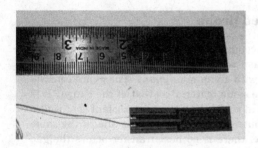

Figure 6.7: *Photograph of a thin film heat flux sensor*

heat flux q from this data if there are 40 hot and 40 cold junctions in the sensor.

Solution :

Since there are 40 junctions, the total output corresponds to $0.01/40 = 0.00025\ V$ per junction. Assuming that $40\ \mu V$ corresponds to $1\ °C$ this translates to a temperature difference of

$$\Delta T = \frac{0.00025}{40 \times 10^{-6}} = 6.25°C$$

Using $k = 0.2\ W/m°C$ and $\delta = 0.25\ mm = 0.00025\ m$ we get

$$q = \frac{k\Delta T}{\delta} = \frac{0.2 \times 6.25}{0.00025} = 5000\ W/m^2$$

Typically the solar heat flux is $1000\ W/m^2$. The above heat flux is some 5 times larger than solar flux!

6.1.4   Cooled thin wafer heat flux gauge

The operational principle of the thin wafer type cooled heat flux gauge is the same as the thin film gauge above. Temperature drop across the constantan wafer is measured by the differential thermocouple arrangement shown in Figure 6.8 (thicknesses of the gauge elements are highly exaggerated in this figure, for clarity). Note that the temperature gradient is in a direction *parallel* to the applied heat flux. There are two T type thermocouple junctions formed by the constantan wafer sandwiched between the two copper wafers.

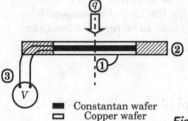

1   Cooled surface
2   Base material
3   Differential thermocouple
    copper lead wires

Figure 6.8: *Sectional view of a thin wafer type cooled heat flux gauge*

6.1.5   Axial conduction guarded probe

This probe (see Figure 6.9) is based on conduction through the probe in a direction *parallel* to the heat flux that is being measured. The gauge consists of a cylinder of a material of known thermal conductivity with an annular *guard*. The guard consists of an outer annular cylinder made of the same material as that of the gauge. It is exposed to the same heat flux and cooled at the back by the same coolant that is also used to cool the probe itself. Since the outer annulus experiences almost the same axial temperature gradient as the probe, there is no heat transfer across the gap between the probe and the guard and hence one dimensional axial conduction takes place in the probe. The temperatures are measured by two embedded thermocouples as indicated in the figure. Fourier law is used to derive the heat flux from the measured temperature difference, the distance between the thermocouples and the known thermal conductivity of the probe material.

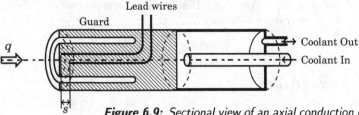

Figure 6.9: *Sectional view of an axial conduction guarded heat flux probe*

The coolant flow may be adjusted using the probe front surface temperature signal to operate the probe with a set front surface temperature. The coolant may be air for low heat flux applications while water may be used as the coolant for high heat flux applications. The axial conduction heat flux probe is conveniently positioned through a hole in the side wall of a furnace for monitoring the wall heat flux. The probe will respond to both convective as well as radiative heat fluxes. The front surface of the probe may be suitably treated for achieving high emissivity.

Example 6.3

Consider an axial conductivity guarded heat flux probe made of an alloy material of thermal conductivity equal to $45W/m°C$. The two thermocouples are placed $1cm$ apart. The incident heat flux is known to be $10^5 \ W/m^2$. The probe has a diameter of $25mm$. Determine the indicated temperature difference ΔT and the rate at which heat (Q) needs to be removed from the back surface of the probe.

Solution :

The data specified are:

Probe diameter: $d = 25 \ mm = 0.025 \ m$

Spacing between thermocouples: $s = 1 \ cm = 0.01 \ m$

Incident heat flux: $q = 10^5 \ W/m^2$

Probe material thermal conductivity: $k = 45 \ W/m°C$

By Fourier law, we have $q = k\dfrac{\Delta T}{s}$ where ΔT is the temperature difference indicated by the thermocouples representing the probe output signal. Hence we have

$$\Delta T = \frac{qs}{k} = \frac{10^5 \times 0.01}{45} = 22.2^\circ C$$

The rate at which heat is to be removed from the back of the probe is nothing but the total heat transfer in to the probe at its front. This is given by the product of the incident heat flux and the probe frontal area. Thus

$$Q = qA = q\frac{\pi d^2}{4} = 10^5 \times \frac{\pi \times 0.025^2}{4} = 49.1W$$

This amount of heat is removable by an air stream.

6.1.6 Slug type sensor

Schematic of a slug type heat flux sensor is shown in Figure 6.10. A mass m of a material of specific heat C is embedded in the substrate as shown. Frontal area A of the slug is exposed to the heat flux to be measured while all the other surfaces of the slug are thermally insulated as indicated. When the incident flux is absorbed at the surface of the slug, it heats the slug, and uniformly so if it is made of a material of high thermal conductivity. The transient temperature response of the slug is related to the incident heat flux and is, in fact, the measured quantity. The heat flux is inferred by either a theoretical model or by calibration.

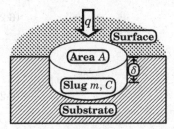

Figure 6.10: *Slug type heat flux sensor*

In the ideal case with no heat loss the equation for the temperature of the slug is given by

$$mC\frac{dT}{dt} = qA \tag{6.20}$$

On integration this will yield the slug temperature as a linear function of time given by

$$T = T_0 + \frac{qA}{mC}t \tag{6.21}$$

where T_0 is the initial slug temperature. Obviously we have to allow the process of heating to terminate by stopping the exposure of the slug to the incident heat flux when the slug temperature reaches its maximum allowable temperature T_{max}.

Response of a slug type sensor with a small heat loss

Consider now the case when there is a small heat leak from the slug sensor. Let the loss be proportional to the temperature excess of the slug with respect to the casing or the substrate temperature T_c. The loss coefficient is given as $K_L W/^\circ C$. Equation 6.20 will now be replaced by

$$mC\frac{dT}{dt} = qA - K_L(T - T_c) \tag{6.22}$$

Let $T - T_c = \theta$ and $\dfrac{K_L}{mC} = \epsilon$. With these Equation 6.22 will be recast as

$$\frac{d\theta}{dt} - \frac{qA}{mC} = -\epsilon\theta \tag{6.23}$$

Since K_L is expected to be small, the parameter ϵ is small. The solution may be sought by expanding it in the form $\theta = \theta^{(1)} + \epsilon\theta^{(2)} + \cdots$. Substituting this in equation 6.23 we get

$$\frac{d\theta^{(1)}}{dt} - \frac{qA}{mC} + \epsilon\frac{d\theta^{(2)}}{dt} + \cdots = -[\epsilon\theta^{(1)} + \epsilon^2\theta^{(2)} + \cdots] \tag{6.24}$$

Collecting terms of same order, the above is replaced by

$$\frac{d\theta^{(1)}}{dt} - \frac{qA}{mC} = 0 \tag{6.25}$$

(Both terms are multiplied by ϵ^0)

$$\frac{d\theta^{(2)}}{dt} = -\theta^{(1)} \tag{6.26}$$

(Both terms are multiplied by ϵ^1)
These equations are solved to get

$$\theta^{(1)} = \frac{qA}{mC}t \tag{6.27}$$

$$\theta^{(2)} = -\frac{qA}{mC}\frac{t^2}{2} \tag{6.28}$$

These assume that $T_0 = T_c$. Thus the response of the slug follows the relation

$$\boxed{\theta = \frac{qA}{mC}t - \epsilon\frac{qA}{mC}\frac{t^2}{2} = \frac{qA}{mC}t\left[t - \frac{K_L}{mC}\frac{t^2}{2}\right]} \tag{6.29}$$

Thus a slight amount of nonlinearity is seen in the response of the slug.

Example 6.4

A slug type of sensor is made of a copper slug of $3mm$ thickness. The specification is that the temperature of the slug should not increase by more than $40^\circ C$ during its operation.

- What is the time for which the slug can be exposed to an incident heat flux of $10000\,W/m^2$ when there is negligible heat loss from the slug?

- What is the time for which the slug can be exposed to an incident heat flux of 10000 W/m^2 when there is small heat loss from the slug specified by a loss coefficient of $K_L = 50 \, W/^\circ C$?

Note: Make all calculations based on unit area of slug exposed to the incident heat flux.

Solution :

No heat loss case:

Properties of Copper are taken from hand book of thermo-physical properties. Density $\rho = 8890 \, kg/m^3$, specific heat $C = 398 \, J/kg^\circ C$.

Mass of slug on unit frontal area basis is

$$m = \rho \delta = 8890 \times 0.003 = 26.67 \, kg/m^2$$

The maximum temperature rise during operation is $\Delta T_{max} = 40^\circ C$

If there is negligible heat loss the temperature increases linearly and the maximum exposure time is given by Equation 6.21, by substituting ΔT_{max} in place of $T - T_0$. Thus

$$t_{max} = \frac{mC\Delta T_{max}}{q} = \frac{26.67 \times 398 \times 40}{10000} = 42.46 \, s$$

Heat loss case:

When the heat loss is taken into account, what happens is worked out below. We have, from the given data

$$\epsilon = \frac{K_L}{mC} = \frac{50}{26.67 \times 398} = 0.0048 \, m^2/s$$

and

$$\frac{qA}{mC} = \frac{10000 \times 1}{26.67 \times 398} = 0.942$$

With these, we get, using Equation 6.29

$$\theta = \Delta T_{max} = 40 = 0.942 \left[t_{max} - 0.0047 \frac{t_{max}^2}{2} \right]$$

This may be rewritten as

$$\theta = \Delta T_{max} = 40 = 0.942 \times t_{max} \left[1 - 0.0047 \frac{t_{max}}{2} \right]$$

For the last term on the right hand side we may conveniently substitute the t_{max} value obtained earlier, ignoring the heat loss. Thus we have, solving the above equation for t_{max}

$$t_{max} \approx \frac{40}{0.942 \times (1 - 0.0047 \times \frac{42.46}{2})} = 47.2 \, s$$

With heat loss taken in to account the slug gauge is useful over a slightly longer time!

The effect of heat loss discussed in Example 6.4 is brought out graphically in Figure 6.11. The temperature increase in the slug departs from the linear behavior (dashed line in Figure 6.11) when there is no heat loss and follows the non-linear variation shown (by the full line) in Figure 6.11. Curves such as those shown in Figure 6.11 may be drawn for other heat flux values and used to estimate the incident heat flux from the time temperature excess plot.

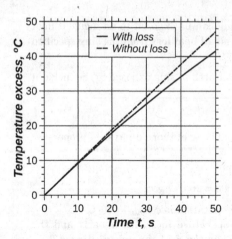

Figure 6.11: *Slug response with and without heat loss for q = 10000 W/m²*

6.1.7 Slug type sensor response including non uniformity in temperature

In section 6.1.6 we have described the use of a slug type sensor for measurement of heat flux. It was assumed that the slug behaves as a lumped system. This meant that there is no spatial variation of temperature within the slug. In practice, however, it is likely that the temperature of the slug is measured at its back face (that is embedded in the body). This temperature will certainly lag behind the front surface temperature. Also the back surface temperature varies linearly with time for a certain duration, starting with a time delay with respect to the front surface temperature. Heat conduction theory provides the theoretical model for explaining these. We shall give below only the results without proof.

As the slug is not at a uniform temperature across its thickness there is a time lag between the front surface and rear surface temperature responses. Accordingly we have to correct the heat flux inferred from that obtained based on the rear temperature response. If q_a and q_i are respectively the actual (incident on the front surface of the slug) and the inferred (from the rear temperature of the slug), we have

$$q_i = q_a \left[1 - 2e^{-\frac{\pi^2 \alpha t}{\delta^2}} \right] \tag{6.30}$$

where α is the thermal diffusivity of the material of the slug. The grouping $\frac{\alpha t}{\delta^2}$ is non-dimensional and is referred to as the Fourier number. If the maximum allowable front surface temperature rise is ΔT_{max}, we have, for the optimum thickness of the slug δ_{opt}, the formula

$$\delta_{opt} = \frac{k \Delta T_{max}}{1.366 q_a} \tag{6.31}$$

The period Δt_{linear} over which the back surface temperature is linear (when a constant incident flux is present on the front surface of the slug) is given by the formula

$$\Delta t_{linear} = \frac{0.366}{\alpha} \left(\frac{k \Delta T_{max}}{q_a} \right)^2 \qquad\qquad (6.32)$$

Example 6.5

A slug type of sensor is made of a stainless steel (SS 304) slug of optimum thickness. The specification is that the temperature of the slug should not increase by more than $80°C$.

1. What is the optimum thickness of the slug if the front surface temperature should be limited to $80°C$ when the heat flux incident on its surface is $10000\ W/m^2$?
2. What is the useful time over which the temperature of back surface of the slug may be used to estimate the heat flux?

Note: Make all calculations based on unit area of slug exposed to the incident heat flux.

Solution :

The properties of SS 304 are taken from data hand book and are given below. Density: $\rho = 7900\ kg/m^3$, specific heat: $C = 477\ J/kg°C$ and thermal conductivity: $k = 14.9\ W/m°C$

The incident heat flux on the front face of the slug has been specified as $q_a = 10^5\ W/m^2$

The maximum allowed temperature of the front face of the slug is $\Delta T_{max} = 80°C$

1. Using equation 6.31, the optimum slug thickness is calculated as

$$\delta_{opt} = \frac{14.9 \times 80}{1.366 \times 10^5} = 0.00873 \approx 0.009\ m = 9\ mm$$

The thermal diffusivity of SS 304 is calculated as

$$\alpha = \frac{k}{\rho C} = \frac{14.9}{7900 \times 477} = 3.95 \times 10^{-6}\ m^2/s$$

2. We now use Equation 6.32 to determine the time for which the slug back surface temperature is linear as

$$\Delta t_{linear} = \frac{0.366}{3.95 \times 10^{-6}} \left(\frac{14.9 \times 80}{10^5} \right)^2 = 13.15\ s$$

We may now calculate the ratio of heat fluxes, using Equation 6.30, as

$$\frac{q_i}{q_a} = 1 - 2 \times e^{-\pi^2 \times \frac{3.95 \times 10^{-6} \times 13.15}{0.009^2}} = 0.997$$

The estimated heat flux based on back surface temperature of the slug will then be given by

$$q_i = 0.997 \times q_a = 0.997 \times 10^5 = 99500\ W/m^2$$

6.1.8   Thin film heat flux gauge - Transient operation

In section 6.1.3 we have described a thin film sensor that operates under the steady sate. We now describe a thin film sensor that operates in the transient mode that is capable of measuring extremely short heat flux events that are characteristic of high speed flows. In supersonic and hypersonic wind tunnels the flow duration may be from a few μs to a few ms and the heat flux has to be estimated within the duration of the test. The basic theoretic model is one dimensional transient heat transfer in a semi-infinite medium that is treated in most books on heat transfer. Again we shall make use of the results form conduction theory without offering proof.

The principle of operation of a thin film heat flux gauge may be understood by referring to Figure6.12. Assume that a constant heat flux q is incident on a thin platinum film of thickness δ mounted on a thick substrate. The platinum film is of negligible thickness (usually a few μm thick) and is assumed to take on a uniform temperature, at any instant of time, equal to the surface temperature of the substrate. For times short compared to the time required for the thermal disturbance to propagate through the thickness of the substrate the substrate may be assumed to be a semi-infinite solid. The transient response of the surface and an interior location are given by the solution of the one dimensional heat equation available from any book on heat transfer.

Surface temperature response:

$$\Delta T(0,t) = \frac{q}{k}\sqrt{\frac{4\alpha t}{\pi}} = 1.128\frac{q\sqrt{\alpha}}{k}\sqrt{t} \tag{6.33}$$

Here $\Delta T(0,t)$ is the increase in temperature of the platinum film with respect to the initial temperature of the platinum film as well as the substrate.

Temperature response at a depth x inside the substrate:

$$\Delta T(\eta,t) = \frac{q}{k}\left[\sqrt{\frac{4\alpha t}{\pi}}e^{-\eta^2} - x\,\mathrm{erfc}(\eta)\right] \tag{6.34}$$

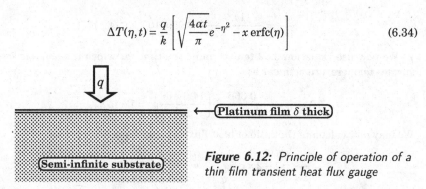

Figure 6.12: *Principle of operation of a thin film transient heat flux gauge*

In the above $\eta = x/\sqrt{\alpha t}$ is a non-dimensional composite variable, α is the thermal diffusivity of the substrate material and k is the thermal conductivity of the substrate material. The complementary error function $\mathrm{erfc}(x)$, a tabulated function makes its appearance in the equation. The response is thus determined by the *thermal properties of the substrate material*. The substrate material may be chosen so as to give the desired response. The temperature of the surface may be inferred by monitoring the electrical resistance of the platinum foil treating it as an RTD. This is usually done by passing a constant direct current through the foil (a few mA) and monitoring the potential difference across the sensor. Figure 6.13 shows the response when different substrate materials are chosen (properties of these materials are given in Table 6.1) and subject to a constant heat flux of 10^5 W/m^2 at the surface. In each case the heat flux sensor is at an initial temperature of 30 $°C$.

Table 6.1: *Typical substrate material thermal properties*

Material	$\rho, kg/m^3$	$C, J/kg°C$	$k, W/m°C$ at 25°C	$\alpha, m^2/s$
MACOR*	2520	790	1.46	7.33×10^{-7}
Corning PYREX 7740	2230	754	1.005	5.98×10^{-7}
Quartz	2600	819	3.0	1.41×10^{-6}

*MACOR is a machinable ceramic that withstands temperatures up to 1000°C

The substrate is normally an electrically insulating material with low thermal properties to obtain a sizable surface temperature response. The surface film materials are typically like those shown in Table 6.2.

Table 6.2: *Properties of common film materials*

Material	$\rho, kg/m^3$	$C, J/kg°C$	$k, W/m°C$	$\alpha, m^2/s$
Platinum	21500	130	70	2.5×10^{-5}
Nickel	8900	450	84	2.1×10^{-5}
Copper	8900	380	397	11.74×10^{-5}

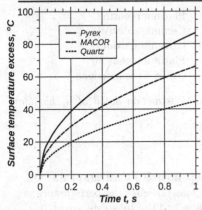

Figure 6.13: *Surface temperature response with three different substrates given in Table 6.2. Incident flux is q = 10^5 W/m^2.*

The choice of the substrate thickness depends on the duration for which the sensor needs to be used and the material of the substrate. This is gleaned by looking at the ratio of temperature at a depth to that at the surface as shown in Figure 6.14. This ratio is in fact given by

$$\frac{\Delta T(x,t)}{\Delta T(0,t)} = e^{-\eta^2} - \sqrt{\pi}\eta \cdot \text{erfc}(\eta) \qquad (6.35)$$

If this ratio is taken as 0.01 for deciding the duration of operation, Figure 6.14 indicates that it need be only 0.6 *mm* for test duration of 50 *ms*. The figure has been made for quartz as the substrate material. Thin film gauges are normally used in transient experiments that last from a microsecond to possibly a few milliseconds, typical of aerospace applications.

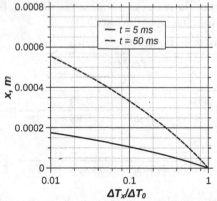

Figure 6.14: Thickness of sub-strate of a thin film gauge with quartz substrate. Incident flux is $q = 10^5 \ W/m^2$.

The construction detail of a typical thin film gauge is shown in Figure 6.15. The platinum film is deposited on a Zirconium oxide substrate of suitable thickness. The electrical connections are through constantan leads embedded in the substrate.[2]

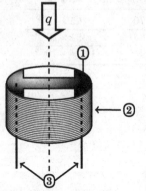

1. Platinum coating
2. Ceramic substrate (zirconium oxide)
3. Platinum leads for electrical connections

Figure 6.15: Construction detail of a thin film gauge

Example 6.6

A thin film of platinum of size $10 \times 2 \ mm$ is deposited on a substrate of MACOR. The thickness of the platinum film is $0.01 \ mm$. Discuss the performance of this as a heat flux gage. Assume that the film and the substrate are at $20°C$ initially and the maximum temperature of the film is not to exceed $200°C$ when it is exposed to a flux of $1 \ MW/m^2$.

Solution :

[2]For more details of this sensor the reader may visit the web site
http://www.swl.rwth-aachen.de/en/industry-solutions/thin-films/

The surface heat flux has been specified as $q = 1\ MW/m^2 = 10^6 W/m^2$. The initial temperature is $T_0 = 20°C$ and hence the surface temperature increase, at the end of the useful interval t_{max} of the gage, is $\Delta T(0, t_{max}) = 200 - 20 = 180°C$.

The substrate is MACOR having a thermal diffusivity of $7.33 \times 10^{-7}\ m^2/s$ and thermal conductivity of $1.46\ W/m°C$. The useful time interval is then given, using Equation 6.33 as

$$t_{max} = \left[\frac{k\Delta T(0, t_{max})}{q} \frac{1}{1.128\sqrt{\alpha}}\right]^2$$

or

$$t_{max} = \left[\frac{1.46 \times 180}{10^6} \frac{1}{1.128\sqrt{7.33 \times 10^{-7}}}\right]^2 = 0.074 s = 74 ms$$

We now calculate the ratio of temperature at depth x inside the substrate to the surface temperature at time t = 74 ms. The calculation uses Expression 6.35, using tabulated values of complementary error function. The data is then plotted as shown in Figure 6.16

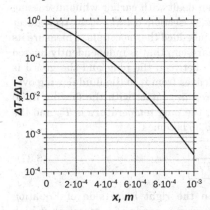

Figure 6.16: Temperature at various depths within the substrate at $t = t_{max} = 74\ ms$

It is seen from Figure 6.16 (compare with Figure 6.14) that the value of the temperature ratio is 0.01 at a depth of 0.000675 m or 0.675 mm, at a time of 75 ms. Hence the substrate need only be this thick for the present application.

6.2   Measurement of heat transfer coefficient

Heat transfer coefficient is an important quantity that is needed for design of thermal systems. It is usually introduced through the so called "Newton's law of cooling". This law postulates that the heat flux from/to a surface is linearly related to the difference in temperature between a surface and a moving fluid in contact with it. The proportionality constant h (which has dimensions of $W/m^2°C$) is called the heat transfer coefficient. Thus, we have

$$h = \frac{q_s}{T_s - T_f} \qquad\qquad (6.36)$$

where q_s is the heat flux *from* the surface, T_s is the surface temperature and T_f is the fluid temperature. We see that the estimation of h requires the measurement of the three quantities, two temperatures and one heat flux, that enter on the right hand side of Equation 6.36.

6.2.1 Film coefficient transducer

In the relevant literature the heat transfer coefficient is also referred to as the film coefficient and hence the name of this transducer. The heat transfer coefficient is visualized via a conductive layer of the fluid called the "film" next to the surface that is assumed to be stationary. The film coefficient is the conductance of this fluid layer.

A possible arrangement for the measurement of heat transfer coefficient is the film coefficient transducer whose schematic is given in Figure 6.17. This transducer makes use of the guard principle that has been dealt with earlier while discussing the axial conduction type heat flux probe (see section 6.1.5). The transducer consists of an electrically heated plug with an embedded thermocouple to measure its temperature. The plug is surrounded by a guard cup that is independently heated electrically. When the temperature difference between the plug and the guard is zero, all the heat that is supplied to the plug leaves from it to the fluid flowing over its surface. If the wetted frontal area (the area in contact with the flowing fluid) of the plug is S, the electric heat input is Q_e, the plug temperature is T_p and the temperature of the fluid is T_f, then Equation 6.36 becomes

$$h = \frac{Q_e}{S(T_p - T_f)} \qquad\qquad (6.37)$$

Note that all the quantities that appear on the right hand side of Equation 6.37 are measurable quantities. The sensing plug is made of a material of high thermal conductivity such that it is very nearly isothermal throughout and hence, in principle, we may measure its temperature at any convenient location on it.

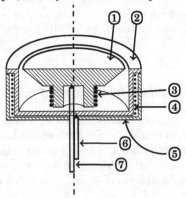

1. Sensing plug
2. Guard cup
3. Sensing plug heater
4. Guard cup heater
5. Casing
6. Sensing plug thermocouple
7. Guard cup thermocouple

Figure 6.17: *Schematic of a film coefficient transducer (section view)*

6.2.2 Cylindrical heat transfer coefficient probe

A variant that is very often used in practice is the cylindrical probe shown in Figure 6.18. In many applications we desire the mean heat transfer coefficient from the surface of a tube or pipe immersed in a flowing medium. The experiment is conducted in the so called "cold" mode wherein the fluid is at or near room temperature and the probe is heated to a higher temperature. The principle of this probe is no different from the film coefficient gauge shown in Figure 6.17. The heat transfer area is $S = \pi D L$, the heat input is Q_m and the heat transfer coefficient is given by

$$h = \frac{Q_m}{\pi D L (T_m - T_f)} \tag{6.38}$$

We may account for radiation by subtracting

$$Q_r = S \sigma \varepsilon (T_m^4 - T_{bkg}^4) \tag{6.39}$$

from Q_m and evaluating the heat transfer coefficient as

$$h = \frac{Q_m - Q_r}{S(T_m - T_f)} \tag{6.40}$$

In the above ε is the emissivity of the surface of the probe and T_{bkg} is the effective temperature of the background with which the probe exchanges heat by radiation. Both these also need to be measured.

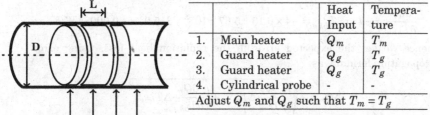

		Heat Input	Temperature
1.	Main heater	Q_m	T_m
2.	Guard heater	Q_g	T_g
3.	Guard heater	Q_g	T_g
4.	Cylindrical probe	-	-
Adjust Q_m and Q_g such that $T_m = T_g$			

Figure 6.18: Schematic of a cylindrical heat transfer coefficient probe

Example 6.7

A cylindrical heat transfer probe of diameter 12.5 mm is made of rough polished copper with a surface emissivity of $\varepsilon = 0.15$. The main heater is 25 mm wide and the two guard heaters are each 12 mm wide. In a certain forced flow experiment in air, the heat supplied to the main heater has been measured to be $2.67 \pm 0.02\,W$ when the surface temperature was $42.7 \pm 0.2°C$. The air temperature was measured at $20.4 \pm 0.2°C$. The effective background temperature is found to be the same as the air temperature. Estimate the convection heat transfer coefficient and its uncertainty.

Solution :

The nominal values of all the quantities specified are written down as:

$$D = 12.5\,mm = 0.0125\,m; L = 25\,mm = 0.025\,m; \varepsilon = 0.15$$

$$T_m = 42.7°C = 315.9\,K; T_f = T_{bkg} = 20.4°C = 293.6\,K$$

The heat input to the main heater is $Q_m = 2.67\,W$ The surface area involved in heat transfer is calculated as

$$S = \pi \times 0.0125 \times 0.025 = 0.00098175\,m^2$$

The heat loss by radiation is estimated, using Equation 6.39 as

$$Q_r = 0.00098175 \times 5.67 \times 10^{-8} \times 0.15 \times (315.9^4 - 293.6^4) = 0.021\,W$$

The convection heat transfer coefficient may now be estimated using Equation 6.40 as

$$h = \frac{2.67 - 0.021}{0.00098175(42.7 - 20.4)} = 120.99 W/m^2\,°C$$

We assume that only the temperatures and heat input are subject to error since nothing is mentioned about the errors in the geometrical parameters. Measured temperatures are subject to errors of $\Delta T = 0.2\,°C$. Hence the temperature difference is subject to an error of $\Delta(T_m - T_{bkg}) = 0.2\sqrt{2} = \pm 0.2 \times 1.414°C = 0.283°C$ (this may easily shown by using the error propagation formula). In addition we shall assume that the background temperature is subject to the same error as the air temperature. The error is to be estimated in the radiant heat loss. Only temperature measurement errors are assumed to be important. The influence coefficients are calculated as:

$$\frac{\partial Q_r}{\partial T_m} = 4\varepsilon\sigma S T_m^3 = 4 \times 0.15 \times 5.67 \times 10^{-8} \times 315.9^3 = 0.00105\,W/K$$

$$\frac{\partial Q_r}{\partial T_{bkg}} = -4\varepsilon\sigma S T_{bkg}^3 = -4 \times 0.15 \times 5.67 \times 10^{-8} \times 315.9^3 = -0.00084\,W/K$$

Hence we get the expected error in the radiation heat loss, using error propagation formula, as

$$\Delta Q_r = \pm\sqrt{\left(\frac{\partial Q_r}{\partial T_m}\Delta T_m\right)^2 + \left(\frac{\partial Q_r}{\partial T_{bkg}}\Delta T_{bkg}\right)^2}$$

or

$$\Delta Q_r = \pm\sqrt{(0.00105 \times 0.2)^2 + (-0.00084 \times 0.2)^2} = \pm 0.0003\,W$$

With $\Delta Q_m = \pm 0.02\,W$ we have

$$\Delta(Q_m - Q_r) = \pm\sqrt{\Delta Q_m^2 + \Delta Q_r^2} = \pm\sqrt{0.02^2 + 0.0003^2} = \pm 0.02\,W$$

It is thus seen that the uncertainty in the estimated value of h is (using logarithmic differentiation)

$$\Delta h = \pm h\sqrt{\left[\frac{\Delta(Q_m - Q_r)}{Q_m - Q_r}\right]^2 + \left[\frac{\Delta(T_m - T_{bkg})}{T_m - T_{bkg}}\right]^2}$$

or

$$\Delta h = \pm 120.99 \times \sqrt{\left[\frac{0.02}{2.67 - 0.021}\right]^2 + \left[\frac{0.283}{42.7 - 20.4}\right]^2} = \pm 1.77\,W/m^2\,°C$$

Thus the heat transfer coefficient is specified as $h = 120.99 \pm 1.77\,W/m^{2°}$.

Concluding remarks

Measurement of heat flux is important in many applications with time scales range from a few milliseconds to hours and hence different types of gages are needed and have been discussed in this chapter. In industrial applications measurement of heat transfer coefficient is an important activity. Applications are to be found in furnaces, heat exchangers, ovens etc. Transient applications require thin film sensors while more robust gages are useful for steady heat flux mesurement applications.

Exercise II

II.1 Temperature measurement

Ex II.1: Use library resources and write a note on International Practical Temperature Scale of 1968 (IPTS68) and International Temperature Scale of 1990 (ITS90). Bring out the major differences between these two.

Ex II.2: Temperature of a gas of known composition may be measured by acoustic thermometry. Search the recent literature and prepare a short bibliography on this topic.
Discuss any one of the references from this bibliography in full detail.

Ex II.3: The following data pertains to measurement of the temperature of a certain freezing metal using a constant volume gas thermometer. The pressure at triple point and the corresponding pressure at the freezing point of the metal are tabulated below:

p_{tp}, mm Hg	15.6	45.7	112.3
p_t, mm Hg	112.5	358.0	890.2

Determine the freezing point of the metal according to the ideal gas scale. Make a plot to indicate how the intercept is estimated from a quadratic fit to data in the form of pressure ratio vs. pressure at the triple point.

Ex II.4: Following readings were taken with a medium precision constant volume gas thermometer. The bulb of the thermometer was immersed in hot water at a constant temperature for these measurements. The amount of gas in the bulb was systematically varied to gather data.

p_{tp}, mm Hg	40	60	80	101	120
p_t, mm Hg	51.2	76.7	102.1	128.7	152.6

Estimate the temperature of water bath using the above data.
Hint: Fit straight line to data of pressure ratios. Use the precision of slope to specify an error bar to the temperature derived by you.

Ex II.5: We would like to use K type thermocouple to measure temperature in the range 30 to $1000°C$ with the reference junction at the ice point. It is desired that the resolution should be $\pm 1°C$. Specify the range and least count of a voltmeter for this purpose.
Make use of thermocouple reference tables for answering this problem.

Ex II.6: A K type thermocouple is used to measure the temperature of a system that is known to be $850°C$. The potentiometer binding posts are known to be at $31°C$ act as the reference junction. What is the mV reading indicated by the potentiometer? If an observer takes the mV reading and calculates the temperature assuming the reference temperature to be the ice point what will be temperature error?
Make use of thermocouple reference tables for answering this problem.

Ex II.7: Figure II.1 shows the way connections have been made for measuring temperature using K type thermocouple. What is the correct temperature if the output is 16 mV?

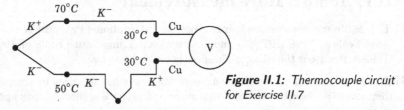

Figure II.1: *Thermocouple circuit for Exercise II.7*

Ex II.8: A millivolt recorder is available with a range of 0 to 10 mV and an accuracy of 0.5% of full scale. A copper-constantan thermocouple pair with a reference junction at the ice point is used with the recorder. Determine the temperature range and the accuracy of the installation.
Note: For a copper-constantan thermocouple 10 mV corresponds approximately to $213°C$ and the Seebeck coefficient is approximately $40\mu V/°C$.

Ex II.9: For a certain thermocouple the following table gives the thermoelectric output as a function of temperature $t°C$. The reference junction is at the ice point.

$t,°C$	10	20	30	40	50
E, mV	0.397	0.798	1.203	1.611	2.022

$t,°C$	60	70	80	90	100
E, mV	2.436	2.850	3.266	3.681	4.095

The manufacturer of the thermocouple recommends that the following interpolation formula should be used for calculating E as a function of T:

$$E(t) = 0.03995t + 5.8954 \times 10^{-6}t^2 - 4.2015 \times 10^{-9}t^3 + 1.3917 \times 10^{-13}t^4$$

How good is the polynomial as a fit to data? Base your answer on all the statistical parameters and tests you can think of. Obtain the Seebeck coefficient at 50 and 100 $°C$.

Ex II.10: Table below gives data for a T type thermocouple. It has been decided to approximate the thermocouple Seebeck emf by the relation $E = 0.03865t + 4.15018E - 5t^2$. What is the error in use of the approximation at the lowest and highest temperatures in the range shown in the table?

$t°C$	50	55	60	65	70	75
E, mV	2.036	2.251	2.468	2.687	2.909	3.132

$t°C$	80	85	90	95	100
E, mV	3.358	3.585	3.814	4.046	4.279

Above thermocouple is used as a differential thermocouple to measure the temperature difference between inlet and outlet to a small heat exchanger. The differential thermocouple gave an output of 0.125 mV. What is the temperature difference? Base your answer on the Seebeck coefficient at 75°C. What is the Peltier emf for the T type thermocouple in the above problem at $t = 75°C$?

A thermopile consisting of 40 cold and 40 hot junctions is made using T type thermocouple wires. The hot junctions are all maintained at 75°C while the cold junctions are all at a temperature of 50°C. What will be the output of the thermopile in mV?

Ex II.11: In a certain calibration experiment the following data was obtained:

$t°C$ (standard)	22.5	46.8	64.7	78.8	93.3	103.5	112
$t°C$ (candidate)	22.8	46.5	64.2	79.2	91.9	102.9	113.2

Comment on the distribution of error between the candidate and the standard. Can you specify a suitable error bar within which the candidate is well represented by the standard?

Ex II.12: Write a note on the constructional details of wire wound and film type Platinum Resistance Thermometers. Comment on their relative merits.

Ex II.13: The resistance of a certain sensor varies with temperature as $R(t) = 500(1 + 0.005t - 2 \times 10^{-5}t^2)$, where the resistance $R(t)$ is in Ω and the temperature t is in $°C$. Determine the resistance and the sensitivity at 100°C.

Ex II.14: A four wire PRT100 is fed a current of 5 mA by suitable external power connected using two of the four lead wires. A digital voltmeter is connected across the RTD by using the other two lead wires. The RTD is exposed to an ambient at 65°C. What will be the voltage reading? If the RTD has a dissipation constant of 0.001 $W/°C$ what will be the voltage reading? Base your answers taking in to account the Callendar correction.

Ex II.15: A thermistor code named 20 A is in the form of a glass coated bead 0.75 mm diameter. Its resistance at three different temperatures are given below:

$t, °C$	0	5.6	50
R, Ω	8800	3100	1270

Determine β for this thermistor. If resistance is measured with a possible error of $\pm 100\,\Omega$, what is the error in estimated t when the nominal value of resistance is $1500\,\Omega$?

Redo the problem using the Steinhart - Hart model.

Ex II.16: A thermistor has a resistance of $100\,\Omega$ at $20°C$ and a resistance of $3.3\,\Omega$ at $100°C$. Determine the value of β for this thermistor. Determine also the ratio of resistance of the thermistor when the temperature changes from0 to $50°C$. What are the sensitivities of the thermistor at these two temperatures. Based on the above, make suitable comments, in general, on the use of thermistors.

Ex II.17: A typical thermistor has the following specifications as given by the manufacturer.

Nominal Resistance at $25°C$	$10\,k\Omega$
β value ($25 < t < 85°C$)	$3480\,K$
Dissipation Constant	$0.4\,mW/K$
Heat capacity	$1.3\,mJ/K$
Thermistor Diameter	$0.8 \pm 0.1\,mm$
Thermistor Length	$1.4 \pm 0.4\,mm$

Determine: (a) Ratio of resistance of thermistor R_{25}/R_{85} and (b) Temperature error when the thermistor carries a current of $1\,mA$ and has a resistance value of $1500\,\Omega$.

Ex II.18: The resistance of a certain temperature detector is given by the data given below:

Temperature, $°C$	0	12.7	24.1	46.2	63.8	76.9	98.6
Resistance, Ω	2757	2556	2354	1955	1691	1464	1054

Estimate the temperature when the resistance is a) 1500, b) 2000 and c) $2600\,\Omega$. You are expected to develop a relation between the temperature and resistance and use it to get the answers.

Ex II.19: A thermistor is connected in series with a variable resistance R_s such that the current through the thermistor is limited to $1\,mA$ when the battery voltage is $9\,V$ and the thermistor is at room temperature of $20°C$ (see Figure II.2). The thermistor is characterized by a resistance of $2\,k\Omega$ at $20°C$ and a β value of $3000\,K$. Determine the value of the series resistance when the thermistor is at $20°C$. The thermistor is exposed to an environment at $40°C$. The variable resistance is to be adjusted such that the current is again $1\,mA$. What is the series resistance?

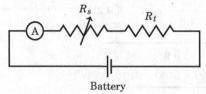

Figure II.2: *Thermistor circuit for Exercise II.19*

Ex II.20: For a certain thermistor $\beta = 3100K$ and the resistance at $70°C$ is $998 \pm 5\Omega$. The thermistor resistance is $1995 \pm 4\Omega$ when exposed to a temperature environment that is to be estimated. Estimate the temperature and its uncertainty.

Ex II.21: Thermistor manufacturer has provided the following data on a special order thermistor.
a) Resistance at ice point: $R_0 = 2050\,\Omega$
b) "Beta" value for the thermistor: $\beta = 3027\,K$
It was decided to verify this data by conducting a carefully conducted "calibration" experiment using a precision constant temperature bath capable of providing an environment within $\pm 0.02°C$ of the set value. The following data was obtained from this experiment:

Temperature $°C$	Resistance Ω
22.3	890
46.7	404

The measured resistance values are within an error band of $\pm 0.5\Omega$. Based on the experimental data comment on the "outcome" of the calibration experiment.

Ex II.22: The thermistor of Example II.21 is connected in series with a standard resistor of $1000\,\Omega$. A well regulated dc power of $9\,V$ is connected across the resistor - thermistor combination. The thermistor is subject to change in temperature in the range $5 - 80°C$. Plot the output across the standard resistor in this range.

Ex II.23: A thermistor is in the from of glass coated bead $0.75\,mm$ in diameter. Its resistance at two different temperatures are given as $R(t = 0°C) = 8800\,\Omega$ and $R(t = 50°C) = 1270\,\Omega$. Determine β for this thermistor. If resistance is measured with an error of $\pm 10\,\Omega$ what is the error in temperature when the nominal value of R is $1600\,\Omega$? What is the sensitivity of the thermistor at this temperature?

Ex II.24: The resistance of a certain semi conductor material varies with temperature according to the relation $R(T) = 1200e^{[3100(\frac{1}{T} - \frac{1}{273.15})]}$ where the resistance is in Ω and the temperature is specified in K. At a certain temperature the resistance has been measured as $600 \pm 10\,\Omega$. Determine the temperature and its uncertainty. Make a plot of R vs T in the range $273.15 < T < 500K$.

Ex II.25: A thermistor follows the resistance temperature relationship given by $R(T) = 1500e^{[3000(\frac{1}{T} - \frac{1}{273.15})]}$ where the resistance is in Ω and temperature is in K. This thermistor is connected in series with a resistance of $500\,\Omega$. A DC power supply of $1.5\,V$ is connected across the resistance -thermistor combination. A voltmeter measures the voltage across the series resistor. What is the voltage reading when the thermistor is at the ice point? What is the indicated voltage when the thermistor is at room temperature of $30°C$? If the voltmeter used for measuring the voltage across the fixed resistance has a resolution of $1\,mV$ what is the resolution in temperature at $30°C$?

Ex II.26: An RTD with $\alpha = 0.0034/°C$ and $R = 100 \ \Omega$ at the ice point has a dissipation constant of 0.04 $W/°C$. This RTD is used to indicate the temperature of a bath at $50°C$. The RTD is part of bridge circuit with all other resistances of $100 \ \Omega$ each. The battery connected to the bridge provides $6 \ V$. What voltage will appear across the RTD? Take into account the self-heating of the RTD due to the current that flows through it. When is the effect of self-heating more significant, at $t = 0°C$ or at $t = 100°C$? Explain.

Ex II.27: A thermistor has a resistance of $1000 \ \Omega$ at $20°C$ and a resistance of $33 \ \Omega$ at $100°C$. Determine the value of β for this thermistor. Determine also the ratio of resistance of the thermistor when the temperature changes from 0 to $50°C$. The above thermistor is connected in parallel with a resistance of $500 \ \Omega$ and the parallel combination is connected in series with a resistance of $500 \ \Omega$. A DC power supply of $9 \ V$ is connected across the combination. If the thermistor is exposed to a process space at $40°C$ what is the voltage across it? What is the temperature error if the dissipation constant is known to be $10 \ mW/°C$?

Ex II.28: A certain incandescent solid is at a temperature of $1200 \pm 5 \ K$. The total emissivity of the object is known to be 0.85 ± 0.02. Estimate along with the uncertainty the brightness temperature of the object.

Ex II.29: Estimate the actual temperature of an object whose brightness temperature has been measured using a vanishing filament pyrometer to be $1000°C$. The spectral emissivity of the object at 0.665 μm is 0.55. If the emissivity is subject to an error of 0.5% what is the corresponding error in the estimated actual temperature?

Ex II.30: The actual, brightness and color temperatures of a target are respectively given by $1185 \ K$, $1136 \ K$ and $1212 \ K$. The brightness temperature is at an effective wavelength of 0.66 μm. The color temperature is based on a second wavelength of 0.45 μm. Determine the emissivities at the two wavelengths.

Ex II.31: Estimate the actual temperature of an object whose brightness temperature has been measured using a vanishing filament pyrometer to be $850°C$. The transmittance of the pyrometer optics is known to be 0.95 ± 0.01. The spectral emissivity of the object at $0.66\mu m$ is 0.58. If the emissivity is subject to an error of $\pm1.5\%$ what is the corresponding error in the estimated actual temperature?

Ex II.32: The brightness temperature of a metal block is given as $950°C$. A thermocouple embedded in the block reads $1032°C$. What is the emissivity of the surface? The pyrometer used in the above measurement is a vanishing filament type with an effective λ of 0.65 μm. Assuming that the thermocouple reading is susceptible to an error of $\pm5°C$ while the brightness temperature is error free determine an error bar on the emissivity determined above.

Ex II.33: The emissive power of a black body in W/m^2 is given by the formula $E_b(T) = 5.67 \times 10^{-8}T^4$ where T is in Kelvin. A certain black body has the temperature measured at $1100 \pm 25 \ K$. Specify the emissive power along with its uncertainty. Express the uncertainty in physical units as well as in percentage form. If the uncertainty in the temperature is reduced by a factor of two what is the corresponding reduction in the uncertainty in the emissive power?

Ex II.34: Brightness temperature of a target is estimated, using a vanishing filament pyrometer operating at $0.66\,\mu m$ wavelength, to be $1245 \pm 10\,K$. Target temperature was measured independently using an embedded precision thermocouple as $1400 \pm 2\,K$. What is the effective target emissivity and its uncertainty?

Ex II.35: Total emissivity of a certain surface varies with temperature according to the relation $\varepsilon(T) = 0.068 + 0.00012\,T$ where $300 < T < 1200\,K$. The brightness temperature of such a surface has been measured using a thermal detector as $T_B = 700\,K$. What is the actual temperature of the surface?
(Hint: Iterative solution is required)

Ex II.36: Repeated measurements, using a vanishing filament pyrometer, by different operators indicates a mean brightness temperature $900\,K$ with a standard deviation of $5\,K$ while an embedded thermocouple gives a reading of $946\,K$ with an uncertainty of $\pm 2\,K$. The transmittance of the pyrometer is known to be 0.95. If the effective wavelength of the pyrometer is $0.655\,\mu m$ what is the emissivity of the target? What is its uncertainty?

Ex II.37: If the volume of mercury in the bulb of a thermometer is $200\,ml$ and the diameter of the capillary is $0.08\,mm$, how much will the mercury column move for a $1°C$ change in the temperature of the bulb? The difference in the cubical expansion of mercury and glass is given as $0.0002°C^{-1}$.

Ex II.38: A bimetallic strip $40\,mm$ long is made of yellow brass and Monel bonded together at $30°C$. The thickness of both the materials is $0.3\,mm$. The bimetallic strip is in the form of a cantilever beam and is used for "on - off" control with a set point of $75°C$. Calculate the deflection at this temperature of the free end for a temperature increase of $0.5°C$ above the set point.

II.2   Transient temperature measurement

Ex II.39: An experiment is conducted to determine the time constant of a K type thermocouple as under. We prepare a beaker of boiling water, insert the thermocouple into it, and allow it to reach equilibrium with the boiling water. Then we quickly remove it to a stand so that it is cooled by air moving across the junction by convection. The emf generated by the thermocouple is amplified with a DC amplifier with a gain of 100 and is then recorded, on a chart recorder at a speed of $25\,mm/s$ during the cooling period. From this record the following data has been gathered.

$T°C$	85	75	65	55	45	35
$t\,s$	0.350	0.600	0.937	1.438	2.175	3.250

Determine the time constant of the system. Also estimate the initial temperature of the thermocouple. Comment on this value. Assume that air temperature remains constant at $25°C$.

Ex II.40: A thermocouple is attached to a thin rectangular sheet of copper of thickness 0.5 mm and extent 50×150 mm. The copper sheet is initially at a temperature of $75°C$ and is suddenly exposed to still air at $15°C$. How long should one wait for the thermocouple to read $50°C$? It may be assumed that the entire plate cools uniformly with time and the heat loss is by natural convection. Assume that the 150 mm edge is vertical. The heat transfer coefficient may be based on an average plate temperature of $62.5°C$. Use suitable natural convection correlation from a book on Heat Transfer.

Ex II.41: A first order system is subjected to a ramp input given by $T_f = 20 + Rt$ where R = 1,2,5, 10 $°C/s$. The initial temperature of the probe is 30 $/circC$. Consider two cases with $\tau = 2$ s and 10 s and for these two cases draw graphs showing input as well as output responses of the system.

Ex II.42: Consider a first order system subjected to a periodic input. The initial temperature of the system is T_0 =30 $°C/s$. The amplitude of the periodic input is $T_a = 10$ $°C/s$, with $\omega\tau$ = 1,2,5. Plot the response of the system as a function of t/τ for all these cases.

Ex II.43: A first order system is subject to a ramp input given by $T_\infty = 20+5t$ where temperature is in $°C$ and t is in s. The initial temperature of the probe is $30°C$ and the time constant of the probe is $\tau = 2$ s. Determine the probe temperature at $t = 25$ s. What is the response of the probe if $t \gg \tau$? Explain your answer.

Ex II.44: The time constant of a first order thermal system is given as 30 s. The uncertainty in the value of the time constant is given to be ±0.5 s. The initial temperature excess of the system over and above the ambient temperature is $25°C$. It is desired to determine the system temperature excess and its uncertainty at the end of 30 s from the start.

Hint: System temperature excess variation is exponential.

Ex II.45: A certain first order system has the following specifications:
- Material: copper shell of wall thickness 1 mm, outer radius 6 mm
- Fluid: Air at $30°C$
- Initial temperature of shell: $50°C$

How long should one wait for the temperature of the shell to reach $40°C$? Assume that heat transfer is by free convection. Use suitable correlation from a heat transfer text to solve the problem.

Ex II.46: In order to determine the time constant of a temperature sensor an experiment was conducted by subjecting it to a step input. The ambient temperature remained constant at $40°C$ throughout the experiment. Temperature time data was collected by a suitable recorder at $\dfrac{1}{4}$ s intervals as shown in the table.

$t\,s$	0	0.25	0.5	0.75	1	1.25	1.5	1.75	2
$T°C$	75.28	69.65	64.82	60.90	57.73	55.54	53.44	51.08	49.31

$t\,s$	2.25	2.5	2.75	3	3.25	3.5	3.75	4	
$T°C$	47.67	46.87	46.05	44.39	43.89	43.40	42.61	43.40	

What is the time constant of the sensor?

II.3 Thermometric error

Ex II.47: A temperature sensor is in the form of a long cylinder of diameter $2\,mm$. It is placed in a duct carrying air at a temperature of $300°C$ moving with a speed of $2\,m/s$ such that the flow is normal to the axis of the cylinder. The duct walls are known to be at a temperature of $200°C$. What is the error in the indicated temperature if the emissivity of the sensor surface is 0.65 and conduction along the sensor is ignored?

If the sensor has an effective thermal conductivity of $12\,W/m°C$ and is immersed to a length of $25\,mm$ what will the thermometer error when the conduction error is included? What is the minimum depth of immersion if the thermometric error should not exceed $1°C$?

Make a plot of thermometric error as a function of surface emissivity of the sensor.

Ex II.48: A temperature sensor is in the form of a long cylinder of diameter $1.5\,mm$. It is placed in a duct carrying air at a temperature of $200°C$ moving with a speed of $5\,m/s$ such that the flow is normal to the axis of the cylinder. The duct walls are known to be at a temperature of $180°C$. What is the error in the indicated temperature if the emissivity of the sensor surface is 0.80 and conduction along the sensor is ignored? Air properties given below may be made use of:

Density $= 0.6423\,kg/m^3$, Dynamic viscosity $= 2.848 \times 10^{-5}kg/m\,s$, Thermal conductivity $= 0.0436\,W/m°C$, Prandtl number $= 0.680$.

Ex II.49: a) A thermocouple pair consists of $0.25\,mm$ diameter wires of copper and constantan placed $0.75\,mm$ apart by encapsulating the two wires in a sheath material of thermal conductivity equal to $1\,W/m°C$. The external dimensions of the sheath are $L_1 = 1.25\,mm$ and $L_2 = 0.75\,mm$. The thermocouple is inserted in a solid up to a depth of $15\,mm$ and is so installed that there is perfect contact between the sheath and the surface of the hole. The temperature indicated by the thermocouple is $T_t = 125°C$ when the exposed leads of the thermocouple is convectively cooled by ambient medium at $T_\infty = 35°C$ via a heat transfer coefficient of $h = 4.5\,W/m^2°C$. Estimate the temperature of the solid.

b) If the solid temperature is maintained at the value you have determined in the above part what should be the depth to which the thermocouple is to be inserted if the thermometer error should be less than $0.5°C$?

Ex II.50: A thermocouple may be idealized as a single wire of diameter $0.5\,mm$ of effective thermal conductivity of $25\,W/m°C$. The insulation may be idealized as a layer of thickness $0.5\,mm$ and thermal conductivity of $0.4\,W/m°C$. The thermocouple is attached to a disk of $12\,mm$ diameter and is exposed to a flowing fluid at a temperature of $300\,W/m°C$ and a heat transfer coefficient of $67\,W/m^2°C$. The disk has a surface emissivity of 0.2. The disk is also able to see a background which is at $250°C$. The thermocouple lead is taken out of the back of the disk and is exposed to an ambient at $30°C$ and a convective heat transfer coefficient of $7.5\,W/m^2°C$. Assuming that the lead wire is very long, determine the thermocouple reading.

Ex II.51: Air is flowing in a tube of diameter $D = 100\ mm$ with a Reynolds number based on tube diameter of $Re = 1.5 \times 10^5$. In order to measure the temperature of the air a thermometer well is installed in the pipe, normal to its axis. The thermometer well material has a thermal conductivity of $45\ W/m°C$, has an ID of $3\ mm$ and an OD of $4.5\ mm$ and has a depth of immersion of $50\ mm$. Determine the true air temperature if a thermocouple attached to the well bottom indicates a temperature of $77°C$ while a thermocouple attached to the pipe wall indicates a temperature of $56°C$. Assume that the thermal conductivity of air is $0.03 W/m°C$ and the Prandtl number of air is 0.7.

Ex II.52: A certain temperature measurement system has the following specifications:
Material: copper shell of wall thickness $1\ mm$, outer radius $6\ mm$. Thermocouple is attached to the shell and is made of wires of very small radii such that lead wire conduction may be ignored.
Fluid: Air at $65°C$ surrounds the spherical shell. The surface of the shell has an emissivity of 0.65.
Determine the temperature indicated by the thermocouple if the shell interacts radiatively with a background at $25°C$.
Assume that heat transfer from air to the shell is by free convection. Use suitable correlation from a heat transfer text to solve the problem.

Ex II.53: A thermometer well has an OD of $6\ mm$ and an effective area conductivity product of $6 \times 10^{-5}\ Wm/°C$. The well is attached to the walls of the duct at $90°C$. Air at a temperature of $100°C$ is flowing through the duct at an average speed of $1\ m/s$. Axis of the thermometer well is normal to the flow direction. The surface of the well is polished metal surface of emissivity of 0.05. What is the thermometer error if the well is $50\ mm$ long?

Ex II.54: A temperature sensor is in the form of a long cylinder of equivalent diameter $2\ mm$. It is placed in a duct carrying air at a temperature of $300°C$ moving with a speed of $2\ m/s$ such that the flow is normal to the axis of the cylinder. The duct walls are known to be at a temperature of $280°C$. Assume that the conductivity of the cylinder is $14\ W/m\ K$.
(a) What should be the length of immersion if the temperature error should be less than $0.5°C$?
Air properties given below may be made use of:
Density= $0.6423\ kg/m^3$, Dynamic viscosity= $2.848 \times 10^{-5}\ kg/m\ s$, Thermal conductivity = $0.0436\ W/m\ K$, and Prandtl number= 0.680.
(b) Assume that the depth of immersion has been chosen as above. The emissivity of the sensor surface is known to be 0.05? How will it affect the thermometric error?

Ex II.55: A mercury in glass thermometer has a diameter of $6mm$ and is placed in a vertical position in a large room to measure the air temperature. The room walls are exposed to the sun and are at $45°C$. The heat transfer coefficient for heat exchange between the room air and the thermometer is $5.5 W/m^{2}°C$. The thermometer may be assumed a gray body with an emissivity of 0.65. What is the true air temperature if the thermometer reads $35°C$?

II.4 Heat flux measurement

Ex II.56: A Gardon gauge is made with a foil of radius $R = 2.3\ mm$ and thickness $\delta = 20\ \mu m$. The foil material has a thermal conductivity of $k = 19.5\ W/m°C$, specific heat of $c = 390\ J/kg°C$ and density of $\rho = 8900\ kg/m^3$. Determine the gauge constant as well as the time constant. What will be the temperature difference between the center of the foil and the periphery when radiant flux of $1000\ W/m^2$ is incident on the foil?

Ex II.57: A Gardon gauge uses a constantan foil of $3\ mm$ diameter and $40\ \mu m$ thickness. The thermal conductivity of the foil material may be taken as $22\ W/m°C$. Determine the gauge constant assuming that the Seebeck coefficient for Copper-Constantan pair is $40\ \mu V/°C$. It is desired to make the gauge twice sensitive as compared to above by suitable choice of the foil diameter-thickness combination. Mention what would be your choice of the diameter-thickness combination?

What is the differential temperature generated in the two cases if the incident heat flux is $10^5\ W/m^2$?

Ex II.58: A Gardon gauge of diameter $2.7\ mm$ is $20\ \mu m$ thick. It is exposed to an incident heat flux of $1.2\ W/cm^2$. What is the temperature difference developed by the gauge? If the diameter of the gauge is uncertain to an extent of 2% and the thickness is uncertain to an extent of 5%, what will be the uncertainty in the output when exposed to the heat flux mentioned above?

Ex II.59: A Gardon gauge of diameter $2.7\ mm$ is $20\ \mu m$ thick. It is exposed to an incident heat flux of $1.2\ W/cm^2$. What is the temperature difference developed by the gauge? What is the time constant of the gauge? If the diameter of the gauge is uncertain to an extent of 2% and the thickness is uncertain to an extent of 5%, what will be the uncertainty in the output when exposed to the heat flux mentioned above, if the heat flux itself is uncertain to the extent of 2.5%? What is the uncertainty in the time constant of the gauge?

Ex II.60: A slug type sensor has a slug of mass $0.005\ kg$ made of a material of specific heat equal to $300\ J/kg°C$. It is in the form of a short cylinder with a flat frontal area exposed to the incident flux of $1\ cm^2$. The exposed surface of the slug which has an emissivity of 0.86 receives incident radiation flux of $10^5 W/m^2$. How long can the slug be exposed to the incident flux if the temperature of the slug is not to exceed $150°C$? Take in to account only the heat loss by radiation from the exposed surface of the slug.

Ex II.61: A cylindrical heat transfer probe has a diameter of $12.5mm$. The main heater is $20mm$ wide. Suitable guards heaters are provided to prevent heat transfer along the axial direction. In a certain cold experiment the heat input to the main heater was $2W$ and the temperature of the sensor was noted to be at $60°C$. The temperature of the ambient medium was recorded at $32°C$. Estimate the heat transfer coefficient. What is the uncertainty in the estimated value of heat transfer coefficient if the temperatures are subject to errors of $\pm 0.5°C$ and the main heater input input has an uncertainty of 2%?

Ex II.62: An axial conduction heat flux probe has the main probe exposed to heat flux of $10^5\ W/m^2$. The frontal area of the main probe is $8\ mm$. The probe is made of an alloy having a thermal conductivity of 45 $W/m°C$. Two thermocouples are used to sense temperatures within the probe body with an axial separation of $4mm$. The temperature difference indicated is $12°C$. Determine the amount of heat that has to be removed by the cooling system, assuming that the guard is an annulus with an ID of 8.75 mm and an OD of 12.5 mm. Suggest a cooling arrangement for this probe.

Ex II.63: Consult the paper "The Plate Thermometer- A Simple Instrument for Reaching Harmonized Fire Resistance Tests" by Ulf Wickström.[3] Based on this paper explain the principle of a plate thermometer and what it measures.

[3]Fire Technology, Volume 30, Issue 2, May 1994, pp 195-208

Module III

Measurement of Pressure, Fluid velocity, Volume flow rate, Stagnation and Bulk mean temperatures

---

Most engineering applications involve the measurement of quantities that are given in the title to this module. Temperature and pressure are two parameters that govern and affect processes that are dealt with by mechanical and other engineers. Pressure measurement has matured over a long period and instruments are available from manufacturers for every need and in a wide range. Chapter 7 deals with the basic principles of all pressure measuring instruments. Also addressed are such important things as the transient response of pressure measuring devices, their calibration and estimation of errors. Most processes involve the flow of fluids - either gas or liquid. We leave out the measurement of flow of solid materials, a topic that is usually considered under "Mechanical Handling". Chapter 8 deals with the measurement of fluid velocity, presumably a point function, just like temperature. However, the measurement of fluid velocity is complicated by the fact that velocity has both a magnitude and a direction. The emphasis here is to deal with the most common methods and lead the interested reader to specialized literature. Volume flow rate, on the other hand, is a gross entity being concerned with the total flow rate and hence the average or the mean velocity across a section. Several methods that deal with volume/mass flow rate are dealt with in Chapter 9. Chapter 10 deals with the applications of the material that has been covered in the previous chapters to the measurement of temperatures in moving media. These are very important in the flow of non-isothermal fluids and in all process applications including high speed flows of interest in the aerospace field.

11

Measurement of Pressure, Fluid Velocity/Volume flow rate, Separation and bulk mean temperatures

Chapter 7

Measurement of Pressure

Most physical and chemical processes are affected by temperature and pressure. Having considered measurement of temperature in Module I it is logical to consider measurement of pressure as the next topic. Since pressure spans a large range from vacuum pressures in the μbar to very high pressures in GPa it is necessary to device suitable techniques to cover the wide range. Simplest of the techniques useful for measurement of pressure is by manometers. Since these are useful in laboratory practice we discuss manometers in detail. Subsequently we discuss the use of pressure transducers that are based on measurement of displacement of a spring element on application of pressure. Vacuum measurement requires special methods that are discussed towards the end of the chapter.

7.1   Basics of pressure measurement

Measurement of pressure is an important activity in the laboratory and industry since it is one of the most important variables that affects most processes. Many other properties of systems are affected by the operating pressure. For example, in an IC engine cylinder, the pressure varies continuously with time and determines the power output of the engine. Chemical reactions are affected by the operating pressure. In thermal power plants, the operating pressure of the boiler, for example, needs to be closely monitored during operation. The following tell us why it is important to measure pressure:

- Pressure is an important quantity that describes a thermodynamic system
- Pressure is invariably an important process parameter .
- Pressure difference is used many a time as a means of measuring the flow rate of a fluid

Pressure level spans some 18 orders of magnitude from the lowest to the highest pressures encountered in practice. The pressure measuring devices and their ranges are given in Table 7.1.

Table 7.1: *Pressure gauge types and ranges*

No.	Type	Lower limit	Upper limit
1	Ionization gauge	10^{-8}	10^{-3}
2	Pirani gauge	10^{-4}	1
3	McLeod gauge	10^{-6}	1
4	Manometer	10^{-1}	10^4
5	Piezoelectric trans- ducers	10^2	10^6
6	Bellows type gauge	10	10^4
7	Diaphragm gauge	1	10^6
8	Bourdon gauge	1	10^7
9	Resistance gauge	10^4	10^9

Entries in the Table are in *mm* of mercury column
1 *mm* of mercury column is equal to 1 *Torr*
(named after Torricelli)

We give below some useful units of pressure and conversion factors:

- $1\,Pa = 1\,N/m^2$
- $1\,atmosphere$ or $1\,atm = 760\,mm$ mercury column $= 1.013 \times 10^5\,Pa$
- $1\,mm$ mercury column $= 1\,Torr$
- $1\,Torr = 1.316 \times 10^{-3}\,atm = 133.3\,Pa$
- $1\,bar = 10^5\,Pa$

7.2 U - Tube manometer

The simplest of the gauges that is used for measuring pressure is a U - tube manometer shown in Figure 7.1. The U tube which has a uniform bore is vertically oriented and the acceleration due to gravity is assumed to be known. The height *difference h*, between the levels of the manometer liquid in the two limbs of the U - tube is the measured quantity. The quantity to be measured, the pressure, is thus *converted* to the measurement of a length, the height of the liquid column.

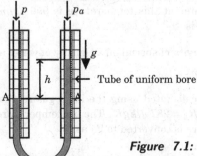

Figure 7.1: *Schematic of a U tube manometer*

The pressure to be measured is that of a system that involves a fluid (liquid or a gas) different from the manometer liquid. Let the density of the fluid whose pressure being measured be ρ_f and that of the manometer liquid be ρ_m. Equilibrium of the manometer liquid requires that there be the same force in the two limbs across the plane AA. We then have

$$p + \rho_f g h = p_a + \rho_m g h \qquad (7.1)$$

This may be rearranged to read

$$\boxed{p - p_a = (\rho_m - \rho_f)gh} \qquad (7.2)$$

Even though mercury is a common liquid used in manometers, other liquids are also used. A second common liquid is water. When measuring pressures close to the atmospheric pressure in gases, the fluid density may be quite negligible in comparison with the manometer liquid density. One may then use the approximate expression

$$\boxed{p - p_a \approx \rho_m g h} \qquad (7.3)$$

The manometer liquid is chosen based on its density with respect to the density of the fluid whose pressure is being measured and also the pressure difference that needs to be measured. In practice *small* pressure differences are measured using water or some organic liquid as the manometer liquid such that the height h measured is sufficiently large and hence measured with sufficient precision.

Example 7.1

(a) A U tube manometer employs special oil having a specific gravity of 0.82 as the manometer liquid. One limb of the manometer is exposed to the atmosphere at a pressure of $740\,mm\,Hg$ and the difference in column heights is measured as $20\,cm \pm 1\,mm$ when exposed to an air source at $25°C$. Calculate the air pressure in Pa and the uncertainty.

(b) The above manometer was carelessly mounted with an angle of $3°$ with respect to the vertical. What is the error in the indicated pressure due to this, corresponding to the data given above?

Solution :

Part (a):

Specific gravity uses the density of water at $25°C$ as the reference. From table of properties, the density of water at this temperature is $996\,kg/m^3$. The density of the manometer liquid is

$$\rho_m = \rho_{\text{special oil}} = \text{Specific gravity of special oil} \times \text{Density of water at } 25°C$$
$$= 0.82 \times 996 = 816.7\,kg/m^3$$

Density of process fluid, air, is calculated using the ideal gas relation. The gas constant for air is taken as $R_g = 287\,J/kgK$. The air temperature is $T = 273 + 25 = 298\,K$. The air pressure is converted to Pa as

$$p_a = \frac{740}{760} \times 1.013 \times 10^5 = 98634.2\,Pa$$

Air density is thus given by

$$\rho_a = \rho_f = \frac{p_a}{R_g T} = \frac{98634.2}{287 \times 298} = 1.15\,kg/m^3$$

The column height is given to be $h = 20\,cm = 0.2\,m$. The measured pressure differential is then obtained using Equation 7.2 as

$$p - p_a = (816.7 - 1.15) \times 9.8 \times 0.2 = 1598.5\,Pa$$

The uncertainty calculation is straight forward. It is the same as the % uncertainty in the column height i.e. $\Delta h\% = \pm\dfrac{0.001}{0.2} \times 100 = \pm 0.5\%$. The error in the measured pressure difference is thus given by

$$\Delta(p - p_a) = \pm\frac{0.5}{100} \times 1598.5 = \pm 7.99 \approx \pm 8\,Pa$$

Part (b):

The mounting error has been given as $\theta = 3°$ with respect to the vertical. It is clear that the manometer liquid height difference, with mounting error is given by

$$h' = h\cos\theta = 0.2 \times \cos 3° = 0.2 \times 0.0086 = 0.1997\,m$$

Indicated pressure difference is then given by

$$p - p_a = (816.7 - 1.15) \times 9.8 \times 0.1997 = 1596.3\,Pa$$

Note that there is thus a systematic error of $1596.3 - 1598.5 = -2.2\,Pa$ because of mounting error. This is about 25% of the error due to the error in the measurement of h.

7.2.1 Well type manometer

A well type manometer is many a time used instead of a U - tube manometer. The advantage of the well type manometer is that it has a single vertical tube of small bore being provided with a large quantity of manometer liquid in a reservoir, referred to as the "well". The height reading is taken using a scale attached to the vertical tube, with respect to an *unchanging* datum which corresponds to the level when the pressure difference between the well side and the tube side is zero. The schematic of a well type manometer is shown in Figure 7.2.

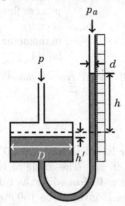

$\rho_f =$ Density of process fluid
$\rho_m =$ Density of manometer liquid

Figure 7.2: *Schematic of a well type manometer*

The dashed line indicates the datum with reference to which the manometer height is measured. The advantage of the well type design is that relatively large pressure differences may be measured with enough manometer liquid being and hence the following holds:

$$h'A = ha \qquad (7.4)$$

This expression simply states that there is no change in the volume of the liquid and hence also the mass. Here A is the well cross section area given by $A = \dfrac{\pi D^2}{4}$ while a is the tube cross section area given by $a = \dfrac{\pi d^2}{4}$. Equation 7.1 is recast for this case as

$$p + \rho_f g(h + h') = p_a + \rho_m g(h + h') \qquad (7.5)$$

Using Equation 7.4 in Equation 7.5, after some rearrangement we get

$$\boxed{p - p_a = (\rho_m - \rho_f)g\left[1 + \frac{a}{A}\right]h} \qquad (7.6)$$

If the area ratio $\dfrac{a}{A}$ is very small compared to unity, we may use the approximate formula

$$p - p_a \approx (\rho_m - \rho_f)gh \qquad (7.7)$$

In case the measured pressure difference is small one may use a well type manometer with inclined tube shown in Figure 7.3. Incompressibility of the manometer liquid requires that $Ah' = aR$. The manometer height is now given by $R\left\{sin\theta + \dfrac{a}{A}\right\}$. Equation 7.6 is replaced by

$$p - p_a = (\rho_m - \rho_f) g \left[\sin\theta + \frac{a}{A} \right] R \qquad (7.8)$$

It is clear that the inclination of the tube amplifies (recall the mechanical advantage of an inclined plane) the measured quantity and hence improves the precision of the measurement.

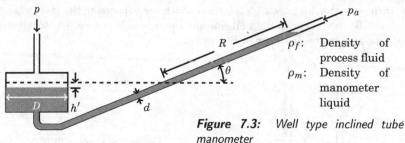

Figure 7.3: *Well type inclined tube manometer*

Example 7.2

(a) In an inclined tube manometer the manometer liquid is water at $20°C$ while the fluid whose pressure is to be measured is air. The angle of the inclined tube is $20°$. The well is a cylinder of diameter $0.05\,m$ while the tube has a diameter of $0.001\,m$. The manometer reading is given to be $150\,mm$. Determine the pressure differential in mm water and Pa. What is the error in per cent if the density of air is neglected?

(b) Determine the error in the measured pressure differential if the reading of the manometer is within $0.5\,mm$ and the density of water has an error of 0.2%. Assume that all other parameters have no errors in them. Neglect air density in this part of the question.

Solution :

Part (a):

Figure 7.3 indicates the notation used in this example. From the given data we calculate the area ratio as

$$\frac{a}{A} = \left(\frac{d}{D} \right)^2 = \left(\frac{0.001}{0.05} \right)^2 = 0.0004$$

Water density at the indicated temperature of $20°C$ is read off a table of properties as $997.6\,kg/m^3$ (pressure is taken as 1 atmosphere for this purpose). The fluid is air at the same temperature and its density is obtained using ideal gas law. We use $p_a = 1.013 \times 10^5\,Pa$, $T = 273 + 20 = 293\,K$, $R_g = 287\,J/kg\,K$ to get

$$\rho_a = \rho_f = \frac{p_a}{R_g T} = \frac{1.013 \times 10^5}{287 \times 293} = 1.205\,kg/m^3$$

The density of air has been calculated at the atmospheric pressure since it is at a pressure very close to it! The indicated pressure difference is then given by (using Equation 7.8)

$$p - p_a = (997.6 - 1.205) \times 9.81 \times (\sin 20° + 0.0004) \times 0.15 = 500.85\,Pa$$

This may be converted to mm of water column by dividing the above by $\rho_m g$. Thus

$$p - p_a = \frac{500.85 \times 1000}{997.6 \times 9.81} = 51.2 \, mm \text{ water column}$$

If we neglect the density of air in the above, we get

$$p - p_a = 997.6 \times 9.81 \times (\sin 20° + 0.0004) \times 0.15 = 501.49 \, Pa$$

The percentage error is given by

$$\Delta p = \frac{501.49 - 500.85}{500.85} \times 100 = 0.12\%$$

Part (b):

The influence coefficients are now calculated. We have

$$\frac{\partial (p - p_a)}{\partial \rho_f} = g \left(\sin \theta + \frac{a}{A} \right) = 9.81 \times (\sin 20° + 0.0004) \times 0.15 = 0.5058$$

$$\frac{\partial (p - p_a)}{\partial R} = \rho_f g \left(\sin \theta + \frac{a}{A} \right) = 997.6 \times 9.81 \times (\sin 20° + 0.0004) = 3363.7$$

The errors have been specified as

$$\Delta R = \pm 0.5 \, mm = \pm 0.0005 \, m; \Delta \rho_f = \pm \frac{0.2}{100} \times 997.6 = \pm 1.995 \, kg/m^3$$

Use of error propagation formula yields the error in measured pressure as

$$\Delta (p - p_a) = \pm \sqrt{(0.5058 \times 1.995)^2 + (3363.7 \times 0.0005)^2} = \pm 1.96 \, Pa$$

7.2.2   Dynamic response of a U tube manometer

Recall the measurement of temperature transients that we considered in Section 4.7. The transient response of the temperature sensor was seen to depend on its own properties as well as the strength of the interaction between the sensor and the process space. In general, similar is the case when one uses a pressure sensor to track time varying pressure. The response time depends on the pressure measuring instrument and the connecting tubes.

We shall look at the transient response of a U - tube manometer based on a simplified analysis. Because the manometer liquid is assumed to be incompressible the total length of the liquid remains fixed at L. We assume that the manometer is initially in the equilibrium position (the liquid level is the same in the two limbs) and the pressure difference Δp is applied across it, for $t > 0$. The liquid column will move and will be as shown in Figure 7.4 at some time $t > 0$. The forces that are acting on the length L of the manometer liquid are:

- Force due to acceleration of the liquid given by

$$F_a = \text{Mass of manometer liquid multiplied by acceleration} = \rho_m A L \frac{d^2 h}{dt^2}$$

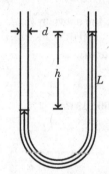

Nomenclature:

ρ_m Density of manometer liquid
μ Dynamic viscosity of manometer liquid
h Manometer height difference at time t
Δp Applied pressure difference
across manometer

Figure 7.4: Transients in a U tube manometer

- Force supporting the change in h given by $F_s = a\Delta p$
- Forces opposing the change:
  - Weight of column of liquid $W = \rho_m a h g$
  - Friction force due to viscosity of the manometer liquid

The symbols have the usual meaning, with a representing the cross sectional area of the U - tube. The viscous force opposing the motion is calculated based on the assumption of fully developed Hagen-Poiseuille (laminar tube) flow. The velocity of the liquid column is expected to be small and the laminar assumption is thus valid. We know from Fluid Mechanics that the pressure gradient and the mean velocity are related, for fully developed laminar flow, as

$$u_m = \frac{dh}{dt} = -\frac{d^2}{32\mu}\frac{dp_f}{dh} = -\frac{d^2}{32\mu}\frac{\Delta p_f}{L}$$

where μ is the dynamic viscosity of the manometer liquid and Δp_f is the friction pressure drop across the moving column of liquid.

In Section 4.7.1 we have looked at the transient response of a temperature sensor and introduced thermal resistance as the reciprocal of the heat transfer coefficient surface area product. This resistance controls or regulates the rate at which heat transfer can take place between the sensor and the surroundings. The input is a temperature change and the response is transfer of heat. The ratio of temperature change to the heat transfer rate, in fact, represents the thermal resistance. In the electrical analogy the temperature change is analogous with potential difference, the heat transfer rate is analogous with the electrical current and the ratio of these two is analogous with the electrical resistance.

In the case of the U - tube manometer the manometer liquid has to move in response to the applied pressure difference. The movement involves mass flow and the rate of mass flow is regulated by the viscous resistance to flow. The analogy may easily be seen in that the ratio of the pressure drop to the mass flow rate represents flow resistance. Hence we define fluid resistance R as the ratio of frictional (viscous) pressure drop (potential difference in electrical circuit or temperature difference in a thermal circuit) to the mass flow rate (current in electrical circuit or heat transfer rate in thermal circuit). We note that the mass flow rate is given by

$$\dot m = \rho_m a u_m = -\rho_m \pi \frac{d^2}{4}\frac{d^2}{32\mu}\frac{\Delta p_f}{L} = -\rho_m \frac{\pi d^4}{128\mu}\frac{\Delta p_f}{L}$$

Hence the fluid resistance due to friction is given by

$$R = -\frac{\Delta p_f}{\dot{m}} = \frac{128\mu L}{\pi \rho_m d^4} \tag{7.9}$$

Note that the resistance involves only the geometric parameters and the liquid properties. The frictional force opposing the motion is thus given by

$$F_f = \dot{m}aR$$

Note that the mass flow rate is itself given by $\dot{m} = \rho_m a u_m = \rho_m a \frac{dh}{dt}$. Hence the frictional force opposing the motion is

$$F_f = (\rho_m a u_m)aR = \rho_m a^2 R \frac{dh}{dt}$$

We may now apply Newton's law to the moving manometer liquid as

$$F_a = F_s - W - F_f$$

Introducing the expressions given above for the various terms, we get

$$\rho_m a L \frac{d^2h}{dt^2} = a\Delta p - \rho_m a g h - \rho_m a^2 R \frac{dh}{dt} \tag{7.10}$$

We may rearrange this equation as

$$\frac{L}{g}\frac{d^2h}{dt^2} + \frac{aR}{g}\frac{dh}{dt} + h = \frac{\Delta p}{\rho_m g} \tag{7.11}$$

This is a second order ordinary differential equation that resembles the equation governing a spring mass dashpot system that is familiar to us from mechanics. The system is thus inherently a second order system. We define a characteristic time given by

$$\tau = \sqrt{\frac{L}{g}} \tag{7.12}$$

(Reader: does it remind you of a simple pendulum?) and damping ratio

$$\zeta = \frac{aR}{2g\tau} \tag{7.13}$$

to recast Equation 7.11 in the standard form

$$\tau^2 \frac{d^2h}{dt^2} + 2\zeta\tau \frac{dh}{dt} + h = \frac{\Delta p}{\rho_m g} \tag{7.14}$$

The above equation may easily be solved by standard methods. The response of the system is shown in Figure 2.74 for three different cases. The system is under-damped if $\zeta < 1$, critically damped if $\zeta = 1$ and over-damped if $\zeta > 1$. When the system is under-damped the output shows oscillatory behavior, exhibits an overshoot (a value more than the input) and the output settles down slowly. In the other two cases the response is monotonic, as shown in the figure. In the over-damped case the response grows slowly to eventually reach the full value.

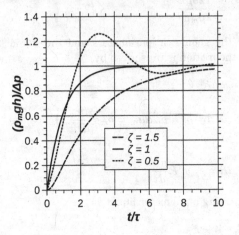

Figure 7.5: *Response of U - tube manometer to step input*

Example 7.3

A U tube manometer uses mercury as the manometer liquid having a density of $13580\ kg/m^3$ and kinematic viscosity of $v = 1.1 \times 10^{-7}\ m^2/s$. The total length of the liquid is $0.6\ m$. The tube diameter is $2\ mm$. Determine the characteristic time and the damping ratio for this installation. Redo the above with water as the manometer liquid. The density and kinematic viscosity of water are $996\ kg/m^3$ and $10^{-6}\ m^2/s$ respectively.

Solution :
Manometer liquid is mercury
The given data is

$$L = 0.6\ m; d = 2\ mm = 0.002\ m \text{ and we take } g = 9.81\ m/s^2$$

$$\rho_m = 13580\ kg/m^3; \quad v = 1.1 \times 10^{-7}\ m/s^2$$

Hence the characteristic time using Equation 7.12 is

$$\tau = \sqrt{\frac{L}{g}} = \sqrt{\frac{0.6}{9.81}} = 0.247\ s$$

The liquid viscous resistance is calculated using Equation 7.9 as

$$R_{Hg} = \frac{128 \times 1.1 \times 10^{-7} \times 0.6}{\pi \times 0.002^4} = 168067.6\ Pa\ s/kg$$

The area of cross section of the tube is

$$a = \frac{\pi \times 0.002^2}{4} = 3.142 \times 10^{-6}\ m^2$$

The damping ratio is then calculated using Equation 7.13 as

$$\zeta = \frac{3.142 \times 10^{-6} \times 168067.6}{2 \times 9.81 \times 0.247} = 0.109$$

The system is under-damped since $\zeta < 1$.
Manometer liquid is water
The manometer liquid properties are changed to

$$\rho_m = 996 \, kg/m^3; \nu = 10^{-6} \, m/s^2$$

The characteristic time remains unchanged since it is purely a function of length of manometer fluid column. The damping ratio changes because of the change in the liquid viscous resistance. With water as the manometer liquid we have

$$R_{H_2O} = \frac{128 \times 10^{-6} \times 0.6}{\pi \times 0.002^4} = 1527887.5 \, Pa \, s/kg$$

The damping ratio is then calculated as

$$\zeta = \frac{3.142 \times 10^{-6} \times 1527887.5}{2 \times 9.81 \times 0.247} = 0.9905$$

The system is very nearly critically damped since $\zeta \approx 1$.

7.3 Bourdon gauge

Bourdon gauges are available to cover a large range of pressures. Bourdon gauges are purely mechanical devices utilizing the *mechanical deformation* of a flattened but bent tube (spring element) that winds or unwinds depending on the pressure difference between the inside and the outside. The motion is against a hair spring torque that prevents any backlash in the movement. A needle attached to the shaft indicates directly the pressure difference. The working principle of the Bourdon gauge is explained with reference to Figures 7.6.

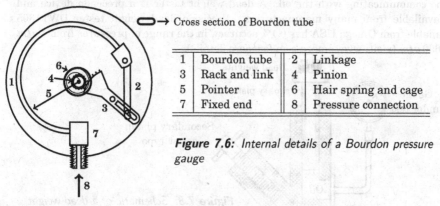

Cross section of Bourdon tube

1	Bourdon tube	2	Linkage
3	Rack and link	4	Pinion
5	Pointer	6	Hair spring and cage
7	Fixed end	8	Pressure connection

Figure 7.6: *Internal details of a Bourdon pressure gauge*

The Bourdon tube is a metal tube of elliptic cross section having a bent shape as shown in the figure. The inside of the tube is exposed to the pressure to be measured. The outside of the Bourdon tube is exposed to a second pressure, usually the local atmospheric pressure. The Bourdon tube is held fixed at one end (the end connected to the pressure source) and the other end is connected by linkages to a spring restrained shaft. A pointer is mounted on the shaft, as indicated in the figure. The needle moves over a circular scale that indicates the pressure directly in appropriate units. The external appearance of a Bourdon gauge is as shown in Figure 7.7. Bourdon gauges are available for measuring pressures higher/lower than the atmospheric pressure. The pressure indicated is referred to as the gauge pressure if it is with respect to the atmospheric pressure outside the Bourdon tube.

Figure 7.7: *Photograph of a Bourdon gauge*

7.3.1 Dead weight tester

Bourdon gauges may be calibrated by using a dead weight tester, schematic of which is shown in Figure 7.8. The weight tester consists of an arrangement by which a piston may be allowed to float over a liquid (usually oil), under internal pressure and a force in the opposite direction, imposed on the piston by weights placed as indicated in the figure. The oil pressure is changed by screwing in the piston. The pressure is calculated as the weight placed on the piston divided by the cross section area of the piston (the piston is to be oriented with its axis vertical!). The gauge under test experiences the same pressure by being connected to a side tube communicating with the oil. A dead weight tester is a precision device and is available from many manufacturers. Precision dead weight tester DWT1305 available from Omega USA has 0.1% accuracy, in the range of pressures from $1\,atm$ to $650\,atm$ (visit: www.omega.com for more details).

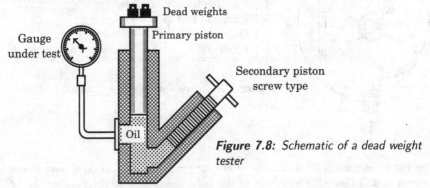

Figure 7.8: *Schematic of a dead weight tester*

7.4 Pressure transducers

Here we discuss several pressure transducers that are grouped together, for convenience. These are:

- Pressure tube with bonded strain gauge
- Diaphragm/Bellows type transducer
- Capacitance pressure transducer
- Piezoelectric pressure transducer

7.4.1 Pressure tube with bonded strain gauge

The principle of a pressure tube with bonded strain gauge may be understood by referring to Figure 7.9. The pressure to be measured is communicated to the inside of a tube of suitable wall thickness, one end of which is closed with a thick plate. Because the end plate is very thick it undergoes hardly any strain. The tube, however, experiences a hoop strain due to internal pressure (indicated by the arrows in the figure). A strain gauge mounted on the tube wall experiences the hoop strain. The quantity to be measured, the pressure, is thus converted to the measurement of a change in *length* of a strain gauge element. The strain gauge element itself responds to the strain in a material on which it *is* attached. The principle of operation of a strain gauge is explained below.

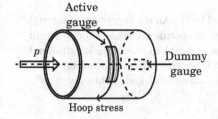

1. Active gauge responds to hoop stress due to internal pressure
2. Dummy gauge is used for temperature compensation

Figure 7.9: *Pressure tube with bonded strain gauge*

Strain gauge theory

Strain gauge consists of a resistance element whose resistance is a function of its size. A wire subject to strain undergoes a deformation as indicated in Figure 7.10.

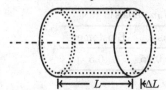

—— Original shape, area of cross section A

···· Changed shape, area of cross section $A - \Delta A$

Figure 7.10: *Deformation of a wire element*

Electrical resistance of a wire of Length L, area of cross section A and made of a material of specific resistance ρ is given by

$$R = \frac{\rho L}{A} \tag{7.15}$$

Logarithmic differentiation yields

$$\frac{dR}{R} = \frac{d\rho}{\rho} + \frac{dL}{L} - \frac{dA}{A} \tag{7.16}$$

Equation 7.16 represents the fractional change in the wire resistance due to fractional changes in the specific resistance, the length and the area of cross section due to an imposed deformation on the wire. The specific resistance is a strong function of temperature but also a weak function[1] of pressure. Note that the second term on the right hand side of Equation 7.16 represents the longitudinal strain and the third term represents the fractional change in the cross sectional area of the wire. If the wire is of circular cross section of radius r, we have $A = \pi r^2$ and hence

$$\frac{dA}{A} = 2\frac{dr}{r} \tag{7.17}$$

Recall from mechanics of solids that Poisson ratio ν for the material is defined as the ratio of lateral strain to longitudinal strain. Hence we have

$$\frac{dA}{A} = 2\frac{dr}{r} = -2\nu\frac{dL}{L} \tag{7.18}$$

Combining these, Equation 7.16 becomes

$$\frac{dR}{R} = \frac{d\rho}{\rho} + (1+2\nu)\frac{dL}{L} = \frac{d\rho}{\rho} + (1+2\nu)\epsilon_l \tag{7.19}$$

In the above the longitudinal strain is represented by ϵ_l. An incompressible material has Poisson ratio of 0.5. In addition if the change in specific resistance due to strain is negligible the fractional change in resistance is equal to twice the longitudinal strain. The ratio of fractional resistance change of the wire to the longitudinal strain is called the gauge factor (GF) or sensitivity of a strain gauge and is thus given by

$$\boxed{GF = \frac{1}{\epsilon_l}\frac{dR}{R} = (1+2\nu)} \tag{7.20}$$

The gauge factor is thus close to 2. Poisson ratio for some useful materials is given in Table 7.2.

Table 7.2: *Poisson ratio of some useful materials*

Material	Steel	Concrete	Gold	Glass	Elastomers
Poisson ratio	0.3	0.2	0.42	0.20 - 0.25	0.5

However, in general we may include the change in specific resistance also and introduce an engineering gauge factor such that

$$\boxed{G = \frac{1}{\epsilon_l}\frac{dR}{R} = (1+2\nu) + \frac{1}{\epsilon_l}\frac{d\rho}{\rho}} \tag{7.21}$$

Table 7.3 presents typical gauge factor values with different materials.

---

[1]Specific resistance changes due to longitudinal strain are brought about by the changes in the mobility of the free electrons due to change in the mean free path between collisions.

Table 7.3: Typical gauge materials and gauge factors

Material	Sensitivity (G)
Platinum (Pt 100%)	6.1
Platinum-Iridium (Pt 95%, Ir 5%)	5.1
Platinum-Tungsten (Pt 92%, W 8%)	4
Isoelastic (Fe 55.5%, Ni 36% Cr 8%, Mn 0.5%)[♯]	3.6
Constantan / Advance / Copel (Ni 45%, Cu 55%)[♯]	2.1
Nichrome V (Ni 80%, Cr 20%)[♯]	2.1
Karma (Ni 74%, Cr 20%, Al 3%, Fe 3%)[♯]	2
Armour D (Fe 70%, Cr 20%, Al 10%)[♯]	2
Monel (Ni 67%, Cu 33%)[♯]	1.9
Manganin (Cu 84%, Mn 12%, Ni%)[♯]	0.47
Nickel (Ni 100%)	-12.1

[♯] Isoelastic, Constantan, Advance, Copel, Nichrome V,
Karma, Armour D, Monel, and Manganin are all trade names

Strain gauge construction details

Figure 7.11 shows the way a strain gauge is made. The strain element is a serpentine metal layer obtained by etching, mounted on a backing material that may be bonded on to the surface whose strain needs to be measured. The active direction is as indicated in the figure. When a force is applied normal to this direction the serpentine metal layer opens up like a spring and hence does not undergo any strain in the material (actually the response will be something like 1% of that along the active direction). When the length of the metal foil changes the resistance changes and the change in resistance is the measured quantity. Connecting wires are used to connect the strain gauge to the external circuit.

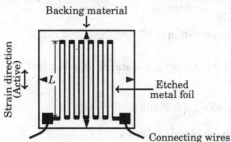

Figure 7.11: Strain gauge construction schematic (not to scale). L is the active gauge length, ◀ = Alignment marks

By making the element in the form of a serpentine foil the actual length of the element is several times the length of the longer side of the element (referred to as gauge length in the figure). The gauge length may vary from 0.2 to 100 mm. For a given strain the change in length will thus be the sum of the changes in lengths of each leg of the element and hence the resistance change will be proportionately larger. Gauges with nominal resistances of 120, 350, 600 and 700 Ω are available. However 120 Ω gauges are very common.

Example 7.4

A typical foil strain gauge has the following specifications:

- Material is Constantan of thickness $\delta = 6\,\mu m$
- Resistance $R = 120\Omega \pm 0.5\%$
- Gauge factor $GF = 2 \pm 2\%$

Such a strain gauge is bonded on to a tube made of stainless steel of internal diameter $d = 6\,mm$ and wall thickness $t = 0.3\,mm$. The tube is subjected to an internal pressure of $p = 10MPa$ gauge. The outside of the tube is exposed to the standard atmosphere. What is the change in resistance of the strain gauge if the Young's modulus of stainless steel is $E = 207\,GPa$? Specify an error bar on this. Assume that all other quantities have negligibly small uncertainties.

Solution :

Step 1 The tube is subject an internal pressure of $p = 10\,MPa$ gauge and hence the pressure difference between the inside and the outside is $\Delta p = 10\,MPa = 10^7\,Pa$.

Step 2 Tube is assumed to be thin and the radius r is taken as the mean radius given by $r = \dfrac{d+t}{2} = \dfrac{0.006+0.0003}{2} = 0.00315\,m$.

Step 3 The Young's modulus of stainless steel is specified to be $E = 207\,GPa = 2.07 \times 10^{11}\,Pa$.

Step 4 The hoop stress developed in the tube is given by

$$\sigma_h = \frac{\Delta p\,r}{t} = \frac{10^7 \times 0.00315}{0.0003} = 1.05 \times 10^8 Pa$$

Step 5 The hoop strain may now be calculated as

$$\epsilon_h = \frac{\sigma_h}{E} = \frac{1.05 \times 10^8}{2.07 \times 10^{11}} = 0.00051$$

For Constantan the maximum allowable elongation is 1% and the above value is hence acceptable.

Step 6 With the gauge factor of $GF = 2$ the change in resistance as a fraction of the resistance of the strain element is

$$\frac{\Delta R}{R} = GF\epsilon_h = 2 \times 0.00051 = 0.00102$$

The change in resistance of the strain gauge is thus given by

$$\Delta R = 0.00102 \times 120 = 0.122\,\Omega$$

Step 7 Uncertainty in the resistance change may be calculated using error propagation formula. With 2% uncertainty in gauge facor and 0.5% uncertainty in gauge resistance, the uncertainty in ΔR is obtained as

$$u_{\Delta R} = \sqrt{2^2 + 0.5^2} = 2.06\%$$

Hence the resistance change has an uncertainty of $\dfrac{2.06 \times 0.122}{100} = 0.003\,\Omega$.

7.4.2 Bridge circuits for use with strain gauges

While discussing resistance thermometers we have already seen how a bridge circuit may be used for making measurements (see section 4.4.3). Since a strain gauge is similar in its operation to an RTD, in that the electrical resistance of the strain gauge element changes with strain, it is possible to use a similar bridge circuit for measuring the strain and hence the pressure. Lead wire compensation is also possible by using a three wire arrangement. For the sake of completeness and to bring out some details which are specific to strain gauges, these will be discussed here.

The basic DC bridge circuit is shown in Figure 7.12. The output voltage V_o is related to the input voltage (or supply voltage) V_s by the relation (derived using Ohm's law and theory of dc circuits)

$$V_o = V_s \frac{R_1 R_3 - R_2 R_4}{(R_1 + R_4)(R_2 + R_3)} \tag{7.22}$$

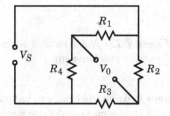

Figure 7.12: *Basic DC bridge circuit for strain measurement.* V_S = *Battery voltage* V_0 = *Gauge output*

If all the resistances are chosen equal (equal to $R = 120\Omega$, for example) when the strain gauge is unstrained, the output voltage will be zero. Let us assume that R_3 is the strain gauge element and all others are standard resistors. This arrangement is referred to as the "quarter" bridge circuit. The other variants such as the half and the full bridge circuits will be discussed at appropriate places. When the strain gauge is strained the resistance will change to $R_3 + \Delta R_3$ where $\Delta R_3 \ll R_3$. The numerator of Equation 7.22 then becomes

$$R_1(R_3 + \Delta R_3) - R_2 R_4 = R(R + \Delta R) - R^2 = R\Delta R$$

The denominator is approximated by

$$(R_1 + R_3)(R_2 + R_4) = (R + R + \Delta R)(R + R) \approx 4R^2$$

since all resistances are equal when the strain gauge is unstrained. We thus see that Equation 7.22 may be approximated by

$$V_o \approx V_s \frac{R\Delta R}{4R^2} = V_s \frac{\Delta R}{4R} \tag{7.23}$$

Using the definition of gauge factor given by Equation 7.20 the above equation may be recast as

$$V_o = V_s \frac{GF}{4} \epsilon_l \tag{7.24}$$

Thus the output voltage is proportional to the strain experienced by the strain gauge. For a gauge with sensitivity of $GF = 2$ the ratio of output voltage to input voltage is equal to half the strain. In practice Equation 7.24 is rearranged to read

$$\epsilon_l = \frac{4V_o}{GFV_s} \tag{7.25}$$

Since the pressure is proportional to the strain, Equation 7.25 also gives the pressure within a constant factor. Note that V_o needs to be measured by the use of a suitable voltmeter, likely a milli-voltmeter. In practice one may use an amplifier to amplify the signal by a known factor and measure the amplified voltage by a suitable voltmeter. Another method is to make the resistance R_2 a variable resistor and null the bridge by using a controller as shown in Figure 7.13.

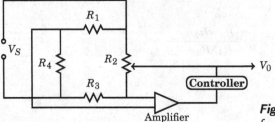

Figure 7.13: Null balanced bridge for strain measurement

The controller uses the amplified off balance voltage to drive the bridge to balance by varying the resistor R_2. The voltage V_o indicated is then a measure of the strain.

In practice, the quarter bridge circuit is actually arranged with one active gauge R_3 and one dummy gauge R_4. Both the gauges respond to ambient temperature variations by identical amount and hence do not disturb the null under "no load". However, the active gauge alone responds to the strain and hence gives a signal whenever the pressure is different from the atmospheric pressure.

An improved bridge may be obtained by having two active gauges R_1 and R_3, and two dummy gauges R_2 and R_4 (Figure 7.12). The active elements may be mounted on the pressure tube, at two different positions, for example at two locations 180° apart along the circumference of the pressure tube. The two dummy gauges may be mounted side by side on the end plate (see Figure 7.9). The bridge circuit is referred to as half bridge circuit. In this case the bridge is driven away from balance by changes in both R_1 and R_3 and hence the output is twice that in the quarter bridge circuit. Equation 7.24is replaced by

$$V_o = V_s \frac{GF}{2} \epsilon_l \tag{7.26}$$

Example 7.5

The supply voltage in the arrangement depicted in Figure 7.12 is $V_S = 5\ V$. All the resistors are equal when the strain gauge is not under load. When the strain gauge is loaded along its axis the output voltage registered is $1.13\ mV$. What is the axial strain of the gauge if the sensitivity is $GF = 2$?

Solution :

The given data is written down as $V_S = 5V$; $V_0 = 1.13\,mV = 1.13 \times 10^{-3}\,V$; $GF = 2$. The axial strain ϵ_l is obtained by using Equation 7.25 as

$$\epsilon_l = \frac{4V_0}{GFV_S} = \frac{4 \times 1.13 \times 10^{-3}}{2 \times 5} = 4.52 \times 10^{-4} = 452\,\mu-\text{strain}$$

where $(1\,\mu-\text{strain is } 10^{-6})$

Example 7.6

Consider the data given in Example 7.4 again. This gauge is connected in a bridge circuit with an input voltage of $V_s = 9\,V$. What is the output voltage in this case?

Solution :

From the results in Example 7.4 the change in resistance has been obtained as $\Delta R = 1.22\Omega$. The supply voltage has been specified as $V_s = 9V$. Hence, using Equation 7.23, we have

$$V_o = V_s \frac{\Delta R}{4R} = 9 \times \frac{1.22}{4 \times 120} = 0.0229\,V = 22.9\,mV$$

Three wire gauge for lead wire compensation

Electrical connections are made as shown in Figure 7.14 while using a strain gauge with three lead wires. Let each lead wire have a resistance of R_L as shown. Also let $R_2' = R_2 + R_L$ and $R_3' = R_3 + R_L$. Equation 7.22 is then recast as

$$
\begin{aligned}
V_o &= V_s \frac{R_1 R_3' - R_2' R_4}{(R_1 + R_4)(R_2' + R_3')} = V_s \frac{R_1(R_3 + R_L) - R_4(R_2 + R_L)}{(R_1 + R_4)(R_2 + R_3 + 2R_L)} \\
&= V_s \frac{(R_1 R_3 - R_2 R_4) + R_L(R_1 - R_4)}{(R_1 + R_4)(R_2 + R_3 + 2R_L)}
\end{aligned}
\tag{7.27}
$$

If all the resistances are identical under no strain condition the second term in the numerator is zero and hence the balance condition is not affected by the lead resistance. Let $R_3 + \Delta R_3$ be the resistance in the strained condition for the gauge. Again we assume that $\Delta R_3 \ll R_3$ and, in addition, we assume that $\Delta R_3 \ll R_L$. With these Equation 7.27 may be approximated by the relation

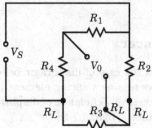

Figure 7.14: *Three wire arrangement for lead wire compensation*

$$V_0 = V_S \frac{R_1R_3 - R_2R_4 + R_1\Delta R_3}{(R_1+R_4)(R_2+R_3+2R_L)} \approx V_S \frac{\Delta R_3}{4R_1\left(1+\frac{R_L}{R_1}\right)} \approx V_S \frac{\Delta R_3}{4R_1}\left(1-\frac{R_L}{R_1}\right) \qquad (7.28)$$

In writing the above we have additionally assumed that the lead resistance is small compared to the resistance of the gauge. Thus Equation 7.24 is recast as

$$V_o = V_S \frac{GF}{4}\left(1-\frac{R_L}{R_1}\right)\epsilon \qquad (7.29)$$

Thus the three wire arrangement provides lead wire compensation but with a reduced gauge factor of $GF' = GF\left(1-\frac{R_L}{R_1}\right)$.

Example 7.7

A three wire strain gauge with a lead resistance of $R_L = 4\,\Omega$ and a gauge resistance under no strain of $R = 120\,\Omega$ is used to measure a strain of $\epsilon_l = 0.001$. What will be the output voltage for an input voltage of $V_S = 5\,V$ if 1. the lead resistance effect is not included? and 2. if the lead wire resistance is included?
The gauge factor GF has been specified to be $GF = 1.9$.

Solution :

1. Lead resistance not included:
The effective gauge factor in this case is the same as $GF = 1.9$. The output voltage with a strain of $\epsilon_l = 0.001$ is obtained by using Equation 7.24 as

$$V_0 = V_S \frac{GF}{4}\epsilon_l = 5 \times \frac{1.9}{4} \times 10^{-3} = 0.00238\,V = 23.8\,mV$$

2. Lead resistance included:
With the lead resistance $R_L = 4\,\Omega$ and the gauge resistance of $R = 120\,\Omega$, the effective gauge factor GF' is obtained as

$$GF' = GF\left(1-\frac{R_L}{R}\right) = 1.9 \times \left(1-\frac{4}{120}\right) = 1.837$$

The output voltage with a strain of $\epsilon_l = 0.001$ is obtained by using Equation 7.29 as

$$V_0 = V_S \frac{GF'}{4}\epsilon_l = 5 \times \frac{1.837}{4} \times 10^{-3} = 0.0023\,V = 23\,mV$$

7.4.3 Diaphragm/Bellows type transducer

Pressure signal is converted to a *displacement* in the case of diaphragm/bellows type pressure gauge. The diaphragm or the bellows acts as a spring element that undergoes a displacement under the action of the pressure. Schematic of diaphragm and bellows elements are shown respectively in Figures 7.15 and 7.16.

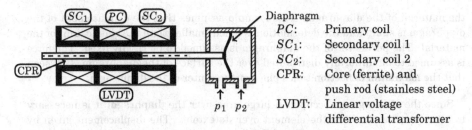

Figure 7.15: *Schematic of "diaphragm" type pressure gauge that uses LVDT for displacement measurement*

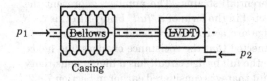

Figure 7.16: *Schematic of "bellows" type pressure gauge that uses LVDT for displacement measurement*

In both cases a Linear Variable Differential Transformer (LVDT) is used as a displacement transducer - a transducer that measures displacement by converting it to an electrical signal. The working principle of LVDT will be discussed later.

Diaphragm pressure gauge with strain gauges

In the case of the diaphragm type gauge one may also use a strain gauge for measuring the strain in the diaphragm by fixing it at a suitable position on the diaphragm (Figure 7.17).

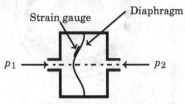

Figure 7.17: *Diaphragm gauge with strain gauge for displacement measurement*

In the case of the bellows transducer the relationship between the pressure difference and the displacement is given by

$$\delta = \frac{(p_1 - p_2)A}{K} \tag{7.30}$$

Here δ is the displacement of the bellows element, A is the inside area of the bellows and K is the spring constant of the bellows element. In the case of the diaphragm element the following formula from strength of materials is useful:

$$\delta(r) = \frac{3(p_1 - p_2)(a^2 - r^2)^2}{16Et^3}(1 - v^2) \tag{7.31}$$

In Equation 7.31, $\delta(r)$ is the deflection at radius r, a is the radius of the diaphragm, t is its thickness, E and v are respectively the Young's modulus and Poisson ratio of the material of the diaphragm. This formula assumes that the displacement of the diaphragm is small and the deformation is much smaller than the elastic limit of the material. The perimeter of the diaphragm is assumed to be rigidly fixed. The force is assumed to be uniformly distributed over the surface of the diaphragm. It is seen that the maximum deflection takes place at the center of the diaphragm.

Since the strain gauge occupies a large area over the diaphragm it is necessary to integrate the strain of the element over its extent. The displacement given by Equation 7.31 is normal to the plane of the diaphragm. Because of this displacement the diaphragm undergoes both radial as well as circumferential strains. As indicated in Figure 7.18 strain gauge elements are arranged so that they respond to both the radial as well as tangential (circumferential) strains. The numbers represent the leads and the strain gauges are connected in the form of a *"full"* bridge. In this case all the four strain gauges in the bridge are active and respond to the appropriate strains. The four elements are so connected that the resistance changes and hence the voltage changes all *add up*. Hence the full bridge circuit has a higher sensitivity as compared to the quarter bridge circuit that was considered earlier in Section 7.4.2. The strain gauges that respond to the radial strain are arranged near the periphery of the diaphragm while those that respond to the tamgemtial strain are placed close to the center of the diaphragm. Terminals are provided for external connections and for compensating resistors. Different strain gauge elements of the bridge respond to radial (ϵ_R) and circumferencial (ϵ_θ) strains as indicated in Figure 7.18. It is apparent that the resistances of *all* the strain gauges *change* due to the flexure of the diaphragm.

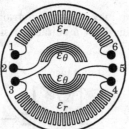

ϵ_r Radial strain

ϵ_θ Circumferencial strain

Figure 7.18: *Strain gauges mounted on a diaphragm gauge (Based on "Strain Gages and Instruments", Tech Note TN-510-1, Document Number: 11060, Revision: 01-Nov-2010, Vishay Precision Group accessible at www.micro-measurements.com)*

The ratio of output to input is then given by the approximate formula (from Vishay Intertechnology, Inc., Manufacturers of Diaphragm pressure gauges)

$$\frac{V_o}{V_s} = 0.75 \frac{(p_1 - p_2)(1 - v^2)}{E} \left(\frac{a}{t}\right)^2 \times 1000 \, mV/V \qquad (7.32)$$

The above formula assumes that the diaphragm is perfectly rigidly held over its perimeter. This is achieved by making the diaphragm an integral part of the body of the pressure transducer. It is also assumed that the sensitivity of the strain gauge is 2. While measuring transient pressures it is necessary that the transients do not have significant components above about 20% of the natural frequency f_n of the diaphragm given by the following expression.

$$f_n = \frac{469t}{a^2} \sqrt{\frac{E}{\rho(1-v^2)}} \qquad (7.33)$$

In the above expression ρ is the density of the material of the diaphragm in kg/m^3. The diaphragm is again assumed to be rigidly clamped along its periphery.

Example 7.8

Steel diaphragm of 15 mm diameter is used in a pressure transducer to measure a maximum possible pressure of 10 MPa. Choose the thickness of the diaphragm such that the transducer gives an output of 3 mV/V at the maximum pressure. Also determine the maximum out of plane deflection of the diaphragm and its natural frequency.

Solution :

Step 1 We take the properties of the steel diaphragm as $E = 207\, GPa$; $v = 0.285$; $\rho = 7830\, kg/m^3$.

Step 2 The given data is written down using appropriate notation.

$$\frac{V_o}{V_s} = 3mV/V; \quad a = \frac{15}{2} = 7.5\,mm; \quad p_1 - p_2 = 10\,MPa = 10^7\,Pa$$

Step 3 We solve for t, the diaphragm thickness from Equation 7.32 and get

$$t = 7.5\sqrt{\frac{0.75 \times 10^7 \times (1-0.285^2) \times 10^3}{2.07 \times 10^{11} \times 3}} = 1.580\,mm$$

Step 4 The maximum out of plane deflection is obtained by putting $r = 0$ in Equation 7.31.

$$\delta(r = 0) = \frac{3 \times 10^7 \times 7.5^4 \times (1-0.285^2)}{16 \times 2.07 \times 10^{11} \times 1.580^3} = 0.107\,mm$$

Step 5 The ratio of the maximum deflection to the diaphragm thickness is

$$\frac{\delta(r = 0)}{t} = \frac{0.107}{1.580} = 0.068$$

This ratio is less than 0.25 and hence is satisfactory.

Step 6 The natural frequency of the diaphragm is obtained using Equation 7.33 as

$$f_n = \frac{469 \times 1.580}{7.5^2} \sqrt{\frac{2.07 \times 10^{11}}{7830 \times (1-0.285^2)}} = 17667.3\,Hz$$

The sensor is thus useful below about 3.4 kHz

Diaphragm/Bellows type gauge with LVDT

A diaphragm/bellows element may be used to measure pressure by accurate direct measurement of the displacement. This may be done by using a Linear Variable Differential Transformer (LVDT) as indicated in Figure 7.15. This figure also shows a schematic of an LVDT. A schematic of electric circuit of LVDT is shown in Figure 7.19. LVDT consists of three coaxial coils with the primary coil injected with an ac current. The coils are mounted on a ferrite core carried by a stainless steel rod. The two secondary coils are connected in electrical opposition as shown in Figure 7.19. One end of the stainless steel rod is connected to the diaphragm/bellows element as indicated in Figure 7.16. When the ferrite core is symmetrical with respect to the two secondary coils the ac voltages induced by magnetic coupling across the two secondary coils V_1, V_2 are equal. When the core moves to the right or the left as indicated the magnetic coupling changes and the two coils generate different voltages and the voltage difference $V_0 = V_1 - V_2$ is proportional to the displacement (see Figure 7.20).

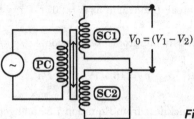

PC: Primary coil
SC1: Secondary coil 1
SC2: Secondary coil 2
V_0:   Electrical output

Figure 7.19: *Electrical circuit schematic of the LVDT*

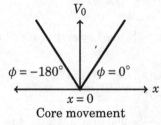

Figure 7.20: *Output of an LVDT excited by alternating current supply. The phase (ϕ) is with respect to the phase of the supply.*

Typical LVDT circuit with signal conditioner is shown in Figure 7.21. The primary input voltage is $1 - 3\ V$ (rms). Full scale secondary voltage is between $0.1 - 3\ V$ (rms). Gain is set between 2 and 10 to increase the output voltage to $10\ V$. The two secondary coils are connected in opposition and sum and difference signals are obtained from the two coils, after demodulating the two voltages. The ratio of these is taken and is output with a gain A as indicated in the figure. The output is independent of the gain settings of the sum and difference amplifiers as long as they are the same. In Example 7.8 the diaphragm undergoes a maximum deflection of $0.126\ mm$ or $126\ \mu m$. As Table 7.4 shows we can certainly pick a suitable LVDT for this purpose.

7.4.4   Capacitance type diaphragm gauge

Schematic of a capacitance type pressure transducer is shown in Figure 7.22. The gap between the stretched diaphragm and the anvil (held fixed) varies with applied

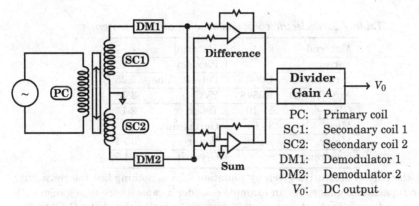

Figure 7.21: Typical LVDT signal conditioner circuit

Table 7.4: Common specifications for commercially available LVDT

Input	Power input is a 3 to 15 V (rms) sine wave with a frequency between 60 to 20,000 Hz (the two most common signals are 3 V, 2.5 kHz and 6.3 V, 60 Hz).
Stroke	Full-range stroke is from 125 μm to 75 mm
Sensitivity	Sensitivity usually ranges from 0.6 to 30 mV per 25 μm und er normal excitation of 3 to 6 V. Generally, the higher the frequency the higher the sensitivity.
Nonlinearity	Inherent nonlinearity of standard units is on the order of 0.5% of full scale.

pressure due to the small displacement of the diaphragm. The displacement of the diaphragm changes the capacitance of the gap. The theoretical basis for this is explained below. Consider a parallel plate capacitor of plate area A, cm^2 and gap

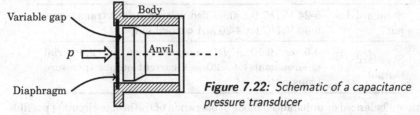

Figure 7.22: Schematic of a capacitance pressure transducer

between plates of x, cm. Let the gap contain a medium of dielectric constant κ. The capacitance C of the parallel plate capacitor is then given in $pico-Farads$ ($1\,pF = 10^{-12}\,F$) by

$$C = 0.0885\kappa\frac{A}{x} \qquad (7.34)$$

The dielectric constants of some common materials are given in Table 7.5. The effect of change in the gap may be seen by differentiating Equation 7.34 with respect to x.

$$\frac{\partial C}{\partial x} = -0.0885\kappa\frac{A}{x^2} = -\frac{C}{x} \qquad (7.35)$$

Table 7.5: Dielectric constants of some common materials

Material	κ	Material	κ
Vacuum	1	Plexiglass	3.4
Air(1 atm)	1.00059	Polyethylene	2.25
Air(10 atm)	1.0548	PVC	3.18
Glasses	5 - 10	Teflon	2.1
Mica	3 - 6	Germanium	16
Mylar	3.1	Water	80.4
Neoprene	6.7	Glycerin	42.5

In fact, the partial derivative given by Equation 7.35 is nothing but the sensitivity S of the capacitance gauge. As an example consider a capacitance type gauge with air as the medium with $A = 1\, cm^2$, $x = 0.3\, mm = 0.03\, cm$. We then have $C = 0.0885 \times 1.00059 \times \dfrac{1}{0.03} = 2.95\, pF$. The sensitivity of this gauge is then given by $S = -\dfrac{2.95}{0.03} = -98.3\, pF/cm$.

Bridge circuit for capacitance pressure gauge

A capacitance gauge requires ac excited bridge arrangement for generating a signal proportional to the displacement. The circuit is hence driven by alternating current input at high frequency, as shown in Figure 7.23.

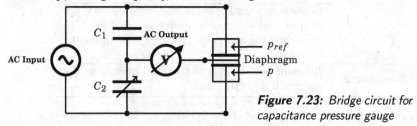

Figure 7.23: Bridge circuit for capacitance pressure gauge

Table 7.6: Electrical input and output of pressure transducers

Standard Input	5-24 V,DC for amplified voltage output transducers 8-30 V,DC for 4-20 mA current output
Standard Output	0-5 or 0-10 V,DC output (used in industrial environments) 4 - 20 mA current output (pressure transmitters)

Either a balanced or unbalanced mode of operation of the bridge circuit is possible. In the balanced mode of operation the variable capacitor is adjusted to bring the bridge back to balance. The position of a dial attached to the variable capacitor may be marked in pressure units. Alternately the output may be related to the pressure by direct calibration. Table 7.6 indicates the typical electrical input output pairs for pressure transducers. In the current output type of instrument 4 mA corresponds to 0 pressure and 20 mA corresponds to full scale (dependent on the range of the pressure transducer). The full scale may be a few mm of water column up to pressures in the MPa range. The scale is linear over the entire range of the transducer. Note that this type of gauge may also be used for measurement of pressures lower than the atmospheric pressure.

7.4.5 Piezoelectric pressure transducer

A piezoelectric force transducer will be dealt in greater detail in section 14.3.3 of Chapter 13. It may also be used for the measurement of pressure. The pressure to be measured compresses the piezoelectric crystal and induces a voltage that is proportional to it. In Figure 7.24 we show the constructional details of a piezoelectric pressure pickup used for measurement of pressure under high temperature condition, such as that prevails inside the cylinder of an internal combustion engine. This transducer has fast response that makes it faithfully follow the transient pressure variations within the engine cylinder. The transducer is mounted on the cylinder using a water cooled bulkhead. The signal is communicated to a suitable voltmeter using coaxial cable via a BNC (Bayonet Neill Concelman connector) connector. The diameter of the piezoelectric crystal is around 3 *mm* in the example shown in the figure.

Figure 7.24: Water cooled piezoelectric pressure pickup

7.5 Measurement of pressure transients

Like temperature transients, pressure transients occur in many process applications. For example, the pressure in the cylinder of an internal combustion engine varies periodically during its operation. The pressure variations are due to the cyclic processes that take in the cylinder. The pressure in a compressed air storage tank changes as the air supply is drawn from it. In some applications one may want to apply time varying pressure for a process to proceed according to a pre determined plan. In all these cases it is essential that one is able to measure the time varying pressure accurately, as it varies. Transient analysis of a U - tube manometer in section 7.2.2 has given us some ideas about these. In light of these, the following should be kept in mind:

- Transient behavior of pressure measuring instruments is needed to understand how they will respond to transient pressures.
- Measurement of transient or time varying pressure requires a proper choice of the instrument and the coupling element between the pressure signal and the transducer.
- Usually process pressure signal is communicated between the measurement point and the transducer via a tube of suitable material.
- The tube element imposes a "resistance" due to viscous friction.

- Viscous effects are normally modeled using laminar fully developed flow assumption (discussed earlier while considering the transient response of a U tube manometer).

Before we proceed with the analysis we shall look at a few general things. These are in the nature of comparisons with what has been discussed earlier while dealing with transients in thermal systems and in the case of the U tube manometer. We shall keep electrical analogy in mind in doing this.

Thermal system

Thermal Capacity defined as C, J/K is the product of the mass M, kg and the specific heat (c J/kgK). With temperature T representing the driving potential, h_e representing the enthalpy, the specific heat capacity is given by

$$c = \frac{\partial h_e}{\partial T} \tag{7.36}$$

Then, we have,

$$C = Mc = \rho V \frac{\partial h_e}{\partial T} \tag{7.37}$$

Thermal resistance is given by $R = \dfrac{1}{hS}$ where h is the heat transfer coefficient and S is the surface area. The time constant is given, using electrical analogy, as

$$\boxed{\tau = \frac{C}{hS} = \frac{Mc}{hS} = \frac{\rho V}{hS} \frac{\partial h_e}{\partial T}} \tag{7.38}$$

Pressure measurement in a liquid system

Driving *potential* is the head of liquid h m. Volume of liquid V m^3 *replaces* enthalpy in a thermal system. Thus the capacitance is defined as

$$C = \frac{\partial V}{\partial h} \tag{7.39}$$

and has units of m^2. We have already seen that the resistance to liquid flow is defined through the relation $R = -\dfrac{dp}{d\dot{m}}$ with p standing for pressure and $\dot{m}$ standing for the mass flow rate. The time constant is hence given by

$$\boxed{\tau = RC = -\frac{dV}{dh} \frac{\partial p}{\partial \dot{m}}} \tag{7.40}$$

Pressure measurement in a gas system

Potential for bringing about change is the pressure p. The *mass of the gas m* replaces the enthalpy in the thermal system. The capacitance is given by

$$C = \frac{dm}{dp} \tag{7.41}$$

With m being given by the product of density ρ and volume V, we have $m = \rho V$. The capacitance depends on the process that the gas undergoes while the pressure transducer responds to the transient. If we assume the process to be polytropic, then

$$\frac{p}{\rho^n} = \text{constant} \tag{7.42}$$

where n has an appropriate value ≥ 1. If, for example, the volume of the gas is held constant, we have

$$C = V \frac{d\rho}{dp} \tag{7.43}$$

For the polytropic process, logarithmic differentiation gives

$$\frac{dp}{p} - n\frac{d\rho}{\rho} \text{ or } \frac{dp}{d\rho} = \frac{\rho}{np} \tag{7.44}$$

With this the capacitance becomes

$$C = \frac{V\rho}{np} \tag{7.45}$$

Assuming the gas to be an ideal gas $p = \rho R_g T$ the capacitance is

$$C = \frac{V}{nR_g T} \tag{7.46}$$

The resistance to flow is again given by the Hagen-Poiseuille law as in the case of a liquid, assuming low speed flow, which is always the case. The time constant is then given by

$$\boxed{\tau = RC = \frac{RV}{nR_g T}} \tag{7.47}$$

7.5.1 Transient response of a bellows type pressure transducer

The bellows element may be considered as equivalent to a piston of area A (in m^2) - spring arrangement as shown in Figure 7.25. Let the bellows spring constant be K (in N/m). Let the displacement of the bellows element be b (in m). Let the bellows gauge be connected to a reservoir of liquid of density ρ (kg/m^3) at pressure P (in Pa). At any instant of time let p (in Pa) be the pressure inside the bellows element. The displacement of the spring is then given by the relation

$$\delta = \frac{\text{Force}}{\text{Spring constant}} = \frac{pA}{K} \tag{7.48}$$

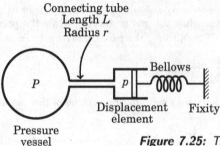

Figure 7.25: Transient in a bellows type gauge

Let the connecting tube between the reservoir and the gauge be of length L (in m) and radius r (in m). If the pressure P is greater than the pressure p, the liquid will flow through the intervening tube at a mass flow rate given by $\dot{m} = \rho A \dfrac{d\delta}{dt}$. We know from the definition of connecting tube resistance that this should equal $\dot{m} = \dfrac{P - p}{R}$. All we have to do is to equate these two expressions to get the equation governing the transient.

$$\rho A \frac{d\delta}{dt} = \frac{P - p}{R} = \frac{P - \frac{K\delta}{A}}{R} \qquad (7.49)$$

The latter part of the equality follows from Equation 7.48. The above equation may be rearranged in the form

$$\frac{A^2 \rho R}{K} \frac{d\delta}{dt} + \delta = \frac{AP}{K} \qquad (7.50)$$

We see that the transient of a bellows type gauge is governed by a first order ordinary differential equation with a time constant given by

$$\boxed{\tau = \frac{A^2 \rho R}{K}} \qquad (7.51)$$

Hence the bellows type gauge is a first order system just like the first order thermal system we considered in section 4.7. We infer from the above that the *capacitance* of a bellows type gauge is

$$\boxed{C = \frac{A^2 \rho}{K}} \qquad (7.52)$$

Example 7.9

A bellows pressure gauge of area $1 cm^2$, spring constant $K = 4.4\ N/cm$ is connected by a $2.5mm$ ID tubing that is $15m$ long in a process application. Measurement is of pressure of water at $30°C$. Determine the time constant of this gauge.

Solution :

The given data for the gauge is written down first using the notation used previously:

$$A = 1\,cm^2 = 10^{-4}\,m^2;\ K = 4.4\,N/cm = 440\,N/m$$

The given data for the connecting tube is:

$$r = \frac{ID}{2} = \frac{2.5}{2} = 1.25 \, mm = 0.00125 \, m; \quad L = 15 \, m$$

Properties of water required are taken from tables. These are

$$\rho = 995.7 kg/m^3; \quad v = 0.801 \times 10^{-6} \, m^2/s$$

Note that $v = \mu/\rho$ represents the kinematic viscosity. The tube resistance is calculated (using Equation 7.9) as

$$R = \frac{8vL}{\pi r^4} = \frac{8 \times 0.801 \times 10^{-6} \times 15}{\pi \times 0.00125^4} = 1.2532 \times 10^7 (m \, s)^{-1}$$

The capacitance is given (using Equation 7.52) by

$$C = \frac{A^2 \rho}{K} = \frac{(10^{-4})^2 \times 995.7}{440} = 2.26 \times 10^{-8} ms^2$$

The time constant is then given by

$$\tau = R \times C = 1.2532 \times 10^7 \times 2.26 \times 10^{-8} = 0.284 \, s$$

7.5.2 Transients in a force balancing element for measuring pressure

Figure 7.26 shows the arrangement in a force balance element. It consists of a diaphragm gauge that is maintained in its zero displacement condition by applying an opposing force that balances the pressure being measured. The required force itself is a measure of the pressure. Let us assume that the pressure being measured is that of a gas like air. If the diaphragm displacement is always in the null the volume of the gas within the gauge is constant. If the pressure P is greater than pressure p mass flow takes place from the pressure source in to the gauge to increase the density of the gas inside the gauge. The time constant for this system is just the one discussed earlier and is given by Equation 7.47. In case the instrument is not the null type (where the diaphragm displacement is zero at all times), the capacity should include both that due to density change and that due to displacement. Since the displacement is very small, it is appropriate to sum the two expressions 7.46 and 7.52 together and write

$$\boxed{C \approx \frac{V}{nR_g T} + \frac{A^2 \rho}{K}} \tag{7.53}$$

The above is valid as long as the transients are small so that the changes in density of the fluid are also small. In many applications this condition is satisfied.

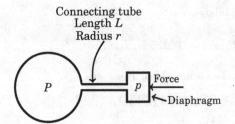

Connecting tube
Length L
Radius r

P

p Force

Diaphragm

Example 7.10

A bellows type pressure gauge of effective area $A = 0.25\ cm^2$, spring constant $K = 100\ N/m$ is connected by a 1.5 mm ID 10 m long stainless tubing in a process application.   The volume of the effective air space in the bellows element is $V = 2\ cm^3$. Determine the time constant if the pressure measurement is a pressure around 5 bar and the temperature is $27°C$. Compare this with the case where the gauge is a null type one.

Solution :

Step 1 We assume that the pressure hovers around 5 $bars$ during the transients. The air properties are taken from appropriate tables under the following conditions:

$$p = 5bar = 5 \times 10^5\ Pa; T = 27°C = 300\ K; R_g = 287\ J/kg\ K$$

Step 2 The density of air is calculated as

$$\rho = \frac{p}{R_g T} = \frac{5 \times 10^5}{287 \times 300} = 5.807\ kg/m^3$$

The dynamic viscosity of air, from table of properties, is $\mu = 1.846 \times 10^{-5}\ kg/m\ s$. The kinematic viscosity is hence given by

$$v = \frac{\mu}{\rho} = \frac{1.846 \times 10^{-5}}{5.807} = 3.179 \times 10^{-6}\ m^2/s$$

Step 3 The capacity is calculated based on Equation 7.53 as

$$C = \frac{\left(0.25 \times 10^{-4}\right)^2 \times 5.807}{100} + \frac{2 \times 10^{-6}}{287 \times 300} = 5.9523 \times 10^{-11}\ m\ s^2$$

In the above we have assumed a polytropic index of 1 that corresponds to an isothermal process i.e. the gas remains at room temperature throughout.

Step 4 The fluid resistance is calculated using Equation 7.9, as

$$R = \frac{8 \times 3.179 \times 10^{-6} \times 10}{\pi \times 0.00075^4} = 2.559 \times 10^8\ (m\ s)^{-1}$$

Step 5 The time constant of the transducer is then given by

$$\tau = R \times C = 2.559 \times 10^8 \times 5.9523 \times 10^{-11} = 0.0152 \, s$$

Step 6 If the gauge operates as a null instrument, the capacitance is given by Equation 7.46 as

$$C = \frac{2 \times 10^{-6}}{287 \times 300} = 2.3229 \times 10^{-11} \, m \cdot s^2$$

The resistance is still given by the value evaluated earlier. Hence the time constant for null mode operation is

$$\tau = R \times C = 2.559 \times 10^8 \times 2.3229 \times 10^{-11} = 0.0059 \, s$$

7.6 Measurement of vacuum

Pressure below the atmospheric pressure is referred to as vacuum pressure. The gauge pressure is negative when we specify vacuum pressure. Absolute pressure, however, is certainly positive! It is possible to use, U - tube manometers, Bourdon gauges for the measurement of vacuum as long as it is not *high* vacuum. In the case of the U - tube with mercury as the manometer liquid we can practically go all the way down to -760 *mm* mercury but there is no way of using it to measure vacuum pressures involving a fraction of a *mm* of mercury (also referred to as *Torr*). A U - tube manometer can thus be used for *rough* measurement of vacuum. Figure 7.27 shows how it is done. A Bourdon gauge capable of measuring vacuum pressure will have the zero (zero gauge pressure) somewhere in the middle of the dial with the pointer going below zero while measuring vacuum, and the pointer going above zero, while measuring pressures above the atmosphere. Again a Bourdon gauge can not be used for measurement of *high* vacuum.

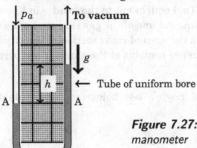

Figure 7.27: *Measurement of vacuum by a U tube manometer*

Now we discuss those instruments that are useful for measuring moderate and high vacuum ranging from, say 0.1 *mm* mercury to 10^{-6} *mm* mercury. Such low pressures (absolute pressures)- or *high* vacuum pressures - are used in many engineering applications and hence there is a need to measure such pressures.

7.6.1   McLeod gauge

McLeod gauge is basically a manometric method of measuring a vacuum pressure that is useful between 0.01 and 100 μm of mercury column. This range translates to absolute pressure in the range of 0.001 to 10 $Pa.$ or 10^{-8} to 10^{-4} bar. The principle of operation of the McLeod gauge is described by referring to Figure 7.28.

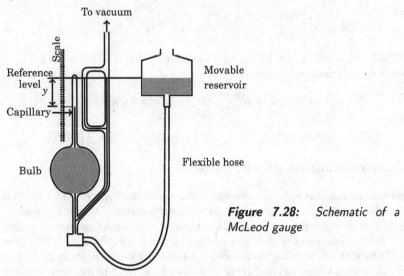

Figure 7.28:   *Schematic of a McLeod gauge*

A known volume (V) of the gas at the vacuum pressure (p), given by the volume of the capillary, the bulb and the bottom tube up to the opening, is trapped by lowering the movable reservoir down to the appropriate extent. It is then slowly raised till the level of the manometer liquid (usually mercury) in the movable reservoir is in line with the reference level marked on the stem of the forked tube. This operation compresses the trapped gas to a pressure (p_c) equivalent to the head y indicated by the manometer as shown. The corresponding volume of the gas is given by the clear volume of the capillary $V_c = ay$ where a is the area of cross section of the capillary. The gauge is exposed to the ambient and hence remains at the ambient temperature throughout this operation.

Since the entire process is, isothermal Boyle's law holds (assuming ideal gas behavior) and hence we have

$$pV = p_c V_c = p_c a y \tag{7.54}$$

Manometer Equation 7.2 gives

$$p_c - p = y \tag{7.55}$$

Note that all pressures are in mm of mercury in the above and y is also in mm. The area of cross section a is in mm^2 and the volume V is in mm^3. We eliminate p_c from Equations 7.54 and 7.55 to get

$$p = \frac{ay^2}{V - ay} \approx \frac{ay^2}{V} \qquad (7.56)$$

The approximation is valid *if the initial volume* V is *much greater than* the final volume ay.

Example 7.11

A McLeod gauge has a volume of 100 ml and a capillary of 0.5 mm diameter. Estimate the pressure indicated by a reading of 25 mm of mercury. What will be the error if the approximate formula is made use of?

Solution :

Step 1 Volume of trapped gas $V = 100\ ml = 100 \times 10^3 = 10^5\ mm^3$

Step 2 Capillary bore is $d = 0.5mm$ and hence the area of cross section of the capillary is $a = \pi \dfrac{d^2}{4} = \pi \times \dfrac{0.5^2}{4} = 0.196\ mm^2$

Step 3 The gauge reading is $y = 25\ mm$

Step 4 Using accurate formula given by first part of Equation 7.56, we have

$$p = \frac{0.196 \times 25^2}{10^5 - 0.196 \times 25} = 0.000122724\ Torr$$

Step 5 If we use the approximate formula (latter part of Equation 7.56), we get

$$p \approx \frac{0.196 \times 25^2}{10^5} = 0.000122719\ Torr$$

Thus there is negligible error in the use of the approximate formula.

Step 6 The answer may be rounded to 1.227 $mTorr$.

7.6.2 Pirani gauge

Another useful gauge for measuring vacuum is the Pirani gauge. The temperature of a heated resistance *increases* with a reduction of the background pressure. This is due to the reduction of rate of heat transfer from the filament to the environment. The rise in temperature changes its resistance just as we found in the case of RTD (see Section 4.4). As shown in Figure 7.29 the Pirani gauge consists of two resistances connected in one arm of a bridge. One of the resistors is sealed in after evacuating its container while the other is exposed to the vacuum space whose pressure is to be measured. Ambient conditions will affect both the resistors alike and hence the gauge will respond to only the changes in the vacuum pressure that is being measured. The Pirani gauge is calibrated such that the imbalance current of the bridge is directly related to the vacuum pressure being measured. The range of this gauge is from 1 μm mercury to 1 mm mercury that corresponds to $0.1 - 100\ Pa$ the absolute pressure range.

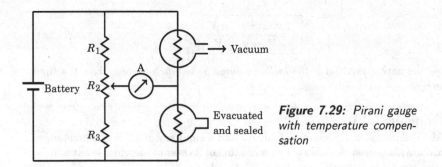

Figure 7.29: *Pirani gauge with temperature compensation*

7.6.3 Ionization gauge

Very high vacuum pressures (higher the vacuum, lower the absolute pressure) are measured using an ionization gauge. Electrical schematic of such a gauge is shown in Figure 7.30 along with the constructional details of an ionization gauge.

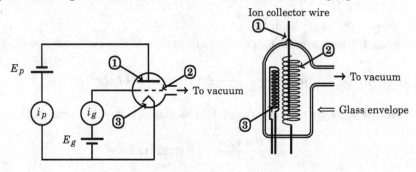

1. Plate, 2. Grid, 3. Filament

Figure 7.30: *Ionization gauge - (a) circuitry and (b) schematic*

The ion gauge is similar to a triode valve which was used in radio receivers built with vacuum tubes before the advent of transistors. The gauge consists of a heated cathode that emits electrons. These electrons are accelerated towards the grid, basically a wire screen, which is maintained at a potential positive with respect to the cathode. The high speed electrons suffer collisions with the molecules of the residual gas in the bulb (that is exposed to the vacuum space) and ionizes these gas molecules. The electrons move towards the grid and manifest in the form of grid current i_g. The positive ions move towards the plate that is maintained at a potential negative with respect to the grid. These ions are neutralized at the plate and consequently we have the plate current i_p. The pressure in the bulb is then given by

$$p = \frac{1}{S}\frac{i_p}{i_g} \tag{7.57}$$

Here S is a proportionality called the ionization gauge sensitivity. S has a typical value of $200\,Torr^{-1}$ or or $2.67\,kPa^{-1}$. The gauge cannot be used when the pressure is greater than about $10^{-3}Torr$ since the filament is likely to burn off. In practice the following is what is done.

The vacuum space is provided with different pressure measuring devices such as those listed below and used at appropriate conditions mentioned therein:

- U tube mercury manometer for indicating that the initial evacuation is complete, usually achieved by a rotary pump or a reciprocating pump.
- A Pirani gauge to measure the reduced pressure as the higher vacuum is achieved by a diffusion pump or a turbo-molecular pump coupled with a cold trap.
- After the pressure has reduced to below $10^{-6} Torr$ the ion gauge is turned on to measure the pressure in the high vacuum range.

For more details the reader should refer to specialized books dealing with high vacuum engineering.

7.6.4 Alphatron gauge

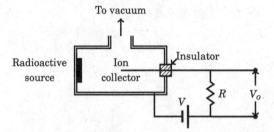

Figure 7.31: Schematic of an Alphatron gauge

In the Alphatron gauge ionization of the residual gas is brought about by alpha particles that are emitted by a radioactive material. The gauge is similar to a common smoke detector in its operation. Since there is no hot cathode as in the case of ion gauge considered earlier, the gauge may be exposed to the atmospheric pressure without any fear of losing a filament! The rate of ion production is dependent on the residual pressure in the gauge, and this pressure is the same as the vacuum pressure that is being measured. A collector electrode is maintained at a positive potential as shown in Figure 7.31. The ion current produces a potential drop V_o across the load resistor and this is indicative of the pressure. The output is fairly linear over the entire range from $1\ mTorr$ to the atmospheric pressure or 0.1 to $10^5\ Pa$.

Concluding remarks

Measurement of pressure spans a wide range from sub atmospheric to very high pressures several thousand times the atmospheric pressure. Different sensors are thus necessary to cover the range of pressures met with in practice. Measurement of pressure transients require an understanding of transient response of pressure sensors. These aspects have been covered in this chapter. Pressure measurement is intimately connected with measurement of flow. This aspect will be discussed in succeeding chapter.

Chapter *8*

Measurement of Fluid Velocity

Applications involving moving fluids are very common in engineering. The fluid velocity is a quantity that directly affects thermal processes such as cooling, heating, chemical reactions and so on. Measurement of fluid velocity is hence very important. Also important, in some applications, a knowledge of the direction of velocity. The present chapter describes different techniques of velocity measurement, intrusive as well as non-intrusive. It also deals with some discussion on the measurement of direction of velocity.

8.1   Introduction

Fluid flow - liquid or gas - occurs in most processes. In applications where the details are important, the fluid velocity needs to be measured as a field variable. A detailed velocity map may be required for optimizing the fluid flow system from the view point of pressure drop and the consequent pumping power requirement. It may also play an important role in heat transfer effectiveness. Hence, in what follows, we look at the measurement of fluid velocity in detail. We will discuss both intrusive and non-intrusive methods, as mentioned below:

- **Velocity map using Pitot tube and Pitot static tube:** Intrusive method, based on fluid flow principles
- **Hot wire anemometer:** Intrusive, based on thermal effect of flow
- **Doppler Velocimeter:** Non-intrusive, based on Doppler effect due to scattering of waves by moving particles
- **Time of Flight (TOF) Velocimeter:** Non-intrusive, based on travel time of sound wave through a moving medium

The size of the probe tends to average the measurement over an area around the point of interest and also introduces a disturbance in the measured quantity by modifying the flow that would exist in the absence of the probe. The size of the probe should be chosen such that there are no 'blockage' effects. The term 'intrusive' refers to this aspect, in the case of intrusive methods! The measurements also are prone to errors due to errors in alignment of the probe with respect to the flow direction. For example, a Pitot tube should be aligned with its axis facing the flow. If the flow direction is not known, it is difficult to achieve this with any great precision. One way of circumventing such a problem is to use a specially shaped probe (reader should consult advanced texts on flow measurement) that is not very sensitive to its orientation with respect to the flow direction.

8.2   Pitot - Pitot static and impact probes

8.2.1   Pitot and Pitot static tube

The basic principle of the Pitot and Pitot static tube is that the pressure of a flowing fluid will increase when it is brought to rest at a stagnation point of the probe. Figure 8.1 shows the streamlines in the vicinity of a blunt nosed body. We assume that, if the flow is that of a gas, like air, the velocity of the fluid is much smaller than the speed of sound in air such that density changes may be ignored. Basically the fluid behaves as an incompressible fluid. The stagnation point is located as shown. Streamlines bend past the body as shown. The pressure at the stagnation point is the stagnation pressure.

- Incompressible flow assumption is valid for liquid flows
- Incompressible flow assumption is also valid for gases if the flow velocity (V) is much smaller than the speed of sound (a)
- The Mach number is defined as $M = V/a$
- Generally density variations are important if $M > 0.3$ or thereabout!

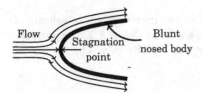

Flow

Stagnation point

Blunt nosed body

Figure 8.1: *Flow of a fluid in the vicinity of a blunt nosed body*

If viscous effects are negligible the difference between the stagnation pressure and the static pressure is related to the dynamic pressure which is related to the square of the velocity. Thus the velocity information is converted to a pressure difference, that may be measured by a pressure measuring device, such as a manometer.

The basic arrangement for measuring fluid velocity using a Pitot tube is shown in Figure 8.2. The Pitot tube consists of bent tube of small diameter (small compared to the diameter or size of the duct) with a rounded nose. The flow is axi-symmetric and in the vicinity of the nose is like the flow depicted in Figure 8.1. The Pitot tube is connected to one limb of a U tube manometer. The other limb of the manometer is connected to a tap made on the tube wall as indicated. The tube tap and the nose of the Pitot tube are roughly in the same plane. It is assumed that the wall tap senses the static pressure p of the fluid while the Pitot tube senses the stagnation pressure p_0 of the fluid. From Bernoulli principle we have (for low speed flow, fluid velocity much less than sonic velocity in the fluid)

$$p_0 - p = \frac{1}{2}\rho V^2 \tag{8.1}$$

The above assumes, in addition, that the Pitot tube is horizontal. In Equation 8.1 ρ is the density (constant in the case of low speed flow) of the fluid whose velocity is being measured. We see that in case of gas flow the temperature also needs to be measured since the density is a function of static pressure and temperature. With ρ_m as the density of the manometer liquid the pressure difference is given by

$$p_0 - p = (\rho_m - \rho)gh \tag{8.2}$$

Combining Equations 8.1 and 8.2, we get

$$V = \sqrt{\frac{2(\rho_m - \rho)gh}{\rho}} \tag{8.3}$$

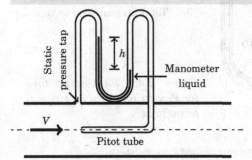

Static pressure tap

h

Manometer liquid

V

Pitot tube

Figure 8.2: *Pitot tube arranged to measure fluid velocity*

An example for the use of a Pitot static tube for the measurement of velocity of an air stream issuing out of a rectangular duct is shown in Figure 8.3. The pressure difference is converted to velocity and is displayed as velocity in *m/s* by the digital micro-manometer.

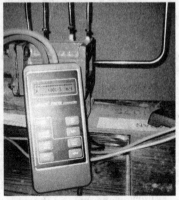

Figure 8.3: *Photograph of a Pitot tube with a digital micro-manometer that displays velocity directly*

A typical application of a Pitot tube is the measurement of velocity of an aircraft. The stagnation pressure is obtained from the Pitot probe mounted on the aircraft wing. The static pressure is sensed by a static pressure hole elsewhere on the aircraft, as for example, on the fuselage. The diaphragm element (we have already seen how pressure can be converted to a displacement, see section 7.4.3) with a dial indicator completes the instrument.

A Pitot static tube or **Prandtl tube**[1] is used very often in laboratory applications. Pitot static tube senses both the stagnation pressure and the static pressure in a single probe. A schematic of a Pitot static probe is shown in Figure 8.4. The proportions of the probe are very important in order to have accurate measurement of the velocity. Typically the probe diameter (D) would be 6 *mm*. The length of the probe would then be around 10 *cm* or more.

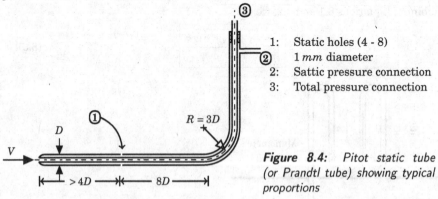

1:   Static holes (4 - 8)
    1 *mm* diameter
2:   Sattic pressure connection
3:   Total pressure connection

Figure 8.4: *Pitot static tube (or Prandtl tube) showing typical proportions*

---

[1]Named after Ludwig Prandtl 1875-1953, German fluid dynamicist.

The static pressure holes are positioned more than $4D$ from the stagation point. A suitable manometer may be connected between the inner and outer tubes to measure the fluid velocity. The probe is inserted through a hole in the side of the duct or wind tunnel, as the case may be.

Example 8.1

A Pitot static tube is used to measure the velocity of an aircraft. If the air temperature and pressure are $5°C$ and $90kPa$ respectively, what is the aircraft velocity in km/h if the differential pressure is $250mm$ water column?

Solution :

We take density of water (manometric liquid) as $999.8 \ kg/m^3$. We calculate first the density of air at $T = 5°C = 278 \ K$ and $p = 90 \ kPa$ using ideal gas relation with gas constant of $R_g = 287 \ J/kg \ K$

$$\rho = \frac{p}{R_g T} = \frac{90 \times 10^3}{287 \times 278} = 1.128 \ kg/m^3$$

The pressure difference given in terms of head of water as $h = 250 \ mm = 0.25 \ m$ may now be converted to SI units as

$$\Delta p = \rho_m gh = 999.8 \times 9.8 \times 0.25 = 2446.8 \ Pa$$

The aircraft speed is thus given by Equation 8.3 as

$$V = \sqrt{\frac{2 \times 2446.8}{1.128}} = 65.87 \ m/s$$

This may be specified in km/hr by noting that $1 \ km/h = 1000/3600 = 1/3.6 \ m/s$.

$$V = 65.87 \times 3.6 = 237.1 \ km/h$$

We shall verify whether the incompressible assumption is good enough. The speed of sound is calculated by assuming that the ratio of specific heats for air is $\gamma = 1.4$.

$$a = \sqrt{\frac{\gamma p}{\rho}} = \sqrt{\frac{1.4 \times 90 \times 10^3}{1.128}} = 334.2 \ m/s$$

The corresponding Mach number is

$$M = \frac{65.87}{334.2} = 0.197$$

Since $M < 0.3$ the incompressible flow assumption is reasonable.

Example 8.2

A Pitot static tube is used to measure the velocity of water flowing in a pipe. Water of density $\rho = 1000 \ kg/m^3$ is known to have a velocity of $V = 2.5 \ m/s$ where the Pitot static tube has been introduced. The static pressure is measured independently at the tube wall and is $2 \ bar$. What is the head developed by the Pitot static tube if the manometric fluid is mercury with density equal to $\rho_m = 13600 \, kg/m^3$?

Solution :

The static pressure that is measured independently is $p = 2 \, bar = 2 \times 10^5 \, Pa$. The dynamic pressure is calculated based on the water velocity and water density as

$$p_d = \frac{\rho V^2}{2} = \frac{1000 \times 2.5^2}{2} = 3125 \, Pa$$

This is also the pressure difference sensed by the Pitot static probe. Hence this must equal $(\rho_m - \rho)gh$. Thus the head developed h is given by

$$h = \frac{p_d}{(\rho_m - \rho)g} = \frac{3125}{(13600 - 1000) \times 9.8} = 0.02531 \, m \text{ or } 25.31 \, mm \text{ mercury column}$$

8.2.2 Effect of compressibility

In Example 8.1 the Mach number of the aircraft is about 0.2. It was mentioned that the incompressible assumption may be acceptable, for this case. Let us, however, look at what will happen if the compressibility effects have to be taken into account. We discuss here subsonic flows ($M < 1$) where it may be necessary to consider the effect of density variations. In such a case we use well known relations from gas dynamics[2] to take into account the density variations. We shall assume that the gas behaves as an ideal gas. The stagnation process (what happens near the nose of the Pitot tube) is assumed to be an isentropic process via which the flow is brought to rest from the condition upstream of the Pitot tube. From gas dynamics we have

$$\frac{p_0}{p} = \left[1 + \frac{\gamma - 1}{2}M^2\right]^{\frac{\gamma}{\gamma - 1}} \tag{8.4}$$

where p_0 is the stagnation pressure, p is the static pressure, M is the Mach number and γ is the ratio of specific heats. Equation 4.53 will have to be replaced by

$$(p_0 - p) = p\left[\left(1 + \frac{\gamma - 1}{2}M^2\right)^{\frac{\gamma}{\gamma - 1}} - 1\right] = (\rho_m - \rho)gh \tag{8.5}$$

We may solve this for M and get

$$M = \sqrt{\frac{2}{\gamma - 1}\left\{\left[1 + \frac{(\rho_m - \rho)gh}{p}\right]^{\frac{\gamma - 1}{\gamma}} - 1\right\}} \tag{8.6}$$

In order to appreciate the above we present a plot of pressure ratio recorded by a Prandtl tube as a function of flow Mach number in Figure 8.5. The figure indeed shows that the incompressible assumption may be alright up to a Mach number of

---

[2]See, for example, H.W. Liepmann and A. Roshko, "Elements of Gas Dynamics", Wiley, NY 1957

about 0.3. This may be appreciated better by looking at the percentage difference between the two as a function of Mach number. Such a plot is shown in Figure 8.6.

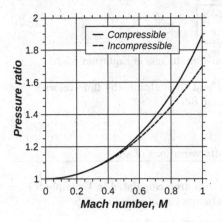

Figure 8.5: *Pressure ratio recorded by a Prandtl tube as a function of Flow Mach number*

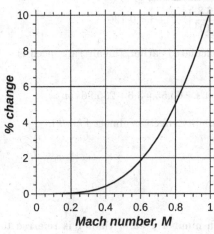

Figure 8.6: *Percent difference in Pressure ratio with compressible and incompressible assumptions recorded by a Prandtl tube as a function of flow Mach number*

Equation 8.4 may be rearranged as

$$M^2 = \frac{2}{\gamma - 1}\left[\left(\frac{p_0}{p}\right)^{\frac{\gamma-1}{\gamma}} - 1\right]$$

Let the pressure difference be defined as $\Delta p = p_0 - p$. Then $\frac{p_0}{p} = 1 + \frac{\Delta p}{p}$. Incompressible flow assumption is expected to be valid if $\Delta p \ll p$. In that case we also have

$$\left(\frac{p_0}{p}\right)^{\frac{\gamma-1}{\gamma}} = \left(1 + \frac{\Delta p}{p}\right)^{\frac{\gamma-1}{\gamma}} \approx 1 + \frac{\gamma-1}{\gamma}\frac{\Delta p}{p}$$

Correspondingly we have

$$M^2 \approx \frac{2}{\gamma - 1}\left[1 + \frac{\gamma-1}{\gamma}\frac{\Delta p}{p} - 1\right] = 2\frac{\Delta p}{\gamma p} = \left(\frac{V}{a}\right)^2$$

With $a^2 = \dfrac{\gamma p}{\rho}$ we then get

$$V = \sqrt{\frac{2\Delta p}{\rho}} \tag{8.7}$$

This expression is identical to the one obtained by the use of Equation 4.52.

Thus we *solve* for the Mach number first and then obtain the flow velocity. We shall look at Example 8.1 again using Equation 8.6.

Example 8.3

Rework Example 8.1 accounting for density variations of air.

Solution :

We make use of the air density and pressure drop calculated in Example 8.1 to calculate the aircraft Mach number as (using Equation 8.6)

$$M = \sqrt{\frac{2}{1.4-1}\left[1 + \frac{2446.8}{90 \times 10^3}\right]^{\frac{1.4-1}{1.4}} - 1} = 0.196$$

With the speed of sound of 334.2 m/s calculated earlier, the aircraft speed is given by

$$V = M \times a = 0.196 \times 334.2 = 65.55 m/s = 65.55 \times 3.6 = 235.96 km/h$$

The error due to the neglect of compressibility effects is almost 1 km/h!

8.2.3   Supersonic flow

Impact probe in supersonic flow

Flow of a gas (for example air) at Mach number greater than 1 is referred to as supersonic flow. The axi-symmetric flow past an impact probe (Pitot tube) in supersonic flow is very different from that in subsonic flow. Typically what happens is shown in Figure 8.7.

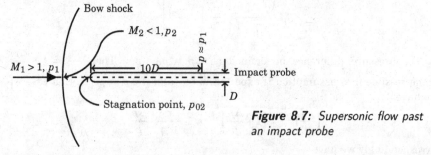

Figure 8.7: *Supersonic flow past an impact probe*

Stagnation process is more complex in the case of supersonic flow. A curved shock (referred to as a bow shock because of its shape) is formed just ahead of the probe and detached from it. As is clear from the sketch the flow across the shock is normal to it just ahead of the stagnation point on the probe. The flow transforms irreversibly to subsonic flow downstream of the shock (state represented by subscript 2) from supersonic condition upstream (state represented by subscript 1) of it. The stagnation process downstream of the shock follows an isentropic process. The following expression can be derived using the appropriate relations for flow across a normal shock followed by an isentropic stagnation process. The appropriate relation is known as Rayleigh Pitot relation and is given by

$$\frac{p_{02}}{p_1} = \frac{\left(\frac{\gamma+1}{2}M_1^2\right)^{\frac{\gamma}{\gamma-1}}}{\left(\frac{2\gamma}{\gamma+1}M_1^2 - \frac{\gamma-1}{\gamma+1}\right)^{\frac{1}{\gamma-1}}} \qquad (8.8)$$

Equation 8.8 relates the ratio of stagnation pressure measured by the impact probe to the static pressure of the flow upstream of the shock to the Mach number of the flow upstream of the shock - the quantity we are out to measure, in the first place. It is found that the static pressure is equl to the upstream static pressure at a location approximately equal to or greater than $10D$ as shown in the figure. Hence we may use a Prandtl tube in supersonic flow with the static pressure tap located about $10D$ downstream of the stagnation point of the probe. A plot of Expression 8.8 (actually the reciprocal of the pressure ratio) is shown in Figure 8.8 and helps in directly reading off M_1 from the measured pressure ratio.

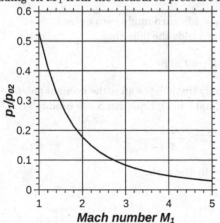

Figure 8.8: *Pressure ratio Mach number relation for supersonic flow*

Wedge probe

A second type of probe that may be used for measurement of supersonic flow is the wedge type probe shown in Figure 8.9. The flow past the wedge is two dimensional with an oblique shock emanating from the apex of the wedge as shown. The flow becomes parallel to the wedge surface downstream of the oblique shock. The static pressure p_2 may be sensed by using a static pressure tap on the surface of the wedge some distance away from the apex. The relationship between the pressures upstream and downstream of the oblique shock is given by

$$\boxed{\frac{p_2}{p_1} = \frac{2\gamma}{\gamma+1}M_1^2\sin^2\alpha - \frac{\gamma-1}{\gamma+1}}\qquad\qquad(8.9)$$

The shock angle α itself is a function of M_1 and θ . Tables relating these are available in books on Gas Dynamics[3]. Alternately one may use the web resource at http://www.aero.lr.tudelft.nl/ bert/shocks.html to calculate these.

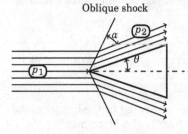

Oblique shock

Figure 8.9: *Wedge probe in supersonic flow*

Example 8.4

A wedge probe has a wedge angle of $\theta = 8°$ and placed in a wind tunnel where the flow is known be at $M_1 = 1.5$. Calculate the static pressure ratio that is expected from the probe.

Solution :

We use the web resource and input $M_1 = 1.5$, turn angle (weak shock) $\theta = 8°$ and $\gamma = 1.4$. The oblique shock calculator yields the following:

$$\alpha = 52.57°, M_2 = 1.208$$

(The entries have been edited to show only three digits after the decimal point and angle with two digits after the decimal point). Equation 8.9 is used to get

$$\frac{p_2}{p_1} = \frac{2\times1.4}{1.4+1}\cdot1.5^2 sin^2 52.57° - \frac{1.4-1}{1.4+1} = 1.489$$

This will be the pressure ratio indicated by the probe.

Cone-cylinder type impact probe

It should be clear by now that a probe in supersonic flow is a device that produces a change in the pressure that may be measured. The wedge probe is an example of this. Another example is a probe having cone-cylinder configuration as shown in Figure 8.9. The probe diameter used was 0.4435 *in* or 11.05 *mm*. Three different probe tips were used, the one shown in the figure being for the measurement of static pressure. Eight static pressure ports are positioned 10 D from the cone shoulder. This design gives accurate flow field static pressure according to the author.

---

[3]For example, see Babu,V 2007, *Fundamentals of Gas Dynamics*, Ane Books, New Delhi

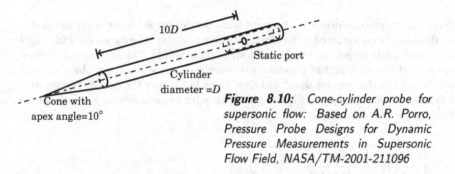

Figure 8.10: Cone-cylinder probe for supersonic flow: Based on A.R. Porro, Pressure Probe Designs for Dynamic Pressure Measurements in Supersonic Flow Field, NASA/TM-2001-211096

8.2.4   Orientation effects and multi-hole probes

The sensitivity to orientation is also a concern while using probes in subsonic as well as supersonic flows. It is also possible that the flow direction may be determined, when it not known, by using this sensitivity. A multiple hole probe (multi-hole probe, for short) is used for this purpose. These probes are precision machined, are a few *cm* long and are capable of being oriented with suitable arrangement for adjusting and measuring the various angles involved in defining the orientation of the probe with respect to the flow direction.

We shall explain the basic principle involved in multi-hole probes. Take, for example, a three hole probe oriented such that the plane passing through the three holes also contains the plane in which the two dimensional flow is taking place. The state of affairs, in general, is as shown in Figure 8.11.

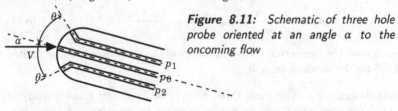

Figure 8.11: Schematic of three hole probe oriented at an angle α to the oncoming flow

We know from fluid flow theory that the pressure is a function of angle the pressure port or hole makes with respect to the direction of flow. Since the axis of the probe is pitched at an angle α as shown, the effective angle made by hole p_1 with respect to the flow direction is $\theta + \alpha$. This is greater than the angle made by hole p_2 with respect to the flow direction given by $\theta - \alpha$. Treating the flow to be potential flow normal to a cylinder, the stagnation pressure remains invariant and hence the total pressure sensed at the three ports are the same. However the velocity varies with θ and hence we have

$$p_\infty + \frac{\rho_\infty V^2}{2} = p_1 + 2\rho_\infty V^2 \sin^2(\theta + \alpha);$$

$$p_\infty + \frac{\rho_\infty V^2}{2} = p_0 + 2\rho_\infty V^2 \sin^2(\alpha);$$

and

$$p_\infty + \frac{\rho_\infty V^2}{2} = p_2 + 2\rho_\infty V^2 \sin^2(\theta - \alpha) \tag{8.10}$$

where subscript ∞ represents the conditions prevailing far away from the probe, the fluid has been assumed to be incompressible and the velocity varies with angle as $2V \sin\theta$ with respect to the forward stagnation point. The pressure difference $p_0 - p_1$ is different from the pressure difference $p_0 - p_2$. When the probe is aligned with zero pitch the two are equal and thus helps to identify the flow direction. Also the pressure difference $p_1 - p_2$ is zero when the pitch angle is zero or when the probe is facing the flow normally.

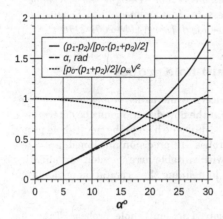

Figure 8.12: *Characteristics of an ideal cylindrical three hole probe*

For small values of α ($\alpha < 10°$) the following approximations apply, as seen from Figure 8.12.

$$\alpha \approx \frac{p_1 - p_2}{p_0 - \frac{(p_1+p_2)}{2}}; \quad \rho_\infty V^2 \approx p_0 - \frac{p_1 + p_2}{2} \tag{8.11}$$

It is also seen from the figure that the useful range of the three hole probe is limited to about $\pm 30°$ for the incident angle α.

A possible design for a cylindrical three hole probe is shown schematically in Figure 8.13. The yaw angle can be easily changed by rotating the cylindrical probe about its own axis as shown in the figure. Apart from the pressures measured with the three holes of the probe, the satatic pressure is also measured using a fourth pressure tap situated at a suitable location. All the pressures are sensed by transducers located within the probe body and the pre-amplified signal is communcated as electronic signal to a computer or data recorder. The probe is calibrated by systematically varying the incident angle α by placing it in a wind tunnel where the direction and magnitude of the flow velocity are known. Calibration takes care of non applicability of the formulae given earlier because of disturbance to flow by the probe itself, viscous effects and also imperfections in the probe head during the manufacturing process. The probe diameter is kept as low as possible, typically a couple of mm. The argument may be extended to a five hole probe. There are two pressure differences, $p_1 - p_2$ and $p_3 - p_4$ that may be monitored. These differences will be functions of two angles, pitch angle α (angle shown in sectional view) and the roll angle β (angle shown in front view) as shown in Figure 8.14.

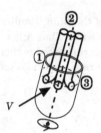

1   Cylindrical probe body with dished end
2   Three pressure tubes
3   Three holes along a circle with 45° angular spacing

Figure 8.13: *One useful type of cylindrical three hole probe*

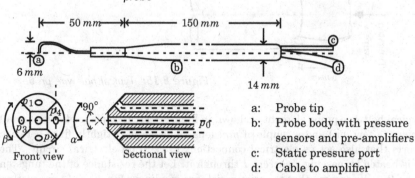

a: Probe tip
b: Probe body with pressure sensors and pre-amplifiers
c: Static pressure port
d: Cable to amplifier

Figure 8.14: *Typical five hole probe with bevelled faces*

Calibration of five hole probe is done by changing the roll and pitch angles systematically in a wind tunnel where the direction of flow is known, collecting the pressure difference data and evolving a relationship between these and the two angles. It is usual to represent the data in terms of pressure coefficients defined by

$$C_p = \frac{\delta p}{\frac{1}{2}\rho V^2} \tag{8.12}$$

where δp is a suitably chosen pressure difference. It is seen that the pressure coefficient C_p is non-dimensional and the dynamic pressure is used as the reference. Each and every multi-hole probe needs to be calibrated because of small variations that are always there between different probes during the manufacturing process. More details about multi-hole probes and their calibration may be obtained from specialized literature.[4]

8.3   Velocity measurement based on thermal effects

8.3.1   Hot wire anemometer

A hot wire anemometer basically involves relating the power that is dissipated by a hot wire to an ambient fluid, to the velocity of the fluid that is moving past it. The larger the velocity of the fluid, larger the heat dissipation from the wire - for

---

[4]D. Telionis and Y. Yang, Recent Developments in Multi-Hole Probe (MHP) Technology, Proceedings of COBEM 2009, Brazil

a fixed wire temperature. Alternately - larger the velocity of the fluid, smaller the wire temperature - for a fixed heat dissipation rate from the wire. Thus a hot wire anemometer is a thermal device and the velocity information is converted to thermal - either temperature change or change in the heat dissipation rate - information.

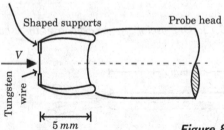

Figure 8.15: *Typical hot wire probe*

Typical hot wire probe element is shown in Figure 8.15. It consists of a very small diameter (a few μm) and a couple of mm long tungsten wire tightly fixed between supports that also act as electrical connections to external electrical circuit. The wire is heated by passing a current I through it. Let the resistance of the tungsten wire be R_w. We assume that the wire resistance is a linear function of temperature and is given by

$$R_w = R_\infty[1 + \alpha(T_w - T_\infty)] \tag{8.13}$$

In the above the subscript ∞ refers to a reference state, α is temperature coefficient of resistance and R_∞ is the wire resistance at the reference temperature. The temperature coefficient is $3.5 \times 10^{-3}/^\circ C$ for Platinum while it is $5.2 \times 10^{-3}/^\circ C$ for tungsten. The heat dissipated by the wire is given by I^2R_w and must equal the heat loss from the wire surface given by $\pi D L h(T_w - T_\infty)$, under steady state. Thus

$$I^2R_w = \pi D L h(T_w - T_\infty) \tag{8.14}$$

In the above expression D is the diameter of the wire, L is the effective length of the wire and h is the heat transfer coefficient. The heat transfer coefficient h (we assume that the hot wire element is like a cylinder in cross flow - flow velocity is normal to the axis of the cylinder) follows the so called King's law [5] which states that

$$\pi D L h = A + B V^n \tag{8.15}$$

Here A, B and index n are constants, usually determined by calibration and V is the fluid velocity. The constant part of the heat transfer coefficient accounts for natural convection and radiation even when the fluid is stationary. The variable part accounts for forced convection heat transfer in the presence of a moving fluid. The power law dependence is the observation due to King. We have seen earlier while discussing heat transfer to a thermometer well (see section 5.4.3) that the power law form appears in the Zhukaskas correlation for flow normal to a cylinder. From Equation 8.13 we have

$$T_w - T_\infty = \frac{R_w - R_\infty}{\alpha R_\infty} \tag{8.16}$$

[5]L.V. King, 1914, "On the Convection of Heat from Small Cylinders in a Stream of Fluid." Phil. Trans. of Roy. Soc. (London), Series A, Vol. 214, No. 14, pp. 373-432.

Substituting these in Equation 8.14, we get

$$I^2 R_w = (A + BV^n)\left(\frac{R_w - R_\infty}{\alpha R_\infty}\right) \tag{8.17}$$

which may be rearranged to read

$$A + BV^n = \frac{I^2 R_w R_\infty \alpha}{R_w - R_\infty} \tag{8.18}$$

Equation 8.18 is the basic hot wire anemometer equation. There are two ways of using the hot wire probe.

1. Constant Temperature or CT anemometer: The hot wire is operated at constant temperature and constant R_w. The current I will respond to changes in the velocity V.
2. Constant Current or CC anemometer: The hot wire current is held fixed. The resistance of the wire responds to change in the velocity V.

8.3.2 Constant Temperature or CT anemometer

A bridge circuit (Figure 8.16) is used for the constant temperature operation of a hot wire anemometer. The series resistance R_s is adjusted such that the bridge is at balance at all times. Thus, if the fluid velocity increases more heat will be dissipated by the hot wire at a given temperature and hence the current needs to increase. This is brought about by decreasing R_s. The opposite is the case when the fluid velocity decreases. We may write the power dissipated by the hot wire element as $I^2 R_w = \dfrac{E^2}{R_w}$. Here E is the voltage across the hot wire that is measured by the voltmeter as shown in Figure 8.16.

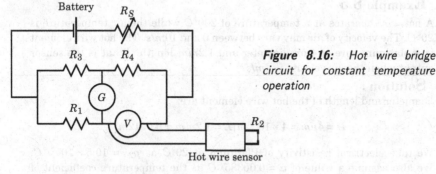

Figure 8.16: *Hot wire bridge circuit for constant temperature operation*

Equation 8.18 then takes the form

$$A + BV^n = \frac{E^2 R_\infty \alpha}{R_w (R_w - R_\infty)} \tag{8.19}$$

We note that, under CT operation, except E all other quantities on the right hand side are constant. We may absorb these in to the constants A and B to write the above equation in the form

$$E^2 = A + BV^n \tag{8.20}$$

Let E_0 be the output of the sensor when $V = 0$. The constant A is then nothing but E_0^2. Hence we have

$$E^2 - E_0^2 = BV^n \text{ or } \left(\frac{E}{E_0}\right)^2 - 1 = \left(\frac{B}{E_0}\right)^2 V^n \tag{8.21}$$

If we take logarithms on both sides, the following equation results:

$$\log\left[\left(\frac{E}{E_0}\right)^2 - 1\right] = \log\left(\frac{B}{E_0}\right)^2 + n\log(V) \tag{8.22}$$

The above expression is a linear representation of the output, obtained by suitable mathematical manipulations. Example 8.5 demonstrates these aspects.

8.3.3 Useful heat transfer correlation

Before proceeding further, we present a useful correlation for the calculation of the heat transfer coefficient for heat transfer from fine wires. The correlation due to Collis and Williams[6] is useful in the Reynolds number range $0.02 \leq Re \leq 44$. We note that the wire diameter in a hot wire sensor is such that this condition is not quite a limitation in the use of the correlation. The correlation itself is given by

$$Nu = \left(\frac{T_w}{T_\infty}\right)^{0.17} \left(0.24 + 0.56Re^{0.45}\right) \tag{8.23}$$

and is valid for air with Prandtl number of 0.7. In the above T_w is the wire temperature and T_∞ is the air temperature. Properties of the medium are calculated at the mean temperature of $T_m = \dfrac{T_w + T_\infty}{2}$. The characteristic length for the calculation of Nu as well as Re is the diameter of the wire D.

Example 8.5

A hot wire operates at a temperature of $200°C$ while the air temperature is $20°C$. The velocity of air may vary between 0 and $10m/s$. The hot wire element is a platinum wire of $4\mu m$ diameter and $1.2mm$ length. What is the sensor output when the air velocity is $4m/s$?

Solution :

Diameter and length of the hot wire element are:

$$D = 4\mu m = 4 \times 10^{-6}m; L = 1.2mm = 0.0012m$$

We take electrical resistivity of Platinum at $20°C$ as $\rho_{20} = 10.5 \times 10^{-8}/°C$. We also assume a value of $\alpha = 0.00385/°C$ as the temperature coefficient of Platinum. Further, we assume that the resistivity of Platinum varies linearly according to the relation $\rho_t = \rho_{20}[1 + \alpha(t - 20)]$ where t is the temperature in $°C$. Since the sensor operates in the CT mode, the wire resistance is constant and is calculated as follows.

$$R_{20} = \frac{\rho_{20}L}{\pi D^2/4} = \frac{10.5 \times 10^{-8} \times 0.0012}{\pi(4 \times 10^{-6})^2/4} = 10.03\,\Omega$$

---

[6]D. C . Collis AND M. J. Williams, "Two-dimensional convection from heated wires at low Reynolds numbers", Journal of Fluid Mechanics, Vol. 6, pp. 357-384, 1959.

$$R_w = R_{20}[1 + \alpha(t - 20)] = 10.03 \times [1 + 0.00385 \times (200 - 20)] = 17.1\,\Omega$$

The heat transfer coefficient is estimated using the correlation due to Collis and Williams. The properties of air are taken at the film temperature of $t_m = \dfrac{200 + 20}{2} = 110°C$. The required properties from air tables are

$$\nu = 24.15 \times 10^{-6}\,m^2/s; \quad k = 0.032\,W/m°C$$

The Reynolds number is calculated, with $V = 4\,m/s$ as

$$Re = \frac{VD}{\nu} = \frac{4 \times 4 \times 10^{-6}}{24.15 \times 10^{-6}} = 0.663$$

Nusselt number is calculated using Equation 8.23 as

$$Nu = [0.24 + 0.56 \times 0.663^{0.45}]\left(\frac{110 + 273}{20 + 293}\right)^{0.17} = 0.738$$

The heat transfer coefficient is then given by

$$h = \frac{Nuk}{D} = \frac{0.738 \times 0.032}{4 \times 10^{-6}} = 5906/;W/m^{2°}C$$

Heat dissipation from the wire when exposed to an stream at $4\,m/s$ is then given by

$$Q = \pi DLh(t - 20) = \pi \times (4 \times 10^{-6} \times 0.0012 \times 5906) \times (200 - 20) = 0.016\,W$$

This must equal $\dfrac{E^2}{R_w}$ and hence the output of the sensor is

$$E = \sqrt{QR_w} = \sqrt{0.016 \times 17.1} = 0.524\,V$$

We make a plot of sensor output as a function of the air velocity in the velocity range from 0.25 to 10 m/s. The plot (left Figure 8.17) indicates that the response is indeed non-linear. However we linearize the output as indicated earlier by making a log-log plot as shown by the right Figure 8.17. We have shown the response at air velocity of 4 m/s corresponding to the case considered in Example 8.5 in the figures.

8.3.4   Constant Current or CC anemometer

In this mode of operation the current through the sensor wire is kept constant by an arrangement schematically shown in Figure 8.17. As the sensor resistance changes the series resistance is adjusted such that the total resistance and hence the current remains fixed. The potential drop across the sensor thus changes in response to a change in its resistance. This voltage is amplified, if necessary, before being recorded.

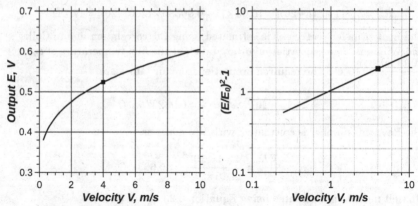

Figure 8.17: *Actual and linearized response of hot wire sensor of Example 8.5 in CT mode*

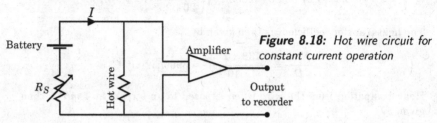

Figure 8.18: *Hot wire circuit for constant current operation*

Referring to Figure 8.17 we see that the output of the circuit E is proportional to IR_w. Since I is constant in CC mode of operation, R_w may be replaced by E, the voltage in the presence of fluid flowing at velocity V normal to the hot wire. Similarly we may replace R_∞ by E_∞. We also represent the output of the sensor when the velocity of the fluid is zero as E_0. With these, Equation 8.18 takes the form

$$A + BV^n = \frac{\alpha E E_\infty}{I(E - E_\infty)} \tag{8.24}$$

We also have

$$A = \frac{\alpha E_0 E_\infty}{I(E_0 - E_\infty)} \tag{8.25}$$

Equation 8.25 helps in obtaining E_∞ as

$$E_\infty = \frac{AE_0}{A + \frac{\alpha E_0}{I}}$$

Introduce this in Equation 8.24 to get

$$A + BV^n = \frac{\alpha}{I} \frac{E\left[\frac{AE_0}{A + \frac{\alpha}{I}E_0}\right]}{\left[E - \left(\frac{AE_0}{A + \frac{\alpha}{I}E_0}\right)\right]} = \frac{\alpha}{I} \frac{AEE_0}{E\left[A + \frac{\alpha}{I}E_0\right] - AE_0} \tag{8.26}$$

The denominator on the right hand side of Equation 8.26 may be rewritten as

$$E\left[A + \frac{\alpha}{I}E_0\right] - AE_0 = (E - E_0)\left(A + \frac{\alpha}{I}E_0\right) + \frac{\alpha}{I}E_0^2$$

Then Equation 8.26 may be recast as

$$K_1(E - E_0) + K_2 = K_3 \frac{(E - E_0)}{A + BV^n} + \frac{K_4}{A + BV^n} \qquad (8.27)$$

where the constants appearing are given by

$$K_1 = A + \frac{\alpha}{I} E_0, \, K_2 = \frac{\alpha}{I} E_0^2, \, K_3 = \frac{AE_0 \alpha}{I}, \, K_4 = \frac{AE_0^2 \alpha}{I}$$

Equation 8.27 may finally be rearranged as

$$E_0 - E = \frac{a_1}{1 + a_2 V^n} \qquad (8.28)$$

where

$$a_1 = \frac{K_2 B}{K_1 A - K_3}, \, a_2 = \frac{K_1 B}{K_1 A - K_3}$$

The output response thus involves two constants a_1, a_2 that may be determined by suitable calibration. Response of a hot wire anemometer operating in the CC mode is typically like that shown in Figure 8.19.

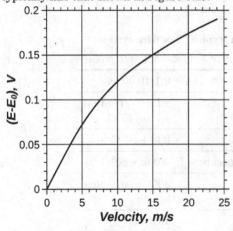

Figure 8.19: Output of a typical micro machined hot wire probe in the CC mode. Length of wire $L = 100 \, \mu m$, Resistance at room temperature = 2000 Ω.

8.3.5 Practical aspects

Hot wire probes are very delicate and need to be handled very carefully. Rugged probes are made using a film of Platinum deposited on a substrate rather than using a wire. Two such designs are shown in Figures 8.20 and 8.21 below.

The probe shown in Figure 8.20 is cylindrical while the probe shown in Figure 8.21 is in the shape of a wedge. A thin film of Platinum is used as the sensor. It is usually protected by a layer of alumina or quartz. The substrate provides ruggedness to the film while the protective coating provides abrasive resistance. The active length of the sensor is between the gold plated portions. The leads are gold wires (Figure 8.20) or gold film (Figure 8.21). In the case of wire type probes it is advisable not to expose the wire to gusts of fluid. The sensor must be handled gently without subjecting it to excessive vibration.

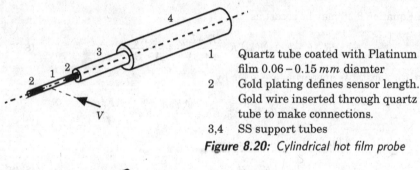

1   Quartz tube coated with Platinum
     film 0.06 – 0.15 mm diamter
2   Gold plating defines sensor length.
     Gold wire inserted through quartz
     tube to make connections.
3,4 SS support tubes

Figure 8.20: Cylindrical hot film probe

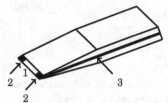

1   Alumina or quartz substrate coated with
     platinum film $0.1 \times 1\, mm$ on each face of bevel
2   Gold plating for connection
3   Gold film lead

Figure 8.21: Hot film wedge probe

Table 8.1 shows the specifications of a "typical" hot wire system.

Table 8.1: Specifications of a typical hot wire probe

Flow Range	0.1-10.00 m/s
Accuracy	±1% reading +5 to +45°C
Operating temperature	0 to + 50°C
Probe tip operating temperature	-20 to + 80°C
Probe length	300 mm plus handle
Probe Diameter	13 mm
Dimensions	$140 \times 79 \times 46\, mm$
Weight	250 g

8.3.6   Measurement of transients (velocity fluctuations)

Hot wire anemometer is routinely used for the measurement of velocity fluctuations that occur, for example, in turbulent flows. The time constant of the sensor plays an important role in determining the frequency response of the hot wire. Take for example the hot wire that was considered in Example 8.5. Suppose the mean velocity at a certain location in the flow is 4 m/s. The fluctuating component will be superposed on this mean velocity. Let us assume that the fluctuation is small compared to the mean. The time constant *may* then be based on the conditions that prevail at the mean speed.

The heat transfer coefficient was obtained as $h = 5906 \, W/m^{2\circ}C$. Thermal conductivity of Platinum is $k = 69 \, W/m^{\circ}C$. With the diameter of the hot wire $D = 4 \, \mu m$ as the characteristic dimension, the Biot number is $Bi = \dfrac{hD}{k} = \dfrac{5900 \times 4 \times 10^{-6}}{69} = 0.00028$. The hot wire element may be treated as a first order lumped system since $Bi \ll 1$. The time constant may now be estimated. The density and specific heat of Platinum are taken as $\rho = 21380 \, kg/m^3$ and $C = 134 \, J/kg^{\circ}C$. The volume to surface area ratio for the wire is $\dfrac{V}{S} = \dfrac{\pi D^2 L/4}{\pi D L} = D/4 = 10^{-6} \, m$. The first order time constant is then obtained as

$$\tau = \frac{\rho C}{h} \frac{V}{S} = \frac{21380 \times 134}{5906} \times 10^{-6} = 0.0005 \, s \tag{8.29}$$

The corresponding bandwidth may be calculated as

$$f = \frac{1}{2\pi\tau} = \frac{1}{2 \times \pi \times 0.0005} = 328.1 \, Hz \tag{8.30}$$

This may not be good enough for measurements of transients that occur in turbulent flows. Literature reports hot wire probes with time constant as low as $2 \, \mu s$.[7] Hot wire anemometer system AN-1003 supplied by A A Lab Systems USA has a frequency response in excess of around $10 \, kHz$.[8]

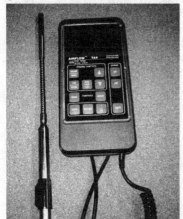

Figure 8.22: *A thermal anemometer probe with display unit*

A typical thermal anemometer probe useful for measuring steady air velocity is shown in Figure 8.22. The sensor is connected to a display unit that is battery operated. The particular model shown is useful for measuring air speed up to $5 \, m/s$.

8.3.7 Directional effects on hot wire anemometer

In the above we have assumed that the flow velocity direction was known and the hot wire probe axis was perpendicular to this direction. In practice we may not

---

[7] see for example F. Jiang, Y. C. Tai, C. H. Ho, and W. J. Li. A Micro-machined Polysilicon Hot-Wire Anemometer. Solid-State Sensor and Actuator Workshop, Hilton Head, SC, pp. 264-267, 1994.

[8] http://www.lab-systems.com/products/flow-mea/an1003/techspec.htm

know the flow direction and hence it would be necessary to look at the effect of flow direction with respect to the hot wire axis. Consider two dimensional flow where the flow is taking place in the $x - y$ plane and the flowing fluid has U and V as the components of the velocity along the x and y directions, respectively. Assume that a single hot wire is positioned such that the wire lies on the $x - y$ plane but makes an angle α with the y axis, as shown in Figure 8.23. The two prongs are of different lengths and are so as to orient the wire at the desired angle. The flow velocity is

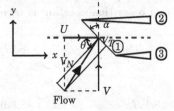

1:   Hot wire
2:   Long prong
3:   Short prong

Figure 8.23: *Velocity components for flow past a hot wire placed at an angle to flow*

along the direction indicated by the large arrowhead. We may resolve the resultant velocity in to a normal component U_N and a tangential component U_T as indicated in the figure. These two component are easily seen to be given by

$$V_N = U \cos \alpha + V \sin \alpha; \quad V_T = V \cos \alpha - U \sin \alpha \qquad (8.31)$$

The normal component of velocity is effective in cooling the wire and hence is also represented as the effective velocity. Heat transfer from the wire due to the tangential component is much smaller than that due to the normal component. Hence the output of the hot wire corresponds to the effective velocity $U_{eff} = U_T = U \cos \alpha + V \sin \alpha = \dfrac{U + V}{\sqrt{2}}$ if $\alpha = 45°$.

Consider now a crossed hot wire arrangement shown in Figure 8.24. Both hot wires and the velocity components are in a single plane. Of curse the two wires are separated by a very small distance such that they do not touch each other.

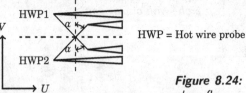

HWP = Hot wire probe

Figure 8.24: *Crossed hot wire probe in plane flow*

Assuming that the angle $\alpha = 45°$ the output of the two hot wires are related to the respective effective velocities given by

$$U_{eff,1} = \frac{U + V}{\sqrt{2}}, \quad U_{eff,2} = \frac{U - V}{\sqrt{2}} \qquad (8.32)$$

We see that the velocity U is proportional to $E_1 + E_2$ while the velocity component V is proportional to $E_1 - E_2$ where the E's represent the hot wire output voltages. We will thus be able to obtain the two velocity components and hence obtain the direction as $\theta = \tan^{-1} \left(\dfrac{V}{U} \right)$.

In three dimensional flows one needs to use a three wire system so that all three velocity components are estimated. For details the reader may refer to specialized texts on hot wire anemometry.

8.4 Doppler Velocimeter

8.4.1 The Doppler effect

As mentioned earlier this is basically a non intrusive method of velocity measurement. This method of velocity measurement requires the presence of scattering particles in the flow. Most of the time these are naturally present or they may be introduced by some means. The Doppler effect is the basis for the measurement. Doppler effect may be used with waves of different types - electromagnetic waves: visible, IR, micro wave - and ultrasonic waves. We explain the Doppler effect by referring to Figure 8.25.

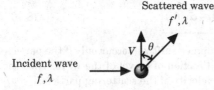

Figure 8.25: *Scattering of radiation by a moving particle*

Let a particle move with a velocity V in a medium in which the wave propagates at speed c. The incident wave is characterized by wavelength λ and frequency f and is incident along a direction that is normal to the velocity V of the scattering particle. We make observation of the wave scattered along a direction that makes an angle θ with respect to the direction of V, as shown. Incident wave wavelength and frequency are related through the well know relation

$$c = f\lambda \text{ or } f = \frac{c}{\lambda} \qquad (8.33)$$

It has been observed that the frequency of the scattered wave is *different* from the frequency of the incident wave because of the motion of the scattering particle. The incident wave travels at the wave speed c while the scattered wave travels at an enhanced speed due to the component of the particle velocity that is added to it. The wavelength does not change. The frequency of the wave alone undergoes a change. Hence, we have, for the scattered wave

$$f' = \frac{c + V\cos\theta}{\lambda} \qquad (8.34)$$

The difference between the two frequencies is referred to as the Doppler shift. It is given by

$$f_D = f' - f = \frac{c + V\cos\theta}{\lambda} - \frac{c}{\lambda} = \frac{V\cos\theta}{\lambda} \qquad (8.35)$$

The general case where the wave is incident along a direction that makes an angle θ_i and is scattered along a direction that makes an angle θ_s with the direction of the particle velocity is shown in Figure 8.26. Note that the incident and scattered waves need not be in the same plane even though Figures 8.25 and 8.26 have been drawn on this basis.

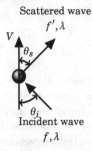

Figure 8.26: *General case of scattering of an incident wave by a moving particle*

The Doppler shift in this case is easily seen to be

$$f_D = \frac{V}{\lambda}(\cos\theta_s - \cos\theta_i) \tag{8.36}$$

It is clear from these expressions that the Doppler shift will occur only if the particle has *non-zero velocity* component along the direction of travel of the wave. Also the Doppler shift is directly proportional to the velocity of the scattering particle.

We take an example of light wave of $\lambda = 0.68\ \mu m$. The speed of light in air may be taken as that in vacuum as $c = 3 \times 10^8\ m/s$. The light wave being considered is actually laser light from a Helium-Neon laser. The frequency of the light wave is

$$f = \frac{c}{\lambda} = \frac{3 \times 10^8}{0.68 \times 10^{-6}} = 4.412 \times 10^{14}\ Hz$$

Consider the scattering of this wave by a particle traveling with a velocity of $V = 10\ m/s$ as shown in Figure 8.25. The Doppler shift is given by

$$f_D = \frac{V\cos\theta}{\lambda} = \frac{10\cos\theta}{0.68 \times 10^{-6}} = 14.706\cos\theta\ MHz$$

The Doppler shift follows a cosine distribution. Doppler shift of $14.7\ MHz$ occurs respectively for $\theta = 0$ (forward scatter) and $\theta = \pi$ (backward scatter). The Doppler shift is zero when $\theta = \pi/2$.

8.4.2   Ultrasonic Doppler velocity meter

The first device we shall consider is one that uses ultrasound (sound waves above $20\ kHz$ and typically $100\ kHz$) as the beam that undergoes Doppler shift due to a moving particle. These are in very common use in medicine, specifically for blood flow measurements in patients. The ultrasound waves are produced by a piezoelectric crystal driven by an ac source at the desired frequency. This transducer may be used as a transmitter as well as a receiver. The construction details of a transducer are shown in Figure 8.27. The transducer consists of a small cylindrical piece of a piezoelectric crystal that is mounted in a casing with a backing of absorbing material. The crystal itself is backed by tungsten loaded resin as shown in the figure. When sound waves are incident on the crystal the pressure variations generate voltage variations across the crystal (receiver mode). These are communicated through lead wires. In case ac power is fed through the leads the crystal oscillates and produces ultrasound waves (transmitter mode).

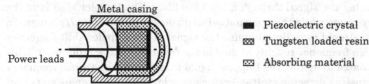

Figure 8.27: *Sectional view of a piezoelectric transducer*

Now we consider the arrangement shown in Figure 8.28. The illustration is that of measurement of velocity of a fluid in a pipe using the Doppler effect. Two transducers are fixed on the walls as indicated. One of the transducers is a transmitter of ultrasonic waves and the second acts as a receiver for the scattered ultrasound wave. The illustration shows the transmitter and receiver to be in contact with the flow. Clamp on transducers are also possible wherein the transducers are mounted outside but in contact with the pipe wall. The latter arrangement is preferred with corrosive fluids and when it is not desirable to make holes in the pipe. In this case the ultrasound passes through the pipe wall and undergoes a change in direction that will have to be taken into account.

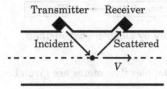

Figure 8.28: *Measurement of fluid velocity by Doppler shift method*

We assume that the particle or some naturally occurring disturbance like air bubbles in the flow, which is moving at the *same* speed as that of the fluid, scatters the incident wave. Figure 8.29 shows the schematic of the electronic circuit that is used for signal conditioning. An oscillator drives the transmitter at the desired or chosen ultrasonic frequency. Let the transmitted ultrasound wave be given by $u_t = u_{t0}\cos(\omega t)$. The scattered wave then is given by $u_r = u_{r0}\cos[(\omega + \omega_D t) + \phi]$. Here u represents the signal in appropriate unit, subscripts t and r represent respectively the transmitted and received waves, ω represents the frequency of the source and ω_D is the Doppler shift and ϕ represents the phase shift. With reference to the circuit shown in Figure 8.29 the u's may be taken in electrical units such as V or mV. ω is the circular frequency that is related to the frequency f by the relation $\omega = 2\pi f$. If we multiply the transmitted and reflected signals together, we get

$$s = u_t u_r = u_{t0} u_{r0}\cos(\omega t)\cos[(\omega + \omega_D)t + \phi] \tag{8.37}$$

Using trigonometric identities we may recast Equation 8.37 as

$$s = \frac{1}{2} u_{t0} u_{r0}[\cos\{(2\omega + \omega_D)t + \phi\} + \cos\{\omega_d t + \phi\}] \tag{8.38}$$

The signal s thus contains a high frequency component at roughly twice the input frequency (usually the Doppler shift is much smaller than the transmitted frequency) and a low frequency component at the Doppler shift frequency. Our interest lies with the latter and hence the former is removed by extracting the latter by demodulation followed by passing the signal through a low pass filter. The Doppler shift is in the audio frequency range while the transmitted frequency is in the $100\,kHz$ range. In summary, the scattered radiation contains the signal at the Doppler shift frequency riding over a high frequency roughly at double the frequency of the input wave. This wave is demodulated to extract the Doppler signal that varies in the audio frequency range. A zero crossing detector, multi-vibrator and a low pass filter yields an output that is proportional to flow. Signal waveforms at various positions in the circuit are also shown in the figure.

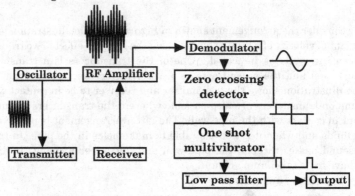

Figure 8.29: *Measurement of fluid velocity by Doppler shift method*

Typical performance figures for an ultrasonic Doppler flow meter are given below:

- Over the flow range 0 to 15 m/s a repeatability of 1% of full scale
- For small pipes with well mixed slurries linearity of 2% for $Re > 10^5$
- Measurements with clamp on meters is possible up to $120°C$

Example 8.6

Consider an ultrasound system with a transmitted frequency of $f = 100\,kHz$. Let the typical velocity of the fluid be $V = 1\,m/s$. Assume that the incident and transmitted waves are almost collinear and the particle is moving in the direction of the incident wave. Discuss the output.

Solution :

Let the speed of the wave in the medium be $a = 350\,m/s$ (speed of sound in air). The wavelength of the incident wave is then given by

$$\lambda = \frac{c}{f} = \frac{350}{100 \times 10^3} = 0.0035\,m$$

In the arrangement specified in the problem the incident angle $\theta_i = 0$ and scattering angle is $\theta_s = \pi$. The Doppler shift is then given (using Equation 8.36 as

$$f_D = -\frac{2V}{\lambda} = -\frac{2 \times 1}{0.0035} = -571.4\,Hz$$

The signal thus consists of approximately 571 Hz low frequency component riding over a wave at 200 kHz. Clearly the Doppler shifted signal is in the audio range and may in fact be heard as a 'hum' if the signal is connected to a speaker via an amplifier.

8.4.3 Laser Doppler velocity meter

In fluid mechanics research the Laser Doppler Velocimeter (LDV) is used for detailed mapping of fluid flow fields. As the name implies the radiation that is used is a highly coherent laser source. The advantage of the laser based measurement is that the beam may be focused to a very small volume thus measuring almost truly the velocity at a point. Since the focused beam is intense enough scattered energy is available to make measurements with adequate signal to noise ratio. Also high sensitivity photomultiplier tubes (PMT) are available that operate with very feeble scattered light. Two possible ways of operating an LDV viz.

1. the fringe based system
2. reference beam method

are described below.

Fringe system

Schematic of a fringe based system is shown in Figure 8.30. A laser source provides highly coherent light of fixed wavelength of very narrow line width. Two beams traveling parallel to each other are obtained by passing it through a partial mirror (beam splitter in figure) and then reflecting one of the beams off a mirror. These two beams are focused at a point within the flow medium as shown. A particle or some disturbance passing through the region close to the focal point will scatter energy as shown. At the focal point the two beams form an interference pattern that is in the form of a fringe system shown as an inset in the figure. When a particle in the medium passes through this region the scattered radiation intensity is modulated by the fringes. If the particle moves at a speed V across the fringe system, and if the fringe spacing is d, the frequency of the "burst" signal is given by

$$f = \frac{V}{d}$$

It can be shown that the fringe spacing is given by

$$d = \frac{\lambda}{2\sin\left(\frac{\theta}{2}\right)}$$

where λ is the wavelength of the laser radiation and θ is the angle between the two beams (see Figure 8.30). From these two expressions, we get

$$V = \frac{f\lambda}{2\sin\left(\frac{\theta}{2}\right)} \qquad (8.39)$$

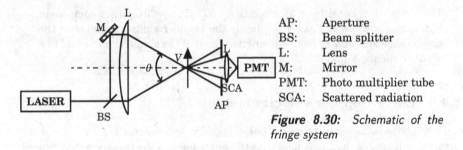

AP:　　Aperture
BS:　　Beam splitter
L:　　Lens
M:　　Mirror
PMT:　Photo multiplier tube
SCA:　Scattered radiation

Figure 8.30: *Schematic of the fringe system*

The aperture helps to discard the main laser beam and collect only the scattered burst of light to be sensed by the photomultiplier tube (PMT). The burst signal (typically like that shown in Figure 8.31 has to be analyzed by suitable signal conditioning system to get the burst frequency and hence the velocity of the particle. The particle (may be smoke particles introduced in to the flow) should have very small size (and hence the mass) so that it moves with the same velocity as the fluid, without any slip. We consider again light wave of $\lambda = 0.68 \, \mu m$. The speed of light in

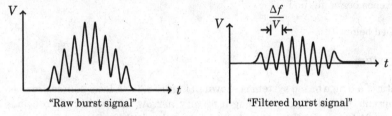

Figure 8.31: *Scattered "burst" signal before and after filtering*

air may be taken as $c = 3 \times 10^8 \, m/s$. Let the two beams cross at an angle of $10°$. The fringe spacing is then given by

$$d = \frac{0.68}{2 \times \sin 5°} = 3.9 \, \mu m$$

Note that the particle should be much smaller than the fringe spacing so that the burst signal is not 'smeared'. Consider a scattering particle traveling with a velocity of $10 \, m/s$. The burst signal has a frequency given by

$$f = \frac{10}{3.9 \times 10^{-6}} = 2.563 \, MHz$$

Reference beam system

The schematic of the reference beam system is shown in Figure 8.32. Here again we cross a reference beam (weakened by a small reflectance of the mirror) and a scattered beam at the focal point of a lens with an aperture. The reference beam is directly collected by the lens and communicated to the PMT. Radiation scattered by the scattered beam within the field of view of the lens is also communicated to the PMT. The PMT responds to the incident intensity which consists of a part that varies at the Doppler shift frequency. This rides over a dc part that is related to the mean intensity of the two beams. Again suitable electronic circuit is required to extract the Doppler shift and hence the particle velocity. Table 8.2 gives the specifications of a typical laser Doppler system.

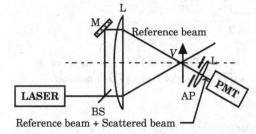

AP:	Aperture
BS:	Beam splitter
L:	Lens
M:	Mirror
PMT:	Photo multiplier tube

Figure 8.32: *Schematic of the reference beam system*

Table 8.2: *Specifications for Canon LV-20Z LDV*

Measurement Range	-200 to 2000 mm/s
Focal Length	40 mm
Depth of focus	±5 mm
Laser spot size	2.4 × 0.1mm (at focal point)
Velocity fluctuation response frequency	0 to 300 Hz
Output Signal Accuracy	less than ±1% of full scale
Doppler pulse output	120 to 1000 kHz
Measurement Certainty	$V < 100 \, mm/s$: ±0.2 mm/s $V > 100 \, mm/s$: ±0.2%
Optical shift frequency output	200 kHz CMOS level
Velocity display	5 digit (mm/s, m/min selectable)
Light source	Semiconductor laser (680 nm)

8.5 Time of Flight Velocimeter

Time of flight refers to the time it takes an acoustic beam to travel through a certain distance in a moving medium. Since the motion of the medium affects the time of flight it is possible to use this as a means of measuring the velocity of the medium. This method does not need the presence of any scattering particles in the flow as in the case of the Doppler shift method. Consider the scheme shown in Figure 8.33. An ultrasound wave pulse (of a short duration) is transmitted from 1 and travels along the fluid and reaches the receiver at 2. Along this path (referred to as the forward path) the sound wave travels at a speed of $a_f = a - V$. The corresponding time of flight is

$$T_{1-2} = \frac{L}{a - V}$$

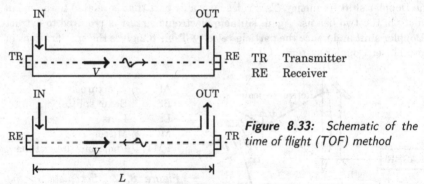

Figure 8.33: *Schematic of the time of flight (TOF) method*

TR	Transmitter
RE	Receiver

However if the sound wave traverses along path 2-1 (return path) the speed of wave is $a_r = a + V$. Correspondingly the time of flight is given by

$$T_{2-1} = \frac{L}{a+V}$$

The difference in the times of flight is given by

$$\Delta T = T_{1-2} - T_{2-1} = \frac{L}{a-V} - \frac{L}{a+V} = \frac{2LV}{a^2-V^2} \tag{8.40}$$

Also the product of the times of flight is given by

$$T_{1-2}T_{2-1} = \frac{L}{a-V} \times \frac{L}{a+V} = \frac{L^2}{a^2-V^2} \tag{8.41}$$

Division of Equation 8.41 by Equation 8.40 then yields the interesting result

$$\frac{T_{1-2}T_{2-1}}{\Delta T} = \frac{L}{2V} \tag{8.42}$$

We notice that the speed of the wave has dropped off and the velocity is given by the ratio of path length and time. Thus we rearrange Equation 8.42 as

$$V = \frac{L}{\frac{T_{1-2}T_{2-1}}{\Delta T}} \tag{8.43}$$

Scheme shown in Figure 8.33 is possible when one is allowed to make a bypass arrangement so that a long straight segment may be arranged as shown. This way it is possible to choose a large L so that accurate measurement of the times of flight and hence of velocity is possible. However, more often, we have the flow taking place in a pipe and the arrangement shown in Figure 8.34 is a feasible arrangement. In this case, as discussed earlier, it is not necessary to make any changes in the piping for making the TOF measurement.

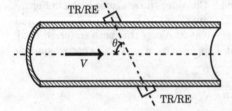

TR:	Transmitter
RE:	Receiver

Figure 8.34: *Clamp on type ultrasonic mean velocity meter*

A variant of the above is when one is allowed to expose the ultrasound transducer to the flowing medium (transducers are said to be wetted). The schematic shown in Figure 8.35 also includes the schematic of the electronic circuit that is needed for making the measurement. In this case the velocity of the fluid has only a component in the direction 1-2 and hence the equations given earlier are to be modified as under. Let the diameter of the pipe be D. The path length is then given by $L = \dfrac{D}{\sin\theta}$.

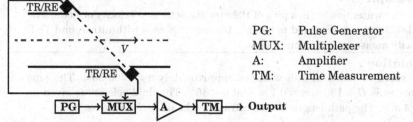

PG: Pulse Generator
MUX: Multiplexer
A: Amplifier
TM: Time Measurement

Figure 8.35: *Wetted type ultrasonic mean velocity meter*

Along the forward path the sound wave travels at a speed of $a_f = a - V\cos\theta$. The corresponding time of flight is

$$T_{1-2} = \frac{\frac{D}{\sin\theta}}{a - V\cos\theta}$$

However if the sound wave traverses along path 2-1 (return path) the speed of wave is $a_r = a + V\cos\theta$ and the time of flight is given by

$$T_{1-2} = \frac{\frac{D}{\sin\theta}}{a + V\cos\theta}$$

The difference between the times of flight is then given by

$$\Delta T = T_{1-2} - T_{2-1} = \frac{\frac{D}{\sin\theta}}{a - V\cos\theta} - \frac{\frac{D}{\sin\theta}}{a + V\cos\theta} = \frac{2DV\cot\theta}{a^2 - V^2\cos^2\theta} \tag{8.44}$$

If we assume that $V \ll a$ (which is almost always the case), the second term in the denominator may be dropped to get

$$\Delta T \approx \frac{2DV\cot\theta}{a^2} \tag{8.45}$$

Thus the difference in times of flight is directly proportional to the velocity. If the angle θ is arranged to be 45°, $\cot\theta = 1$ and the velocity is given by

$$V \approx \frac{a^2\Delta T}{2D} \tag{8.46}$$

Again, we may also take the ratio of the product of times of flight to ΔT to get

$$\frac{T_{1-2}T_{2-1}}{T_{1-2} - T_{2-1}} = \frac{D}{2V\sin\theta\cos\theta} = \frac{D}{V\sin(2\theta)} \tag{8.47}$$

Again, if the angle θ is arranged to be 45°, the above reduces to

$$\frac{T_{1-2}T_{2-1}}{T_{1-2} - T_{2-1}} = \frac{D}{V} \tag{8.48}$$

What is interesting again is that the wave speed has dropped off. In addition the velocity of the fluid is given by the ratio of the diameter of the pipe to the effective time that is measured by the ratio. The velocity measured is the mean velocity across the path traveled by the ultrasound wave in case the fluid velocity varies across the pipe.

Example 8.7

Consider water flowing in a pipe of 100 mm diameter at a velocity of 1 m/s. The angle θ has been arranged to be 45°. Discuss this case with ultrasound TOF velocity measurement point of view.

Solution :

The speed of sound in still water is approximately $a = 1500$ m/s. The pipe diameter is $D = 100$ $mm = 0.1$ m and $\theta = 45°$. The fluid velocity is given as $V = 1$ m/s. The path length for the ultrasound beam is obtained as

$$L = \frac{D}{\sin\theta} = \frac{0.1}{\sin 45°} = 0.141\ m$$

The forward and return transit times are calculated as

$$T_{1-2} = \frac{0.141}{1500 - 1 \times \cos 45°} = 94.325\ \mu s$$

$$T_{2-1} = \frac{0.141}{1500 + 1 \times \cos 45°} = 94.236\ \mu s$$

Thus the time difference is only about

$$\Delta T = T_{1-2} - T_{2-1} = 94.325 - 94.236 = 0.089/mus = 89\ ns$$

Measurement of such small time differences is quite a challenge! This simply means that the signal analysis in the ultrasound TOF method will have to be of high sensitivity and quality making it very expensive! From Equation 8.48, we however, have

$$\frac{T_{1-2}T_{2-1}}{T_{1-2} - T_{2-1}} = \frac{D}{V} = \frac{0.1}{1} = 0.1\ s!$$

We see from Example 8.7 that the transit times are very small and some way of increasing these are desirable. One of these is to use the arrangement shown earlier in Figure 8.33. Another arrangement is to provide for multiple reflections as shown in Figure 8.36. The effective path is increased several fold by this arrangement.

Transmitter   Receiver

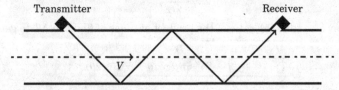

Figure 8.36: Method of increasing path length in TOF measurement

8.5.1 Simultaneous measurement of position and velocity

We can combine TOF and Doppler shift measurements to study the velocity and the position of a scattering particle in a flowing medium. This is referred as pulsed echo technique and is commonly used in medical blood flow measurement applications. The principle of this may be understood by referring to Figure 8.37.

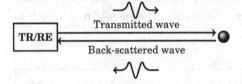

Transmitted wave

TR/RE

Back-scattered wave

Figure 8.37: *Simultaneous position and velocity measurement*

Let us assume that a scatterer is present and is moving as shown in Figure 8.37. A short pulse of ultrasound is transmitted at $t = 0$ and the received pulse is analyzed for both the transit time and the Doppler shift. If the velocity of the scatterer V is small compared to the wave speed a, the transit time τ is related to the position by the simple relation $\tau = \dfrac{2X_s}{a}$. At the same time if the Doppler shift is measured it is directly related to the velocity of the scatterer V, as has been shown earlier. Thus both the position X_s and velocity V may be measured simultaneously. An interesting variant of this is to gate the measurement such that a reflected pulse received after a time delay only is sampled. This will correspond to a certain location of the scatterer. The corresponding Doppler shift will provide the velocity at the location specified by the gated time.

8.5.2 Cross correlation type velocity meter

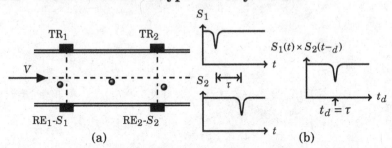

Figure 8.38: *(a) Cross correlation type velocity meter (b) Cross correlation time measurement*

Another method of measurement which converts velocity measurement to a time measurement is the cross correlation type of measurement. The schematic of the arrangement is shown in Figure 8.38(a). Two transmitters and receivers are arranged a distance L apart as shown in the figure. A disturbance that is carried by the fluid at its own speed crosses the first transmitter-receiver pair at some time that may be taken as zero. After a certain delay the same disturbance will pass the second transmitter-receiver pair. This is schematically shown in Figure 8.38(b).

To ascertain the time accurately the signal received by first receiver is multiplied by the signal from the second receiver with a variable time delay τ_d. The product will be substantially zero till the delay time τ_d is equal to the transit time τ. The product signal will show a non-zero blip when the delay time is the same as the transit time of the disturbance. The two signals, under this condition, are said to be highly correlated. The fluid velocity is then given by $V = \dfrac{L}{\tau}$. Thus the velocity measurement has been reduced to its very definition - the ratio of path length to transit time!

Concluding remarks

Measurement of velocity by several techniques has been presented in this chapter. Both intrusive methods - using probes such as the Pitot static, multiple hole probes, hot wire or hot film probe - and non-intrusive methods - such as ultrasonic or laser Doppler, have been discussed in detail. While a knowledge of magnitude of velocity may be sufficient in many applications direction may also be important in some applications. Determination of the direction of velocity needs multiple probes or multiple hole probes. These have also been discussed in this chapter.

Chapter 9

Volume flow rate

This chapter deals with the measurement of volume or mass flow rate of fluids in process applications. Fluid may be incompressible as in the case of most liquids and gases at low velocities or compressible such as gases flowing at high speeds. Flow meters of several types are discussed in this chapter. They may be of the variable area type, drag effect type or positive displacement type. Calibration of flow meters is discussed in some detail since it forms an integral part of flow measurement activity.

9.1   Measurement of volume flow rate

Process applications invariably involve the measurement of volume or mass flow rate of a fluid - gas (such as air) or liquid (such as water). Volume flow rate is expressed in units of m^3/s, m^3/h, l/min, l/s or ml/s. The magnitude of the flow rate will decide the unit that is appropriate. It is usual to represent very small volume flow rates in ml/s and very large volume flow rates in m^3/h. Mass flow rate is represented in kg/s in SI units. However, if the mass flow rate is very small we may represent it in g/s, mg/s or $\mu g/s$. Very large mass flow rates are usually represented in t/h where $1\,t = 1\,tonne = 1000\,kg$.

In a flow system the volume flow rate represents the volume of the process fluid that passes a station in a unit of time. In a very simple way we see that two measurements are needed: the volume that crosses the station and the time it takes this volume to cross the station. It is seldom that we can measure these two quantities and then obtain the volume flow rate as a ratio of the two. If the volume flow is relatively slowly changing it may be possible to do the measurement this way. However, when the volume flow rate changes with time we have to use other methods.

These methods are based on measuring some secondary quantity that varies systematically with the volume flow rate. In this sense the methods are akin to those we used in thermometry for estimating the temperature of a system. It was seen in Module II, Chapter2 that temperature is seldom directly measurable! In the case of volume flow rate the most common approach is to look at the pressure drop due to the flow and relate these two in as *unique* a way as possible. The pressure drop is, in fact, the measured quantity while the volume flow rate is the estimated quantity. The measurement process is complicated by issues such as the temperature, the static pressure, friction and so on.

Some of the methods of measurement of volume/mass flow rate available are:

- Variable area type flow meters
- Drag effect type flow meter
- Other types of flow meters such as:
  1. Positive displacement type flow meters
  2. Vortex shedding type flow meter
  3. Turbine type flow meter

We discuss each one of these in detail in this section.

9.2   Variable area type flow meters

9.2.1   Principle of operation

All variable area flow meters introduce a change in the area available for flow and hence bring about a *change* in the pressure of the flowing fluid. This pressure change or pressure difference is measured by a suitable differential pressure transducer to infer the volume or mass flow rate of the fluid.

To understand the principle of operation of variable area flow meters we look at one dimensional flow of an incompressible medium (density of the medium ρ is constant) in a duct of non-uniform area as shown in Figure 9.1. The principle is similar to that of the Pitot or Pitot static tube where the velocity information was converted to a pressure change (see Module III, Chapter5).

We shall ignore the effect of viscosity, in this analysis.

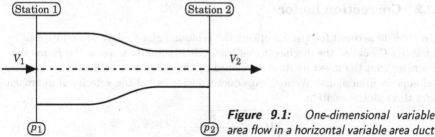

Figure 9.1: *One-dimensional variable area flow in a horizontal variable area duct*

Let the flow areas be A_1 and A_2 respectively at stations 1 and 2. The fluid velocity is uniform across the flow area in one-dimensional flow. The flow velocity changes from V_1 at station 1 to V_2 at station 2 because of the area change. In order to conserve mass (conserve volume in the case of an incompressible fluid) we should have the same volume flow rate across each of the two stations. Thus

$$A_1 V_1 = A_2 V_2 \tag{9.1}$$

According to Bernoulli's principle the pressure p and the velocity V are related through the relation

$$\frac{p_1}{\rho} + \frac{V_1^2}{2} = \frac{p_2}{\rho} + \frac{V_2^2}{2} \tag{9.2}$$

From Equation 9.1 we have $V_2 = \dfrac{A_1}{A_2} V_1$. Substitute this in Equation 9.2 and rearrange to get

$$(p_1 - p_2) = \frac{\rho}{2}(V_2^2 - V_1^2) = \frac{1}{2}\rho V_1^2 \left[\left(\frac{A_1}{A_2}\right)^2 - 1 \right] \tag{9.3}$$

We may solve for the velocity at station 1 as

$$V_1 = \frac{1}{\sqrt{\left(\frac{A_1}{A_2}\right)^2 - 1}} \sqrt{\frac{2(p_1 - p_2)}{\rho}} \tag{9.4}$$

Thus we see that the velocity of the fluid at station 1 is related to the pressure drop between the two stations and the area ratio. We may multiply the above by A_1 to get the volume flow rate Q and further by the density ρ to get the mass flow rate $\dot{m}$. Thus we have:

$$Q = \frac{A_1}{\sqrt{\left(\frac{A_1}{A_2}\right)^2 - 1}} \sqrt{\frac{2(p_1 - p_2)}{\rho}} = \frac{A_2}{\sqrt{1 - \left(\frac{A_2}{A_1}\right)^2}} \sqrt{\frac{2(p_1 - p_2)}{\rho}} \tag{9.5}$$

and

$$\dot{m} = \rho Q = \frac{A_2}{\sqrt{1 - \left(\frac{A_2}{A_1}\right)^2}}\sqrt{2\rho(p_1 - p_2)} \tag{9.6}$$

9.2.2   Correction factor

In order to account for viscous effects the relations given above are multiplied by a quantity C_d called the discharge coefficient. It is actually a correction factor that recognizes that the pressure drop developed by the variable area flow is not all due to change in area alone. We also introduce a factor called the velocity of approach factor through the relation

$$M = \frac{1}{\sqrt{1 - \left(C_c \frac{A_2}{A_1}\right)^2}} \tag{9.7}$$

where C_c is a factor equal to the ratio of flow area at vena contracta to the orifice area, in the case of orifice plate with the downstream tap at the vena contracta. In the case of flow nozzle as well as the venturi (see discussion below) this factor is equal to 1. With these, Equation 9.5 is recast as

$$Q = C_d A_2 M \sqrt{\frac{2(p_1 - p_2)}{\rho}} \tag{9.8}$$

Often the product of the coefficient of discharge and the velocity of approach factor is referred to as the flow coefficient K and the above equation is recast as

$$Q = K A_2 \sqrt{\frac{2(p_1 - p_2)}{\rho}} \tag{9.9}$$

$$\dot{m} = K A_2 \sqrt{2\rho(p_1 - p_2)} \tag{9.10}$$

The coefficient of discharge and the flow coefficient depend on the type of variable area device that is being used as well as the flow Reynolds number. These will be discussed later on.

9.2.3   types of variable area flow meters

Three forms of variable area type flow meters are used in practice. They are:

- Orifice plate
- Flow Nozzle
- Venturi

These are illustrated schematically in Figures 9.2 - 9.4.

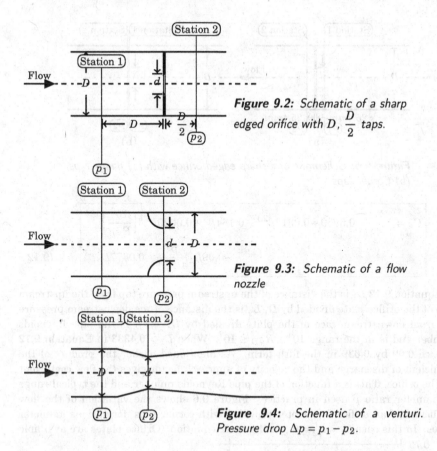

Figure 9.2: Schematic of a sharp edged orifice with $D, \dfrac{D}{2}$ taps.

Figure 9.3: Schematic of a flow nozzle

Figure 9.4: Schematic of a venturi. Pressure drop $\Delta p = p_1 - p_2$.

9.2.4 Orifice plate meter

This is by far the simplest to fabricate. It consists of a thin plate of a suitable material that is mounted between two flanges. The pipe diameter is D while the orifice diameter is d. The ratio of these two is represented as

$$\beta = \frac{d}{D} \tag{9.11}$$

Three different ways of arranging the pressure taps (and hence the choice of stations 1 and 2) are shown in Figures 9.2 and 9.5. Any one of the three arrangements is chosen in practice. Of course, between the pressure taps we can connect, for example, a U tube manometer to measure the pressure difference. Typically the orifice plate thickness may be a few mm and the orifice itself is chamfered to a sharp edge as shown in Figure 9.2. Flange taps are some 25 mm on the two sides of the orifice plate, the corner taps are very close to the flanges and the $D, D/2$ taps are provided as indicated in Figure 9.5. The last arrangement is the most common. The coefficient of discharge of an orifice plate depends on the flow Reynolds number, the β value and the locations of the pressure taps. The coefficient of discharge is given by the Stolz equation.

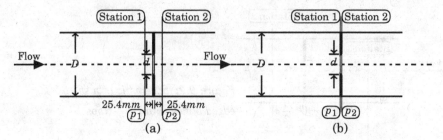

Figure 9.5: *Schematic of a sharp edged orifice with (a) flange taps, (b) Corner taps.*

$$C_d = 0.5959 + 0.0312\beta^{2.1} - 0.184\beta^8 + 0.0029\beta^{2.5}\left[\frac{10^6}{Re_D}\right]^{0.75}$$
$$+0.09L_1\frac{\beta^4}{1-\beta^4} - 0.0377L_2'\beta^3 \qquad (9.12)$$

In Equation 9.12 L_1 is the distance of the upstream pressure tap from the upstream face of the orifice plate divided by D, L_2' is the distance of the downstream pressure tap from downstream face of the plate divided by D. Re_D is the pipe Reynolds number and is in the range $10^4 \leq Re_D \leq 10^7$. When $L_1 > 0.4333$ in Equation 9.12 replace 0.09 by 0.039 in the fifth term. As mentioned earlier the product of the coefficient of discharge and the velocity of approach factor gives the flow coefficient for the orifice. This is a function of the pipe Reynolds number and the typical range of diameter ratio β used in practice. Figure 9.6 shows the variation of the flow coefficient of an orifice flow meter provided with corner taps, for various diameter ratios. In this case,$L_1 = L_2' = 0$, in the Stolz equation. Orifice plates are available

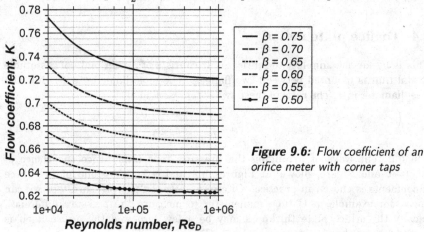

Figure 9.6: *Flow coefficient of an orifice meter with corner taps*

from many manufacturers. Orifice plates are available from, for example, Crane Universal Orifice Plates 5531 E. Admiral Place, Tulsa, Oklahoma 74115-8411. They are made of many materials such as Stainless 304 and 316 that are common; Monel, Hasteloy B and C, Titanium, Nickel and other special materials as special order.

Style 500 plates are manufactured to AGA, ASME, ISA and ANSI specifications and can be used with all Crane orifice fittings and plate holders.

An orifice plate may also be easily machined in a student work shop. Figure 9.7 shows the end view and the cross sectional view indicating the three parameters that characterize it. Note the bevel on the downstream side. Apart from the indicated features one may also provide suitable holes for fixing the orifice plate between two flanges in a flow measurement installation. Suitable gaskets may have to be provided to avoid leaks. An installation using an orifice plate with $D - D/2$ taps is shown

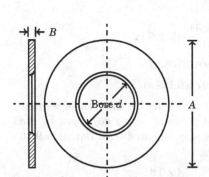

Figure 9.7: *Orifice plate specifications*

Figure 9.8: *Photograph of an orifice plate in use*

in Figure 9.8 along with a water manometer that is used for measuring the head developed by the orifice.

Example 9.1

An orifice plate with corner taps is to be used to measure flow of air at $5\,bar$ and $25°C$. The maximum flow rate is $1\,kg/s$. The minimum flow rate is to be $0.3\,kg/s$. Determine the size of the orifice plate such that the pipe Reynolds number is not less than 10^5. Calculate the differential pressure developed across the orifice plate if $\beta = 0.6$.

Solution :

Air properties are calculated first. We have

$$p_1 = 5bar = 5 \times 10^5\,Pa;\ T_1 = 25°C = 298\,K$$

The density of air is then calculated as

$$\rho = \frac{p_1}{R_g T} = \frac{5 \times 10^5}{287 \times 298} = 5.846\,kg/m^3$$

From air tables the dynamic viscosity at $25°C$ is $\mu = 18.41 \times 10^{-6}\,kg/m\,s$. The kinematic viscosity of air may thus be obtained as

$$v = \frac{\mu}{\rho} = \frac{18.41 \times 10^{-6}}{5.846} = 3.15 \times 10^{-6}\,m^2/s$$

It is specified that the pipe Reynolds number should not be less than 10^5. This means that the pipe Reynolds number should be 10^5 when the mass flow rate of air is $\dot{m}_{min} = 0.3\,kg/s$ (the lower limit). The minimum air velocity is V_{min} such that

$$V_{min}D = Re_D\nu = 10^5 \times 3.15 \times 10^{-6} = 0.315\,m^2/s$$

With mass flow rate of $0.3\,kg/s$ we also have $\dot{m}_{min} = \dfrac{\rho V_{min}\pi D^2}{4}$ and hence

$$V_{min}D^2 = \frac{4\dot{m}_{min}}{\pi\rho} = \frac{4 \times 0.3}{\pi \times 5.846} = 0.06534\,m^3/s$$

By division we then have

$$D = \frac{V_{min}D^2}{V_{min}D} = \frac{0.06534}{0.315} = 0.207\,m$$

Since $\beta = 0.6$, the orifice diameter is at once given by

$$d = \beta D = 0.6 \times 0.207 = 0.124\,m$$

We may now read off the flow coefficient from Figure 9.6 as 0.652 (or we may use the Stolz equation). The differential pressure Δp_{min} developed by the orifice plate between the two corner taps may be found from Equation 9.10 that may be rearranged as

$$\Delta p_{min} = \frac{1}{2\rho}\left[\frac{\dot{m}_{min}}{KA_2}\right]^2 = \frac{1}{2 \times 5.846}\left[\frac{4 \times 0.3}{0.652 \times \pi \times 0.124^2}\right]^2 = 124.9\,Pa$$

This may be converted to head of water (at standard water density of $\rho_w = 1000\,kg/m^3$) as

$$h_{min} = \frac{\Delta p_{min}}{\rho_w g} = \frac{124.9}{1000 \times 9.8} \times 1000 = 12.74\,mm$$

The maximum mass flow rate for this facility is $\dot{m}_{max} = 1\,kg/s$. The velocity in the pipe may be calculated as

$$V_{max} = \frac{4\dot{m}_{max}}{\rho\pi D^2} = \frac{4 \times 1}{5.846 \times \pi \times 0.207^2} = 5.083\,m/s$$

The corresponding pipe Reynolds number is

$$Re_D = \frac{V_{max}D}{\nu} = \frac{5.083 \times 0.207}{3.15 \times 10^{-6}} = 334048$$

The corresponding flow coefficient using the Stolz equation is 0.649. The head developed by the orifice plate is thus given by

$$\Delta p_{max} = \frac{1}{2\rho}\left[\frac{\dot{m}_{max}}{KA_2}\right]^2 = \frac{1}{2 \times 5.846}\left[\frac{4 \times 1}{0.649 \times \pi \times 0.124^2}\right]^2 = 1392.4 Pa$$

This may be converted to head of water (at standard water density of $1000\,kg/m^3$) as

$$h_{max} = \frac{\Delta p_{max}}{\rho_w g} = \frac{1392.4}{1000 \times 9.8} \times 1000 = 142.1\,mm$$

It is thus desirable to use a water manometer of range of some 150 mm of water column. In order to obtain satisfactory performance we may use an inclined tube manometer.

9.2.5 Flow nozzle

The flow nozzle is also an obstruction type meter that is characterized by a higher value of the coefficient of discharge than that in the case of the orifice plate. The flow nozzle consists of a converging nozzle introduced in the pipe. The flow accelerates in the converging part, separates from the wall near the exit with a low pressure separated region close to the nozzle exit. The flow diverges and reattaches to the pipe at some downstream point along the pipe. Figure 9.9(a) shows the schematic of long radius ASME nozzle. Figure 9.9(b) indicates the nomenclature that is used to

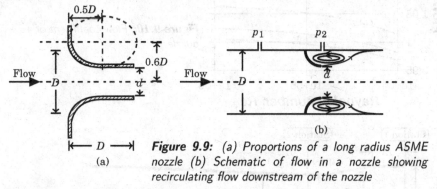

Figure 9.9: (a) Proportions of a long radius ASME nozzle (b) Schematic of flow in a nozzle showing recirculating flow downstream of the nozzle

characterize the nozzle. Also shown is the nature of the flow as it emerges from the nozzle. The flow tends to converge well past the nozzle exit to achieve the smallest diameter for the flow at "vena contracta". The nozzle is normally made by a machining process. The pipe diameter may be from $50\ mm$ to $650\ mm$. The β value may be chosen between 0.2 and 0.8. For pipe Reynolds number in the range 10^4 to 10^7, the coefficient of discharge is well represented by the formula

$$C_d = 0.9965 - 6.53\sqrt{\frac{\beta}{Re_D}} \qquad (9.13)$$

Flow coefficient variation with pipe Reynolds number for long radius flow nozzles is shown plotted in Figure 9.10. We note that the flow coefficient increases with Reynolds number as opposed to a decreasing trend in the case of an orifice plate. The flow coefficient is always more than that for an orifice plate and hence, for a given diameter ratio and pipe Reynolds number, the orifice plate will develop a larger head than a flow nozzle.

9.2.6 Venturi meter

The last type of an obstruction type flow meter is the venturi. It consists of a converging section, a uniform area section and a diverging section as shown schematically in Figure 9.11. Thus it is essentially a converging - diverging nozzle. Since the diverging portion has a gentle slope (or divergence angle) the flow does not separate as it does in the case of an orifice plate or a flow nozzle. This leads to some desirable qualities as we shall see later. However for fixed conditions of diameter, diameter ratio and inlet velocity the venturi will develop the smallest pressure drop between stations 1 and 2, as compared to the flow nozzle or the orifice plate.

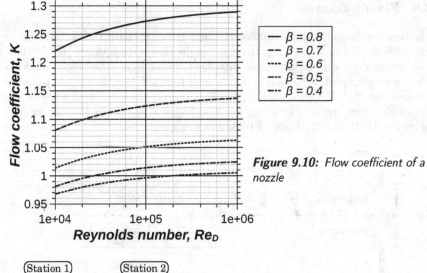

Figure 9.10: *Flow coefficient of a nozzle*

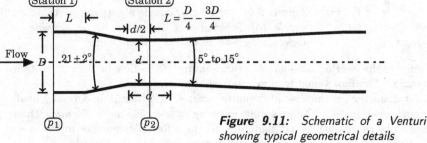

Figure 9.11: *Schematic of a Venturi showing typical geometrical details*

Venturies are fabricated by different methods depending on the size and cost:

Rough casting: Coefficient of discharge is $C_d = 0.984$ with an uncertainty of $\pm 0.7\%$. The following ranges apply: Diameter: 100 - 800 mm; β: 0.3 - 0.75; Re_D: $2 \times 10^5 - 2 \times 10^6$

Machined entrance cone: Coefficient of discharge is $C_d = 0.995$ with an uncertainty of $\pm 1\%$. The following ranges apply: Diameter: 50 - 250 mm; β: 0.4 - 0.75; Re_D: $2 \times 10^5 - 1 \times 10^6$

Rough welded sheet metal: Coefficient of discharge is $C_d = 0.985$ with an uncertainty of $\pm 1.5\%$. The following ranges apply: Diameter: 200 - 1200 mm; β: 0.4 - 0.7; Re_D: $2 \times 10^5 - 2 \times 10^6$

We see that the coefficient of discharge of a venturi is close to unity and its variation is also very small (Figure 9.12). The flow through a venturi is also smooth without the flow separation that is present in both the orifice plate as well as the flow nozzle. Hence the so called "irrecoverable pressure loss" across the venturi is much smaller than in the other two cases, as shown in Figure 9.13. Irrecoverable pressure loss is the loss of pressure across the device between its inlet and outlet. Since the diameter of the variable area flow meters are the same before and after the meter section, one would expect the pressure to get back to the value that existed ahead

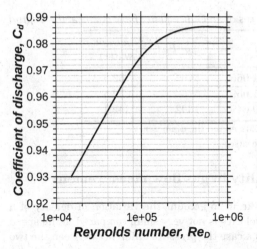

Figure 9.12: *Variation of coefficient discharge of a venturi with pipe Reynolds number*

of the meter. This is not so since there are losses due to friction as well as the recirculating flow that dissipates kinetic energy as heat. Hence the irrecoverable pressure (or head) loss. Pressure difference developed and measured for determining the flow rate is referred here as the head developed and is used as the normalization factor for expressing the irrecoverable head loss. The ordinate in Figure 9.13 is the irrecoverable pressure loss expressed as a percentage of the head developed. This

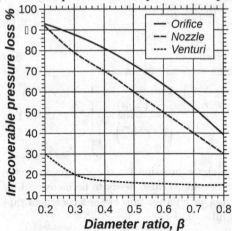

Figure 9.13: *Irrecoverable pressure loss comparisons*

simply means that the installation of a venturi meter does not appreciably affect the flow. Also, since the venturi develops a smaller differential pressure as compared to the orifice plate, for a given diameter ratio, the absolute irrecoverable pressure loss is even more in the case of an orifice plate! Nozzle is slightly better as compared to an orifice plate.

We consider again the data specified in Example 9.1. We make calculations corresponding to the maximum mass flow rate of 1 kg/s by replacing the orifice plate by a long radius flow nozzle and a rough cast venturi. The comparisons are summarized in Table 9.1.

Table 9.1: *Comparisons of performances of an Orifice plate, Flow nozzle and a Venturi*

	C	K	Δp (Pa)	Δh (mm water)
Venturi	0.984	1.055	519.2	52.9
Nozzle	0.988	1.059	515.2	52.5
Orifice	0.605	0.649	1392	142.1

The pipe diameter is 0.207 m and the diameter ratio is $\beta = 0.6$ in all the cases

9.2.7 Effect of compressibility in gas flow measurement

Many a time it is necessary to take in to account the variation of density of a flowing gas medium, like air, if the velocity is not very small compared to the speed of sound in the medium. Also in such a case the pressure difference between the two measuring stations $p_1 - p_2$ is not small compared to p_1. Hence the pressure ratio $\dfrac{p_2}{P_1}$ is not close to unity. The relationship between the flow rate and the head developed becomes more complicated because of the variation in the density of the fluid.

Making use of gas dynamic relations we can show that Equation 9.9 has to be modified by introducing a factor Y known as the expansion factor.

$$Q = YKA_2\sqrt{\frac{2(p_1 - p_2)}{\rho}} \tag{9.14}$$

In Equation 9.14 the expansion factor is given by

$$Y = \sqrt{\frac{p_2}{p_1} \cdot \frac{\gamma}{(\gamma - 1)} \cdot \frac{\left(1 - \left(\frac{p_2}{p_1}\right)^{\frac{\gamma-1}{\gamma}}\right)}{\left(1 - \frac{p_2}{p_1}\right)} \cdot \frac{(1 - \beta^4)}{\left(1 - \beta^4\left(\frac{p_2}{p_1}\right)^{2/\gamma}\right)}} \tag{9.15}$$

The ratio of specific heats, γ appears in the expression for the expansion factor. The pressure ratio, apart from the head developed, also needs to be measured for evaluating the expansion factor. For orifices with flange taps or vena contracta taps the following approximate formula is useful.

$$Y \approx 1 - \left[0.41 + 0.35\beta^4\right] \cdot \frac{(p_1 - p_2)}{\gamma p_1} \tag{9.16}$$

For orifice with flange taps one may use the following approximate expression.

$$Y \approx 1 - \left[0.333 + 1.145(\beta^2 + 0.75\beta^5 + 12\beta^{13})\right] \cdot \frac{(p_1 - p_2)}{\gamma p_1} \tag{9.17}$$

Example 9.2

Consider flow of dry air through an orifice plate with $\beta = 0.5$. The upstream pressure and temperatures are respectively $p_1 = 2\,bar$ and $T_1 = 300\,K$. The mean velocity of the flow has been measured independently in the pipe and is known to be $V_1 = 35\,m/s$. The pipe diameter is $D = 0.06m$ and flange taps are made use of. Determine the head developed by the orifice plate in Pa. Is it necessary to take into account the expansion factor? Explain. Determine the correct Δp developed by the flow meter.

Solution :

The air density in the pipe is determined as

$$\rho_1 = \frac{p_1}{R_g T_1} = \frac{2 \times 10^5}{287 \times 300} = 2.333\,kg/m^3$$

Volume flow rate of air may now be calculated as

$$Q_1 = A_1 V_1 = \frac{\pi \times 0.06^2}{4} \times 35 = 0.09896\,m^3/s$$

The dynamic viscosity of air at this temperature is $\mu_1 = 18.46 \times 10^{-6}\,kg/m\,s$. The kinematic viscosity is thus given by $v_1 = \dfrac{\mu_1}{\rho_1} = \dfrac{18.46 \times 10^{-6}}{2.333} = 7.947 \times 10^{-6}\,m^2/s$. The pipe Reynolds number may be calculated as

$$Re_D = \frac{V_1 D}{v_1} = \frac{35 \times 0.06}{7.947 \times 10^{-6}} = 264250$$

From Figure 9.6 the flow coefficient is estimated as $K = 0.625$. Ignoring the compressibility effect we calculate the head developed by the orifice plate as

$$\Delta p = \frac{\rho_1}{2}\left[\frac{Q_1}{KA_2}\right]^2 = \frac{2.333}{2}\left[\frac{0.09896}{0.625 \times 0.5^2 \times \frac{\pi}{4} \times 0.06^2}\right]^2 = 58530\,Pa$$

The approximate value of the downstream pressure is

$$p_2 = p_1 - \Delta p = 2 \times 10^5 - 58530 = 141470 Pa$$

The pressure ratio is thus given by $\dfrac{p_1}{p_2} = \dfrac{200000}{141470} = 1.414$ which is certainly not very close to unity! We thus conclude that the expansion factor needs to be taken in to account. This we do by calculating Y approximately using the pressure drop estimated by the incompressible formula. Using Equation 9.17 we get

$$Y \approx 1 - [0.333 + 1.145(0.5^2 + 0.7 \times 0.5^5 + 12 \times 0.5^{13})] \cdot \frac{200000 - 141470}{1.4 \times 200000} = 0.865$$

We have taken γ for air as 1.4. We may correct the pressure drop to

$$\Delta p = \frac{\rho_1}{2}\left[\frac{Q_1}{YKA_2}\right]^2 = \frac{2.333}{2}\left[\frac{0.09896}{0.865 \times 0.625 \times 0.5^2 \times \frac{\pi}{4} \times 0.06^2}\right]^2 = 78225\,Pa$$

We may use this value to correct the pressure ratio and hence the expansion factor. The process converges within two iterations to give $\Delta p = 87146\,Pa$.

9.2.8   Sonic orifice or the sonic nozzle

From compressible flow theory it is well known that the velocity of a gas will equal
the velocity of sound at the smallest area (throat) in the case of variable area flow if
the pressure ratio is more than a certain value called the critical pressure ratio. If
the upstream pressure is p_1 and the downstream pressure p_2, the sonic velocity will
be achieved at the throat (or more accurately the vena contracta) if

$$\frac{p_1}{p_2} \geq \left[\frac{2}{1+\gamma}\right]^{-\frac{\gamma}{\gamma-1}} \tag{9.18}$$

Let the throat conditions be indicated by subscript th . Then, we have the mass flow
rate given by $\dot{m} = \rho_{th} A_{th} a_{th}$ where the speed of the fluid is equal to the speed of
sound a_{th}. Under isentropic assumption the conditions at the throat may be written
in terms of those at the upstream to get

$$\dot{m} = \rho_1 a_1 A_{th} \left[\frac{2}{1+\gamma}\right]^{-\frac{\gamma+1}{2(\gamma-1)}} \tag{9.19}$$

Using gas laws this may be recast in the more useful form

$$\dot{m} = p_1 A_{th} \sqrt{\frac{\gamma}{R_g T_1}} \left[\frac{2}{1+\gamma}\right]^{-\frac{\gamma+1}{2(\gamma-1)}} \tag{9.20}$$

Thus all that one has to do is to measure the pressure and temperature in the
upstream pipe to obtain the mass flow rate. Also note that if the upstream pressure
and temperature are held fixed the mass flow rate does not change and hence the
sonic orifice may also be used as a flow controller. Within limits the orifice diameter
may be changed to dial up or down the mass flow rate. This is especially useful
when working between a pressure above the atmosphere and a pressure below the
atmosphere as in vacuum systems. The flow coefficient (K) of a typical sharp-edged
orifice is generally accepted to be 0.61, whereas the flow coefficient of a typical nozzle
is generally accepted as 0.97. Equation 9.20 may be used by multiplying it with the
appropriate K value.

Example 9.3

A sonic nozzle is to be used to measure air flow of $2\,MPa$ and $40°C$. The flow
rate is expected to be 0.2 kg/s. Calculate the throat diameter if critical flow
conditions are assumed to exist there. Take $\gamma = 1.4$ for air.

Solution :

The given data:

$$p_1 = 2 \times 10^6 \, Pa \quad T_1 = 313 \, K \quad \dot{m} = 0.2 \, kg/s$$

Assume a value of 0.97 for the flow coefficient. We can rearrange Equation
9.20 to get the throat area as

$$A_{th} = \frac{\dot{m}}{K p_1 \sqrt{\frac{\gamma}{R_g T_1}} \left[\frac{2}{1+\gamma}\right]^{-\frac{\gamma+1}{2(\gamma-1)}}}$$

$$= \frac{0.2}{0.97 \times 2 \times 10^6 \sqrt{\frac{1.4}{287 \times 313}} \left[\frac{2}{1+1.4}\right]^{-\frac{1.4+1}{2(1.4-1)}}} = 1.1281 \times 10^{-4} \, m^2$$

The corresponding throat diameter may be inferred as

$$d_{th} = \sqrt{\frac{4A_{th}}{\pi}} = \sqrt{\frac{4 \times 0.00010943}{\pi}} = 0.01985 m \text{ or } 12 \, mm$$

Example 9.4

A small sonic nozzle having a diameter of $0.8 \, mm$ is used to measure the flow in a $7.5 \, cm$ diameter pipe. The upstream pressure in the nozzle is varied to suit the flow requirements. The upstream temperature is fixed at $20°C$. The downstream pressure is such that it is always less than the critical pressure. What are the minimum and maximum flow rates if the upstream pressure varies between 0.1 and $1 \, MPa$?

Solution :

Given data:

$$d_{th} = 0.8 \, mm = 0.0008 \, m \quad D = 7.5 \, cm = 0.075 \, m; \quad T_1 = 293 \, K$$

The throat area based on the throat diameter is

$$A_{th} = \frac{\pi d_{th}^2}{4} = \frac{\pi 0.0008^2}{4} = 5.027 \times 10^{-9} \, m^2$$

The mass flow rate is a linear function of upstream pressure as seen from Equation 9.20. Using that equation we have

$$\dot{m}(p_1) = 5.027 \times 10^{-9} \sqrt{\frac{1.4}{287 \times 293}} \left[\frac{2}{1+1.4}\right]^{-\frac{1.4+1}{2(1.4-1)}} = 1.187 \times 10^{-9} p_1 \, kg/s$$

with p_1 given in Pa.

The minimum mass flow rate corresponds to $p_1 = 0.1 \, MPa = 10^5 \, Pa$ and is given by

$$\dot{m}(10^5 \, Pa) = 1.187 \times 10^{-9} \times 10^5 = 1.187 \times 10^{-4} \, kg/s \text{ or } 0.1187 \, g/s$$

The maximum mass flow rate corresponds to $p_1 = 1 \, MPa = 10^6 Pa$ and is given by

$$\dot{m}(10^6 \, Pa) = 1.187 \times 10^{-9} \times 10^6 = 1.187 \times 10^{-3} \, kg/s \text{ or } 1.187 \, g/s$$

Note that the upstream diameter does not enter the calculations. As it happens in this case, the pipe diameter is very large compared to the orifice diameter. Hence the pressure and temperature are effectively the stagnation values. These may be measured using suitable pressure and temperature sensors.

A high pressure tank that has a small leak can be modeled as a flow through a sonic orifice as long as the pressure ratio is greater than about 1.9 or say 2. The leakage flow will keep reducing linearly with the tank pressure as long as this condition is satisfied. Once the pressure ratio goes below this value the leakage flow is fully subsonic.

9.2.9   Selection of variable area flow meters

We have already seen that irrecoverable pressure drop is an important factor in the selection of a particular type of variable area flow meter. The cost of installation depends on the initial cost of the meter as well as the accompanying piping that is involved in the installation. The Venturi meter itself is considerably long and requires, in addition, straight upstream and downstream sections. The straight sections are required so that the standard formulae for the coefficient of discharge may be used, without appreciable error. An Orifice plate requires the least modification in an existing line. Flow nozzle is somewhere between these two cases. Also all these meters have to necessarily be oriented horizontally and enough space must be available for the additional straight lengths ahead of and after the meter. Also, the variable area meters are inherently non-linear. These meters may be designed to cover a wide range of flow rates.

While measuring gas flows it is necessary to include both temperature and pressure sensors in the flow measuring device. If the speeds are high compressibility effects also become important. The same calibration will not, in general, hold for different gases. Corrections will have to be applied if the gas is different from the one with which the meter has been calibrated. Change in pressure levels as well as pressure ratios become important when compressibility effects are significant. The number of measured quantities goes up when these effects are to be taken in to account. Signal conditioning and data collection requires expensive electronics support.

Since there are no moving parts, variable area flow meters are not subject to wear and tear. This is a big advantage.

9.3   Rotameter or Drag effect flow meter

Rotameter is also referred to as a drag effect meter since the volume/mass flow measurement is based on equilibrium of a body (referred to as the "bob") under the action of inertia and drag forces. In fact the drag force experienced by the bob is due to the pressure distribution around the bob as the flow takes place past it. Viscous forces do not have much role in this.

Rotameter consists of a vertical tapered tube through which the flow of the fluid (gas/liquid) takes place. A suitably shaped body (called the 'bob') floats in it as shown in Figure 9.14, in its schematic form. Photograph of a rotameter used in laboratory is shown in Figure 9.15 When the float moves up, the flow area around the float increases. This is because the tube is internally tapered, as indicated in the figure.

9.3.1   Rotameter analysis

Let the mean velocity of the fluid as it flows past the float be V_m . The area over which this flow takes place is the smallest annular area between the bob and the tapered tube. Let the volume of the float be V_b and density of the material of the float be ρ_b. Let the fluid density be ρ_f . The weight of the bob is the downward force given by $\rho_b V_b g$ while there are two upward forces. The first one is the force of buoyancy given by $\rho_f V_b g$ while the second is the drag force that is generally

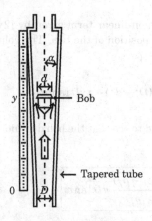

Figure 9.14: *Schematic represen-*   **Figure 9.15:** *Typical Rota meter*
tation of a rotameter   *in the laboratory*

represented by an expression of the form $F_D = S_b C_D \dfrac{\rho_f V_m^2}{2}$. In this expression C_D is
the drag coefficient, dependent on the shape of the bob and S_b is the frontal area of
the bob facing the flow. The vertical position of the float adjusts itself, as mentioned
earlier, by a balance of the downward and upward forces. Thus

$$\rho_b V_b g = \rho_f V_b g + S_b C_D \frac{\rho_f V_m^2}{2} \tag{9.21}$$

If the shape of the bob is properly chosen it is possible to have a C_D that does not
vary much over the useful range of a given Rotameter. For a smooth sphere, for
example, the drag coefficient is more or less constant at 0.4 for $2 \times 10^3 \le Re \le 10^5$.
It becomes clear, at once, that the mean velocity V_m should be a constant since all
other quantities entering Equation 9.21 are fixed! Thus the constant mean velocity
is given by

$$V_m = \sqrt{\frac{2V_b g}{C_D S_b} \cdot \frac{(\rho_b - \rho_f)}{\rho_f}} \tag{9.22}$$

Note that the bob material should have density greater than the fluid density.
Secondly the Rotameter has to be oriented vertically with the diameter of the
tapered tube increasing with y! The area available for the flow is obviously given
by $A_f = \dfrac{\pi \left(d(y)^2 - d^2 \right)}{4}$ where $d(y)$ is the inner diameter of the tapered tube at the
bob position y. Assuming that the taper angle is α we see that $d(y) = d + 2y \tan \alpha$.
With this, and assuming that the taper angle is small, we may write approximately
the flow area as

$$A_f = \frac{\pi}{4}[\{D + 2y \tan \alpha\}^2 - d^2] \approx \frac{\pi}{4}[D^2 - d^2 + 4yD \tan \alpha] = \frac{\pi}{4}[D^2 - d^2] + \pi yD \tan \alpha \tag{9.23}$$

The approximation is based on neglect of the non-linear term given by $[2y\tan\alpha]^2$. Thus the area for flow varies linearly with the position of the bob. The volume flow rate is thus given by

$$Q = V_m A_f = \sqrt{\frac{2V_b g}{C_D S_b} \cdot \frac{\rho_b - \rho_f}{\rho_f}} [\frac{\pi}{4}(D^2 - d^2) + \pi y D \tan(\alpha)] \qquad (9.24)$$

If $D = d$, (some arrangement, however, is needed to see that the bob does not fall into the tube) then the above becomes

$$Q = Ky \quad \text{where} \quad K = \sqrt{\frac{2V_b g}{C_D S_b} \cdot \frac{(\rho_b - \rho_f)}{\rho_f}} \pi D \tan\alpha \qquad (9.25)$$

A scale attached to the tapered tube (usually engraved on the tube itself) may be marked directly in volume flow rate unit, absorbing the proportionality factor K into the scale. Of course, the scale would be correct for a specified fluid at a specified temperature. Correction factors will have to be incorporated for any changes in these. The manufacturer normally supplies the required information. Figure 9.15 shows photograph of a typical rotameter.

Example 9.5

A rotameter is to be used to measure air flow at 7 *bar* and 21°C. The maximum flow rate of 0.0005 *kg/s*, the inlet diameter of the meter is 8*mm* and the length of the meter is limited to 10 *cm*. The bob is constructed such that it has a frontal diameter of 8 *mm* and a volume of 2.5 *ml*. Determine the taper of the tube. The shape of the bob has been chosen such that the drag coefficient is constant at 0.8. The density of the bob material is known to be 50 kg/m^3.

Solution :
We assume that the measured air flow corresponds to

$$\dot{m} = \dot{m}_{max} = 0.0005\, kg/s \quad y = y_{max} = 10\, cm = 0.1\, m$$

The density of air is based on $p_f = 7\, bar = 7\times 10^5\, Pa$ and $T_f = 273 + 21 = 294\, K$ and is

$$\rho_f = \frac{p_f}{R_g T_f} = \frac{7 \times 10^5}{287 \times 294} = 8.406\, kg/m^3$$

The mass flow rate is given by $\dot{m}_{max} = \rho_f Q_{max} = \rho_f K y_{max}$. Hence the Rotameter constant K is

$$K = \frac{\dot{m}_{max}}{\rho_f y_{max}} = \frac{0.0005}{8.406 \times 0.1} = 0.000595\, m^2/s$$

The bob frontal area facing the flow is based on $d = 8\, mm = 0.008\, m$ and is

$$S_b = \frac{\pi \times 0.008^2}{4} = 5.027 \times 10^{-5}\, m^2$$

Also we are given

$$V_b = 2.5\ ml = 2.5 \times 10^{-6}\ m^3 \quad \rho_b = 50\ kg/m^3 \quad C_D = 0.8$$

Equation 9.25 may be rearranged to obtain the taper parameter $\tan \alpha$ as

$$\tan(\alpha) = \frac{K}{\pi D \sqrt{\frac{2V_b g}{C_D S_b} \frac{\rho_b - \rho_f}{\rho_f}}} = \frac{0.000595}{\pi \times 0.008 \sqrt{\frac{2 \times 2.5 \times 10^{-6} \times 9.8}{0.8 \times 5.027 \times 10^{-5}} \cdot \frac{50 - 8.406}{8.406}}} = 0.00964$$

The taper angle is then obtained as

$$\alpha = \tan^{-1}(0.00964) = 0.55°$$

Example 9.6

A rotameter is designed with a tapered tube having a taper angle of $0.8°$. The diameter of the tapered tube is $15\ mm$ at the bottom and is $150\ mm$ long. The bob has a volume of $5\ ml$ and an effective density of $3000\ kg/m^3$. The bob design is such that the drag coefficient is $C_D = 0.4$. What is the maximum flow rate of a liquid of density $\rho_f = 800\ kg/m^3$ that may be measured using this rotameter?

Solution :
Given data:

$$\alpha = 0.8° \quad y_{max} = 0.15\ m \quad d = 0.015\ m \quad V_b = 5 \times 10^{-6}\ m^3$$

$$\rho_b = 3000\ kg/m^3 \quad \rho_f = 800\ kg/m^3 \quad C_D = 0.4$$

The frontal area of the bob may be calculated as

$$S_b = \frac{\pi \times 0.015^2}{4} = 1.7672 \times 10^{-4}\ m^2$$

We may now calculate the gage constant K directly using Equation 9.25 as

$$K = \pi \times 0.015 \times \tan 0.8° \sqrt{\frac{2 \times 5 \times 10^{-6} \times 9.8}{0.4 \times 1.7672 \times 10^{-4}} \cdot \frac{3000 - 800}{800}} = 0.001285\ m^2/s$$

The maximum flow rate that may be measured is then given by

$$Q_{max} = K y_{max} = 0.001285 \times 0.15 = 1.93 \times 10^{-4}\ m^3/s$$

The corresponding mass flow rate is given by

$$\dot{m}_{max} = \rho_f Q_{max} = 800 \times 1.93 \times 10^{-4} = 0.1544\ kg/s = 555.8\ kg/h$$

Drag effect meters are useful for measurement of both gas and liquid flows. For very small gas flows the rotameter may employ a very narrow bore tube with a very small low density float material. The scale on the meter is correct only for the stated pressure and the fluid. Correction factors need to be applied when these are different during use. Rotameters have to be oriented vertically with the inlet at the bottom and the outlet at the top. In those cases where the flow is through horizontal tubes significant modifications in piping may become necessary. These changes in the piping involve introduction of several bends that introduce significant head loss and also increase installation cost.

9.4 Miscellaneous types of flow meters

We describe below the following types of flow meters:

- Positive displacement meters
- Vortex shedding type flow meter
- Turbine flow meter

9.4.1 Positive displacement meters

Various positive displacement meters are available for measuring or controlling delivery of a required amount of fluid. The simplest method one can think of is the collection of a liquid flowing through a system in a unit of time in a measuring jar or bucket to obtain the volume or mass flow rate. A syringe pump (piston moving inside a cylinder of accurate bore at a constant linear speed) is an example of such a meter, that is commonly employed in laboratory practice. It is also used in drug delivery systems in medical practice. A positive displacement meter that uses lobed rotors is shown in Figure 9.16. As is evident each rotation of the rotor will mean a delivery of volume V of the fluid to the outlet. Hence one has to measure the number of rotations of the rotor per unit time and multiply it by V to get the volume flow rate.

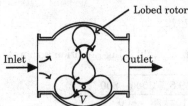

Figure 9.16: *Positive Displacement type with Lobed Rotors*

9.4.2 Vortex shedding type flow meter

It is observed that when a fluid flows past an immersed object, vortices are shed into the flow which gives rise to periodic disturbances behind the flow. Figure 9.17 shows how this can be used as a flow measurement device. Different body shapes that are useful are also shown in the figure. One simply monitors the pressure or velocity fluctuations behind the immersed object and measures the dominant frequency f to get the fluid velocity. The fluctuations in pressure may be monitored by using either a piezoelectric transducer or a strain gauge. It is also possible to use immersed hot film sensor buried in the vortex shedding body to monitor the shedding frequency. The vortices may also be monitored using an ultrasound beam which is interrupted each time a vortex passes across the beam. Vortex shedding frequencies are in the $200-500\,Hz$ range and the frequency responds within about1 $cycle$.

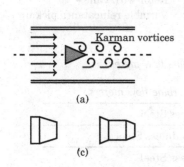

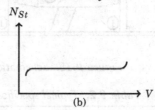

(a)

(b)

(c)

Figure 9.17: (a)Vortex shedding type flow meter (b) Characteristic of vortex shedding meter (c) Different body shapes

With proper choice of the body shape, the Strouhal number, a non-dimensional parameter defined as $N_{St} = \dfrac{fD}{V}$, remains fairly constant over a range of fluid velocities as long as the flow is turbulent. Diameter D is the characteristic length. The fluid velocity is then a linear function of the vortex shedding frequency.

The vortex shedding type meter has the advantage that it is not necessary to calibrate each and every meter. Once the shedder geometry is fixed, one time calibration will suffice, for all meters using the shedder of that fixed geometry. There is no wear and tear of the meter since there are no moving parts. The meter may be used for clean as well as contaminated liquids.

9.4.3   Turbine flow meter

The turbine flow meter (see Figure 9.18) consists of a turbine rotor mounted in a tube through which the flow takes place. Flow straighteners both upstream of and downstream of the rotor provide parallel flow past the meter. The rotational speed of the rotor is directly related to the flow rate. The rotational speed is measured using a variable reluctance (magnetic) pickup. The pickup consists of a permanent magnet around which a coil is wound. The magnetic coupling between the permanent magnet and the coil varies and changes every time a rotor vane passes across it. The change in the magnetic flux coupling induces a voltage across the coil. The frequency of this wave is directly proportional to the rotational speed that is thus proportional to the flow velocity. Turbine flow meters are available for fluids over large pressure and temperature ranges. Typical performance figures for turbine meters are given in Table 9.2.

Turbine flow meters are highly accurate and practically linear over a fairly large range of flow rates. The head loss is moderate and the meter itself is very compact. However these meters are very expensive and the bearings pose service problems. Turbine meters can handle only pure liquids without any contamination. Also these

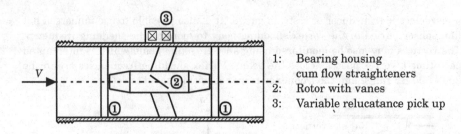

Figure 9.18: *Schematic of a Turbine flow meter*

Table 9.2: *Typical specifications for turbine flow meters*

Body material	316 Stainless Steel
Rotor material	CD4MU Stainless Steel
Rotor support and Bearings	316 Stainless Steel
Rotor shaft	Tungsten Carbide
Operational temperature range	-101 to 177 $^\circ C$
Maximum pressure	As high as 350 bars for some models
Accuracy	1% of reading
Repeatability	0.1%

Supplied by Blancett Flow Meters, Wisconsin, USA

meters are sensitive to flow disturbances, especially to the presence of swirl. Hence flow straigteners are an integral part of the meter.

We conclude the discussion on miscellaneous types of flow meters by summarizing the factors that need to be kept in mind in the choice of a particular flow meter.

9.5   Factors to be considered in the selection of flow meters

Performance considerations: Accuracy, Repeatability, Linearity, Rangeability, Pressure drop, Signal characteristics, Response time, Reliability

Fluid property considerations: Liquid or gas, Operating temperature and pressure, Density, Viscosity, Lubricity, Chemical properties, Surface tension, Compressibility, Presence of other phases,

Economic considerations: Cost of equipment, Installation costs, Operation and maintenance costs, Calibration costs,

Installation considerations: Orientation, Flow direction, Upstream and downstream pipe work, Line (diameter of pipe) size, Servicing requirements, Effects of local vibration, Location of valves, Electrical connections and the nature of power required like ac or dc, Provision of accessories

Environmental considerations: Effect of ambient temperature, Effect of humidity, Signal conditioning and transmission, Pressure effects, Effect of atmosphere, Electrical interference

9.6 Calibration of flow meters

Comparison of the output of an instrument with a standard one of established uncertainty is called calibration. Calibration is required only when the absolute accuracy of a meter is required. This may be either because of prevailing standards or required because of concern over the quality of the measurement. More expensive the fluid being metered higher is the desired accuracy and hence the more important the quality of calibration. Since flow rate is a derived quantity calibration is involved and may require several measurements requiring several instruments.

9.6.1 Methods of calibration

The simplest and cheapest method of calibrating a flow meter is to check its performance against another meter, the calibration of which is traceable to known standards. The reference meter should be *superior* in quality to the meter to be calibrated and the test bench must be so constructed that the location of one does not affect the performance of the other. The available methods of calibration of flow meters are shown in Figure 9.19. We shall describe some of these in detail in what follows.

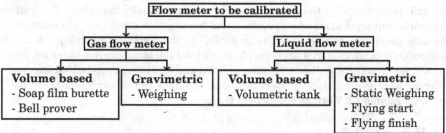

Figure 9.19: *Different methods of calibration of flow meters*

9.6.2 Soap film burette

Very low gas flows at around the atmospheric pressure are measured by meters such as the rotameter or variable area meters that have been described earlier. Calibration of these may be made by the soap film burette technique. Figure 9.20 shows the typical appearance of such a calibration system. Two light source - photocell (light detector) combinations are arranged at two locations along an accurate burette as shown, to demarcate an accurately known volume. The gas flow meter under "test" is placed in series with the burette. The burette has a rubber bulb attached at the bottom as shown in the figure. The gas flow of desired rate is maintained through the burette by supplying the gas through a control valve (not shown in the figure). A soap film is injected into the burette and the gas flow carries it up the burette. Time taken by the soap film to move from the bottom detector to the top detector is the measured quantity. When the soap film crosses the path of the first detector it disturbs the light reaching the photocell and gives a 'blip'. When the soap film crosses the second detector it again produces a 'blip'. A timing device measures the time between these two 'blips' electronically. The volume of the burette between the two detectors is accurately known and hence the volume flow rate is given by the ratio of this volume to the time between the 'blips'.

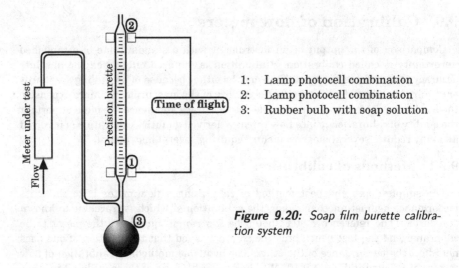

1:   Lamp photocell combination
2:   Lamp photocell combination
3:   Rubber bulb with soap solution

Figure 9.20: *Soap film burette calibration system*

The 'output' of the meter under test may therefore be assigned the volume flow rate that is measured by the soap film burette system. Usually the volume flow rate of gases is converted to standard pressure and temperature conditions. For example one may use 1 standard atmosphere and 25°C as the standard conditions. If the calibration is conducted at a pressure slightly above the atmosphere (otherwise the gas will not flow through the calibration system) and room temperature that may be different from the standard temperature, one will have to use correction based on gas laws. Example 9.7 brings out some of these.

Example 9.7

A calibration run was performed on a rotameter with a gauge constant of $K = 1.3 \times 10^{-4} \ m^2/s$ valid at STP. The bob of this rotameter has a density of $\rho_b = 5 \ kg/m^3$. The reading of the rotameter during the experiment was noted to be $y = 98mm$. The experiment itself was conducted with dry air as the gas at a gauge pressure of $50mm$ of water column and room temperature of $32°C$. The soap film apparatus consists of an accurate burette with the volume between the detectors of $1000 \ cm^3$. The time interval registered by the timing device was $77 \ s$. Analyze the data and discuss the result of the calibration experiment.

Solution :

We first calculate the volume flow rate indicated by the rotameter, correcting for the effect of non STP conditions prevailing during the experiment. From the rotameter analysis presented earlier, the gauge constant is directly related to the factor $\sqrt{\dfrac{\rho_b - \rho_f}{\rho_f}}$ where the symbols have the familiar meanings. The K value specified for the rotameter is based on STP that is, assuming the pressure to be $101325 \ Pa$ and temperature to be $25°C$ or $298 \ K$. Let ρ_f represent the fluid density (air density) at STP and ρ_f' represent the same at the actual non STP conditions. We calculate these two from the following: The density at STP is given by

$$\rho_f = \frac{p}{R_g T} = \frac{101325}{287 \times 298} = 1.185 \, kg/m^3$$

Actual pressure is $50 \, mm$ water column gauge. This is equivalent to $p' = 50 \times 10^{-3} \times 9.8 \times 1000 = 490 \, Pa$ gauge or $p' = 101325 + 490 = 101815 \, Pa$. The temperature is $T' = 32°C = 305 \, K$. The actual density of air is thus given by

$$\rho'_f = \frac{p'}{R_g T'} = \frac{101815}{287 \times 305} = 1.163 \, kg/m^3$$

Let the corrected gauge constant be K'. We have

$$K' = \frac{\sqrt{\rho_b/\rho_f - 1}}{\sqrt{\rho_b/\rho'_f - 1}} K = \frac{\sqrt{5/1.185 - 1}}{\sqrt{5/1.163 - 1}} \times 1.3 \times 10^{-4} = 1.32 \times 10^{-4} \, m^2/s$$

The volume flow rate, Q' corrected for non STP operation is now given by

$$Q' = K'y = 1.32 \times 10^{-4} \times 98 \times 10^{-3} \times 10^6 = 12.9 \, cm^3/s$$

We may now convert this to STP conditions using the principle of conservation of mass, according to which $Q'\rho'_f = Q\rho_f$. Hence

$$Q = \frac{Q'\rho'_f}{\rho_f} = \frac{12.9 \times 1.163}{1.185} = 12.7 \, cm^3/s$$

From the soap film calibrator data we calculate the volume flow rate as below. Volume swept by the soap film is $V = 1000 \, cm^3$. The time taken by the soap film to sweep this volume is $t = 77 \, s$. The volume flow rate is then given by

$$Q'_{Calib} = \frac{1000}{77} = 12.99 \, cm^3/s$$

We have to correct this for non STP effects and convert it to STP volume flow rate Q_{Calib}. This is easily done as

$$Q_{Calib} = \frac{Q'_{Calib}\rho'_f}{\rho_f} = \frac{12.99 \times 1.163}{1.185} = 12.75 \, cm^3/s$$

We may now compare the meter indicated value and the soap film based value, both at STP conditions to get

$$Error = \frac{Q - Q_{Calib}}{Q_{Calib}} \times 100 = \frac{12.7 - 12.75}{12.75} \times 100 = -0.72\%$$

It appears that the gauge constant of the rotameter is fairly accurate and is adequate for laboratory use.

9.6.3 Bell prover system

This method is used as a basic standard for measurement of gas flow rate near ambient pressures. The details of the bell prover system are shown in Figure 9.21. A hollow cylinder is inverted over a bath of sealant liquid which may be water, or to avoid humidity problems, a light oil of low vapor pressure. The cylinder is full of the gas that will be used for testing the flow meter. The cylinder is allowed to fall, in a controlled manner using weight and pulley arrangement shown in the figure. During this fall the cylinder displaces the gas through the flow meter under test.

By timing the rate of fall of the cylinder, and knowing the relation between height change and the volume displaced, the volume flow rate through the meter can be determined. Comparison with the meter reading will provide the required calibration data. As in the previous method non STP corrections need to be incorporated.

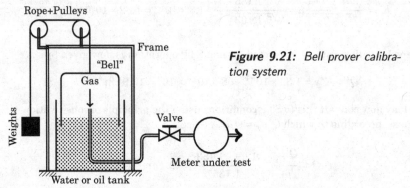

Figure 9.21: *Bell prover calibration system*

9.6.4 Flying start - Flying finish method with static weighing

This method is suitable for the calibration of liquid flow meters. Figure 9.22 gives the schematic of the method.

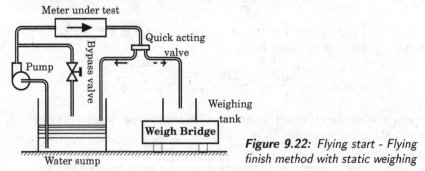

Figure 9.22: *Flying start - Flying finish method with static weighing*

Static weighing means the weighing of liquid that has been collected for a known time and is at rest. A large sump supplies the liquid for conducting the calibration test. The pump has a valve controlled bypass such that the flow rate through the meter under test may be controlled to have a desired value. The liquid is discharged through a two way quick acting valve. The discharge of the liquid is either back to the sump or to the weighing tank. The quick acting valve is operated to divert the liquid, that would be normally discharged to the sump, to the weighing tank. After a measured time interval the valve is operated to send the liquid back to the sump. The mass of the liquid is determined using the weigh bridge. The mass of water collected divided by the duration of collection yields directly the mass flow rate. This may be compared with the reading of the meter under test to provide the required calibration.

Concluding remarks

Volume flow rate or mass flow rate is an often measured quantity in laboratory practice as well as industrial applications. Several methods have been discussed in detail in this chapter. Variable area flow meter or rotameter is a device that is used very often for the measurement of volumetric flow rate of both liquids and gases. The principle of such an instrument has been discussed in sufficient detail. Corrections for operation of a rotameter at other than the design pressure and temperature also has been dealt with. Positive displacement type flow meters also have been presented. The chapter has concluded with a discussion on calibration of flow meters.

Chapter 10

Stagnation and Bulk mean temperature

When a high speed flow is brought to rest isentropically the temperature increases and attains the stagnation temperature because of conversion of kinetic energy of directed motion to internal energy. Proper design of probes is necessary to measure the stagnation temperature in such an application. Heat transfer and fluid flow applications involve fluids in motion. When the fluid velocity and temperature fields are non uniform across a section it is necessary to measure the bulk mean temperature in order to determine the enthalpy flux across the section.

10.1 Stagnation temperature measurement

Earlier we have described various methods of temperature measurement. We have also discussed the measurement of temperature of a moving fluid, essentially at low velocities. However, when flow velocities of a gas are high (as for example, in high subsonic or supersonic flows) the temperature of the gas under stagnation conditions is higher than the temperature of the moving gas because of conversion of kinetic energy to internal energy. The gas temperature takes on a value that is referred to as the stagnation temperature. The stagnation temperature is the same regardless of whether the process through which the stagnation process takes place is reversible or irreversible, as long as it is adiabatic. In many engineering application it is essential to measure the stagnation temperature. We shall consider two different probes that may be used for this purpose.

10.1.1 Shielded thermocouple stagnation temperature probe

Typical construction details of a shielded thermocouple stagnation temperature probe are shown in Figure 10.1. The probe consists of a stagnation chamber with

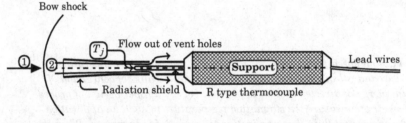

Figure 10.1: *Shielded thermocouple stagnation temperature probe in a supersonic stream (Based on NASA Technical Memorandum 4407, March 1993)*

vent holes, the chamber walls doubling as radiation shield. The leads are taken out through the back as shown. The thermocouple itself is R type. The flow enters normal to the rectangular opening, comes to a halt adjacent to the thermocouple and rejoins the flow after passing through the vent holes. The nature of flow is also indicated schematically in the figure.

The temperature measured by the probe is actually the temperature of the junction T_j. This is not necessarily the stagnation temperature (also referred to as the total temperature T_t). The reason for this is that the lead wire conduction introduces heat loss thus making the process non-adiabatic. Secondly there may be heat loss by radiation from the probe to the cold ambient. However this may be reduced by the radiation shield (the walls of the stagnation chamber). Thirdly the stagnation process involves conversion of kinetic energy to internal energy through the viscous dissipation process. This last mentioned process involves the so called recovery factor that involves a difference between the total temperature and the adiabatic wall temperature. The last quantity is the temperature at a surface exposed to high speed flow when it is perfectly impervious to flow of heat. If the temperature of the gas is T_∞ far away from the probe, the recovery factor is defined as the ratio $\dfrac{T_j - T_\infty}{T_t - T_\infty}$.

It is known that the recovery factor correlates with the parameter $p_{02} T_J^{-1.75}$, where

p_{02} is the total (or stagnation) pressure downstream of the shock. The relationship is obtained experimentally. One may look upon the recovery factor as purely a correction factor for determining the correct value of the stagnation temperature. For more details about the probe types and valuable experimental data the reader is referred to NASA TM 4407.[1]

In a particular design the following were used:

- Radiation shield is made of Pt-20% Rhodium tubing 0.0585"(1.486 *mm*) OD, 0.049"(1.245 *mm*) ID
- Vent holes are 2 in number and are 0.022"(0.559 *mm*) diameter
- Total probe length is 1.22"(30.988 *mm*)
- Area ratio - vent area to entry area - 0.5 to 0.6 for optimum recovery

10.1.2 Dual thin film enthalpy probe

A second type of probe that has been used for measurement of stagnation temperature is the dual thin film probe shown in Figure 10.2.

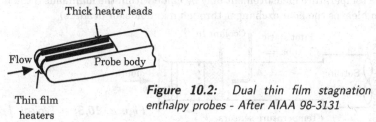

Figure 10.2: *Dual thin film stagnation enthalpy probes - After AIAA 98-3131*

The probe consists of two thin films that are heated externally such that the heating rates are different for the two films. Because of this the two films will run at different temperatures. Let the heat input rates and the respective temperatures be Q_1, T_1 and Q_2, T_2 . From heat transfer theory we know that the following hold:

$$Q_1 = hS_1(T_t - T_1) \tag{10.1}$$

$$Q_2 = hS_2(T_t - T_2) \tag{10.2}$$

Here h is the heat transfer coefficient between the fluid and each film, assumed the same for both the films (the films are close to each other on a common substrate). The surface areas of the films exposed to the fluid are S_1 and S_2 respectively. We may eliminate h and rearrange Equations 10.1 and 10.2 to get

$$T_t = T_1 + \left(\frac{Q_1}{S_1}\right) \frac{(T_2 - T_1)}{\left(\frac{Q_1}{S_1} - \frac{Q_2}{S_2}\right)} \tag{10.3}$$

The estimation of the total temperature thus requires the measurement of two heat rates and two film temperatures. This method circumvents the requirement of knowledge of the recovery factor.

---

[1]C. W. Albertson and W. A. Bauserman, Jr., Total Temperature Probes for High-Temperature Hypersonic Boundary-Layer Measurements, NASA TM-4407, March 1993

10.2   Bulk mean temperature

Consider a typical heat exchanger test rig shown in Figure 10.3. The test rig is basically a low speed wind tunnel with the heat exchanger placed facing the wind. The heat exchanger has two streams taking part in the heat exchange process, air passing through the wind tunnel (a duct) and the coolant passing through the heat exchanger tubes. The example shown is a cross flow heat exchanger, like for example, a radiator used in automotive applications. In order to determine the heat gained by the air flowing in the duct one needs to measure the bulk mean temperature of air just before the heat exchanger and again just after the heat exchanger. The bulk mean temperature is a temperature that is averaged over the flow passage weighted by the local flow velocity. In general, both the local temperature and the local velocity may vary across the flow area. The measurement thus requires the simultaneous measurement of the velocity and temperature fields. Similar measurements are also required for the coolant stream. The coolant may be a liquid like water or antifreeze liquid. The flow rate and bulk temperatures may be measured using appropriate instruments such as a variable area flow meter and RTD. The temperature measurement may be made within the manifolds if the liquid enters and leaves the heat exchanger through manifolds (or headers).

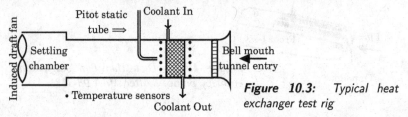

Figure 10.3: *Typical heat exchanger test rig*

Bulk mean temperature may be defined based on concepts borrowed from heat transfer theory. Consider, as an example, flow of a fluid in a circular tube as shown in Figure 10.4. Consider any section across the tube. The temperature of the fluid as well as the velocity is assumed to vary with r, the radial distance from the tube axis. If the density and specific heat of the fluid respectively are ρ and C, the flux of enthalpy $\dot{E}$ across the section of the tube is given by

$$\dot{E} = \int_0^R \rho C T(r)V(r)2\pi r dr \qquad (10.4)$$

Similarly we have the mass flux across the section given by

$$\dot{m} = \int_0^R \rho V(r)2\pi r dr \qquad (10.5)$$

We may write the enthalpy flux as the product of a mean temperature called the bulk mean temperature T_b and the mass flux specific heat product $\dot{m}c$. Thus the bulk mean temperature is given by

$$T_b = \frac{\dot{E}}{\dot{m}C} = \frac{\int_0^R \rho C T(r)V(r)2\pi r dr}{\int_0^R \rho C V(r)2\pi r dr} = \frac{\int_0^R T(r)V(r)rdr}{\int_0^R V(r)rdr} \qquad (10.6)$$

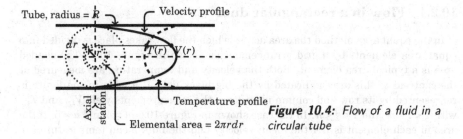

Figure 10.4: *Flow of a fluid in a circular tube*

The last result is appropriate for a fluid with constant properties. The bulk mean temperature is also referred to as the mixing cup temperature since this is the temperature obtained if one were to collect the fluid passing the section and mix it adiabatically. The defining equation involves integration across the section of the tube. In practice integration may not be possible unless the local velocities and temperatures are measured with high spatial resolution. This is possible only with laser based measurements. In normal laboratory practice one uses a Pitot tube for velocity measurement (described in Module III, Chapter5) and thermocouple for temperature measurement. Hence the integrals are replaced by summation as described below. We divide the cross section of the circular tube in to elements of

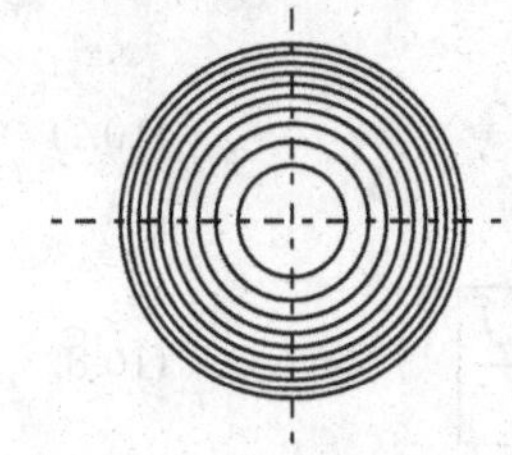

Element i	1	2	3	4	5
R_i/R	0.316	0.447	0.548	0.632	0.707
Element #	6	7	8	9	10
R_i/R	0.775	0.837	0.894	0.949	1.000

R_i=Radius of element i R=Radius of tube

Figure 10.5: *Division of a circle in to 10 elements of equal area*

equal area. For example, if the area of cross section of the tube is divided in to 10 elements of equal area, the bounding circles for the 10 elements are as shown in Figure 10.5. The corresponding radii of the bounding circles are given in the table given as an inset in the figure.

The number of area elements one is able to use in practice depends on the size of the probes (Pitot or Thermocouple). For high spatial resolution one may need to use very small sizes for the probes. Secondly the probes should not disturb the flow and temperature fields to any great extent. Thirdly the importance of the measurement decides the choice of the probe size, the number of area elements to be used and the spatial resolution. It is thus clear that the precision of the estimated value of the bulk mean temperature is dependent on several factors.

The temperature and velocities are measured close to the middle of these area elements (bounded by adjacent circles) and the integrals appearing in Equation 10.6 are replaced by summation. This is explained in greater detail with an example of flow in a duct of rectangular cross section. In view of the axi-symmetric nature of the flow velocity and temperature measurements may be made along any radial direction, in the case of a circular tube.

10.2.1   Flow in a rectangular duct

In the equal area method the area across which the flow takes place is divided into equal area elements by a grid arrangement as shown in Figure 10.6. The shaded area is a typical area element. Both the velocity and temperature are measured at the centroid of this area indicated by the big black dot. If a node in the figure is represented by its row and column identifiers, the measurements yield $V_{i,j}$ and $T_{i,j}$ with $1 < i < m$ and $1 < j < n$. In the case shown in Figure 10.6 $m = 4$ and $n = 5$. If the area of each element is ΔA the mean velocity and the bulk mean temperature are given by

	1	2	3	4	5
1	40.5	38.2	38.3	38.1	39.8°C
	2.98	3.65	3.63	3.61	2.98 m/s
2	39.2	37.8	37.3	37.4	38.9
	3.04	3.73	3.77	3.81	3.04
3	32.4	35.4	35.1	35.0	35.0
	3.03	3.77	3.79	3.81	3.03
4	42.7	38.5	38.9	38.8	39.2
	2.07	3.68	3.64	3.59	2.91

Figure 10.6: *Typical temperature - velocity data*

$$V_m = \frac{\sum_1^m \sum_1^n V_{i,j}\Delta A}{\sum_1^m \sum_1^n \Delta A} = \frac{\sum_1^m \sum_1^n V_{i,j}}{m \times n} \qquad (10.7)$$

$$T_b = \frac{\sum_1^m \sum_1^n V_{i,j} T_{i,j}\Delta A}{\sum_1^m \sum_1^n V_{i,j}\Delta A} = \frac{\sum_1^m \sum_1^n V_{i,j} T_{i,j}}{m \times n \times V_m} \qquad (10.8)$$

One way of accomplishing this is to use an array of thermocouples (or thermocouple rake, as it is referred to) and Pitot tube rake. One may move the rake (a line array) systematically using a precision traverse and hence get the required velocity and temperature readings. The experimental arrangement is as indicated in Figure 10.7. Example 10.1 is based on actual data taken in our laboratory.

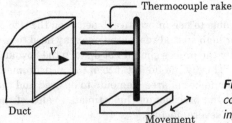

Thermocouple rake

Duct

Movement

Figure 10.7: *Schematic of thermocouple rake arrangement for measuring the bulk mean temperature*

Example 10.1

The data collected in an experiment is given in Figure 10.6 is in the form of a matrix. Temperatures are in $°C$ while the velocity entries are in m/s. Estimate the bulk mean temperature assuming that each grid has an area of $0.125 cm^2$.

Solution :

Table below gives the data and the velocity temperature products in a form useful for computing the bulk mean temperature.

Cell number	$V, m/s$	$T, °C$	$V \times T$	Cell number	$V, m/s$	$T, °C$	$V \times T$
1,1	2.98	40.5	120.69	3,1	3.03	32.4	98.17
1,2	3.65	38.2	139.43	3,2	3.77	35.4	133.46
1,3	3.63	38.3	139.03	3,3	3.79	35.1	133.03
1,4	3.61	38.1	137.54	3,4	3.81	35	133.35
1,5	2.98	39.9	118.90	3,5	3.03	35	106.05
2,1	3.04	39.2	119.17	4,1	2.91	42.7	124.26
2,2	3.73	37.6	140.25	4,2	3.68	38.5	141.68
2,3	3.77	37.3	140.62	4,3	3.64	38.9	141.60
2,4	3.81	37.4	142.49	4,4	3.59	38.8	139.29
2,5	3.04	38.9	118.26	4,5	2.91	39.2	114.07

The bulk mean temperature is then obtained, using Equation 10.8 as

$$T_b = \frac{\sum_1^4 \sum_1^5 V_{i,j} T_{i,j}}{4 \times 5 \times V_m} = \frac{2581.34}{68.4} = 37.74°C$$

Concluding remarks

This has been a short chapter dealing with the measurement of stagnation temperature as well as bulk mean temperature. These are respectively important in high speed flows and in studying transport of heat and mass in internal flows.

Exercise III

III.1 Pressure measurement

Ex III.1: Atmospheric pressure is usually measured by using a barometer. Discuss the constructional details of a barometer and discuss its usefulness.

Ex III.2: A U - tube manometer uses an oil of specific gravity 0.8 as the manometer liquid. One side of the manometer is open to local atmospheric pressure of 730 mm Hg and the difference in column heights is measured to be 200±1 mm when exposed to an air source at 25 $°C$. Standard acceleration due to gravity exists. Calculate the pressure of the air source and its uncertainty in Pa.

Ex III.3: In an inclined tube manometer the sealing liquid has a specific gravity 0.92 while the fluid whose pressure is to be measured is air at 1 atm and 20°C. The angle made by the inclined leg with the horizontal is 12° with a maximum deviation of ± 0.25 °. The well is a cylinder of diameter 0.08 m while the inclined tube has a bore of 0.003 m. Both these are susceptible to a variation of ±0.3%. Determine the pressure difference in Pa when the manometer reading is given as 90±0.75 mm. Express the result also in $Torr$. What is the uncertainty in the indicated pressure?

Ex III.4: In an inclined tube manometer the sealing fluid is a liquid of specific gravity 0.88 while the fluid whose pressure is to be measured is nitrogen at close to 1 atm and 30°C. The angle made by the inclined leg with the horizontal is 15° with a maximum deviation of ± 0.2°. The well is a cylinder of diameter 0.075 m while the inclined tube has a bore of 0.0025m. Both these are susceptible to a variation of ±0.2%. Determine the pressure difference in Pa when the manometer reading is given as 100±1 mm. Calculate the uncertainty in the indicated pressure. Note: The inclined tube is exposed to room air at 1 atm and 30°C.

Ex III.5: A well type manometer has the measurement leg inclined at 14° to the horizontal. The diameter of the measurement column is 3 mm while the well has a diameter of 40 mm. Oil having a specific gravity of 0.85 is used as the manometer liquid. A differential pressure in air at 1 atm is applied such that the displacement of the liquid along the slanted column is 180 mm. What is the correct differential pressure in mm of water and Pa?

351

Ex III.6: A well type inclined tube manometer uses a glass tube of $1.5mm$ ID for the inclined leg while the well has a diameter of $50mm$. The manometer liquid is an oil of density $920kg/m^3$. The manometer is used to measure the differential pressure of air at an average pressure of $1atm$ and $298K$. What is the differential pressure in Pa if the liquid level in the inclined leg is read as $200mm$? The inclination of the leg is $10°$ with the horizontal.

Ex III.7: A mercury U - tube manometer has total length of column of 60 cm in a $1.25\ mm$ diameter tube. Determine the period of oscillation and the damping ratio.

Ex III.8: A U tube manometer has a tube diameter of $3\ mm$. The manometer liquid is water at $30\ °C$. The total length of the water column is $650\ mm$. Obtain the parameters that describe the transients in this system. Identify whether the system is under/over/critically damped. If the diameter of the U tube alone is reduced to $1.5\ mm$ and all other quantities are unchanged what will be your answer regarding the nature of the damping?

Ex III.9: A U-tube manometer uses water as the manometer fluid. The total length of the column of water is $25\ cm$. The U-tube has a uniform bore of $1.8\ mm$. Determine the parameters that govern the transient response of this system. Make a plot of the response of the manometer to a step input.

Ex III.10: A pressure signal is fed through a line having an inside diameter of $2.5\ mm$ and a length of $4\ m$. The line is connected to a pressure transducer having a volume of $3.5\ ml$. Assume that during measurement this volume remains constant. Air at $2\ bar$ and $15°C$ is the transmitting fluid. Calculate the time constant of the system. Is this arrangement suitable for measuring a transient, which is cycling at a) $10\ Hz$ b) $50\ Hz$? Give a brief account of how you arrive at your decision.

Ex III.11: A bellows type pressure gauge of area $2\ cm^2$, spring constant $k = 15$ N/cm is connected by a $3\ mm$ ID tubing $5\ m$ long in a process application. Measurement is of pressure of water of density $1000\ kg/m^3$. What is the time constant?
Suggest ways of reducing the time constant by a factor of 3.

Ex III.12: A pressure signal is fed through a line having an inside diameter of 2 mm and a length of $2.5\ m$. The line is connected to a pressure transducer having a volume of $5\ ml$. Assume that during measurement this volume remains constant. Air at $1.2bar$ and $30°C$ is the transmitting fluid. Calculate the time constant of the system. Is this arrangement suitable for measuring a transient, which is cycling at $100\ Hz$? Justify your answer.

Ex III.13: Density of air is to be estimated by measuring its pressure and temperature. The value of gas constant for air is given to be $287.1\ J/kg \cdot K$. The measured pressure and temperature are: $p = 350 \pm 1\ kPa$; $t = 80 \pm 0.5\ °C$.
Determine the density along with its uncertainty.

Suggest suitable instruments for the measurement of pressure and temperature in the above. What is the basis for your choice? Explain.

Ex III.14: A McLeod gauge is available which has a volume of 100 cm^3. The capillary has a diameter of 0.22 mm. What is the gauge reading when the pressure to be measured is 1 Pa? If the volume of the gauge is uncertain by ± 0.2 cm^3 and the capillary diameter is uncertain to the extent of ± 0.002 mm, what is the uncertainty in the gauge reading?

Ex III.15: A McLeod gauge has a volume of 20 ml. The capillary has a diameter of 0.2 mm. What is the gauge reading when the pressure to be measured is 2 Pa? If the volume is uncertain by $\pm 0.1 ml$ and the capillary diameter is uncertain by ± 0.003 mm, what is the uncertainty in the gauge reading?

III.2 Velocity measurement

Ex III.16: Using library resources discuss "calibration" of multi-hole probes. How would you reduce the data to a useful "formula"?

Ex III.17: An impact probe is to be used to measure a maximum Mach number in air of 6. The static pressure in the stream may be as high as 50 kPa. What will be the upper limit of pressure that is expected?

Ex III.18: An impact tube is used with air flow at 10 °C, 40 kPa and a velocity of 700 m/s. What pressure will be indicated by the probe?

Ex III.19: Consider a hot-wire anemometer working as a fixed wire temperature (and hence its resistance). The wire operates at a temperature of 200 °C while the air temperature is 20 °C. The velocity of the air varies between 0 and 20m/s. The platinum hot wire of 5 μm diameter is 1.5 mm long. What is the sensor output when the air velocity is 8 m/s? Take the resistivity of platinum at 20 °C, $\rho_{20} = 10.5 \times 10^{-8}\Omega m$ and its temperature coefficient $\alpha = 0.00392/^c ircC$.

Ex III.20: A hot wire anemometer is operated in the constant temperature mode. A calibration experiment was performed to determine the relationship between the fluid velocity and the output of the hot wire. The experiment was set up in a small tube of diameter 50mm and the velocity was estimated from an independent measurement using a Pitot static tube. The data is tabulated below:

Velocity, $V, m/s$	1.05	2.10	2.95	3.94	5.06
Hot wire output, E, V	6.755	6.746	6.562	6.558	6.716

Derive by suitable means a useful relation between the output and the velocity. Specify also an error bar.

Ex III.21: Portable thermal anemometers are now available from several manufacturers. One example is "Airflow Instruments" accessible through their web site www.airflowinstruments.co.uk. Make a study of the different models available from them and write a short note on how you would choose an appropriate model for measurement of air velocity in a rectangular duct of 100 × 150mm cross section. The velocity range of interest is 0.05 – 4m/s.

Ex III.22: Hot wire anemometer systems are available from several manufacturers such as A.A Lab Systems, U.S.A. accessible at www.lab-systems.com. Study the specifications of $AN - 1005$ Hot-Wire Anemometry System supplied by them. Comment on its use for turbulence measurement.

Ex III.23: Brief description of LDV has been provided in Chapter 5. Prepare a more detailed note on LDV systems by using the library resources.

III.3   Volume flow rate

Ex III.24: An orifice meter with pressure taps one diameter upstream and one-half diameter downstream is installed in a 55 mm diameter pipe and used to measure a maximum flow of water of $4l/s$ (ρ = 999.5 kg/m^3, $\mu = 9.75 \times 10^{-4} kg/m \cdot s$ at 21 $°C$). For this orifice, $\beta = 0.5$. A differential pressure gauge is to be selected which has an accuracy of 0.25% of full scale and the upper scale limit is to be selected to correspond to the maximum flow rate given above. Determine the maximum range of the differential pressure gauge and estimate the uncertainty in the mass flow measurement at nominal flow rates of 4 and 2 l/s.

Ex III.25: An orifice plate of diameter ratio 0.65 is used to measure the flow of water at room temperature in a 52 ± 1 mm ID pipe. The water flow velocity in the pipe is known to be 3.5 ± 0.05 m/s. Determine the head developed across the orifice plate in Pa and mm of mercury if $D, D/2$ taps are made use of. Also determine the uncertainty in the head developed.

Ex III.26: A sonic nozzle has an orifice diameter of $0.8mm$ and the upstream air temperature remains fixed at $20°C$. Determine the mass flow rate of dry air through the sonic orifice if the upstream pressure varies between 100 and $1000kPa$. Assume that the downstream pressure is always less than the critical pressure.

Ex III.27: An orifice meter is to be used to measure flow of water at 10 bar and 25 $°C$. The maximum flow rate is 1 kg/s. The minimum flow rate is to be 0.3 kg/s. Determine the size of the orifice plate such that the pipe Reynolds number is not less than 10000. Calculate the differential pressure developed across the orifice plate if $\beta = 0.43$. Assume that $D - D/2$ taps are made use of.

Ex III.28: An orifice plate with area ratio of 0.5 is used with $D, D/2$ taps for the measurement of water flow at an upstream absolute pressure of 2 bar. A differential pressure transmitter with $100mm$ water head full scale is used to measure the head developed. If the pipe diameter is 24 mm what is the flow rate if the head developed is 65 mm water column?

Ex III.29: A venturi is designed to measure flow in a tube of 8 cm ID. The flowing fluid is dry air at atmospheric pressure and room temperature of 25 $°C$. The Reynolds number at the throat is to be not less than 2×10^5. Use a β value of 0.62 for the venturi. What is the minimum flow rate that may be measured using this venturi? What is the corresponding pressure drop in mm of water column?

Ex III.30: A venturi is to be used to measure flow of air at 1 *bar* and 60°C. Maximum flow rate is 0.1 m^3/s. The minimum flow rate is to be 0.03 m^3/s. Determine the size of the venturi such that the pipe Reynolds number is not less than 100000. Calculate the differential pressure developed across the venturi if $\beta = 0.55$.

Ex III.31: Redo problem III.30 assuming that the installation is an orifice plate with corner taps. The area ratio for the orifice plate is to be taken as 0.3.

Ex III.32: A venturi is to be used to measure flow of air at 2 *bar* and 25°C. The maximum flow rate is 0.8 *kg/s*. The minimum flow rate is to be 0.4 *kg/s*. Determine the size of the venturi such that the pipe Reynolds number is not less than 150000. Calculate the differential pressure developed across the venturi if $\beta = 0.55$.

Ex III.33: A venturi is to be used to measure flow of air at 5 *bar* and 25 °C. The maximum flow rate is 1 *kg/s*. The minimum flow rate is to be 0.3 *kg/s*. Determine the size of the venturi such that the pipe Reynolds number is not less than 100000. Calculate the differential pressure developed across the venturi if $\beta = 0.6$.

Ex III.34: A rotameter is to be used for measurement of the flow of water at 6 bar and 21°C. The maximum flow rate is 1.2 *kg/s*, the inlet diameter of the meter is 25 *mm*, and the length of the meter is 15 *cm*. The bob is constructed so that its density is 3 times that of water and its volume is 3 *ml*. Calculate the taper for a drag coefficient of 0.6 and determine the meter constant.

Ex III.35: A rotameter is chosen to measure air flow at 25°C and 1 atmosphere pressure. The bob is spherical and has a diameter of 5*mm*. Bob is made of cork with a density of 240kg/m^3 . The total length of the rotameter tube is 100*mm* and the diameter of the tube at its inlet is 5*mm*. The rotameter tube has a taper of 0.6°. What is the gage constant? What is the flow rate when the rotameter reading is 75*mm*? Assume a constant coefficient of drag for the bob of 0.4.

Ex III.36: The rotameter in Exercise III.35 is used in an installation where the air is at a pressure of 50*mm* water column gage and the temperature is 35°C. What is the correction factor that has to be applied to the gage constant calculated there?

Ex III.37: The rotameter in Exercise III.35 is used in an installation where the fluid is Nitrogen at a pressure of 25*mm* water column gage and the temperature is 15°C. What is the correction factor that has to be applied to the gage constant calculated there?

Ex III.38: A rotameter used for measurement of water flow has the following specifications:
Inlet diameter of Rotameter d = 0.025 *m*, length of tube L = 0.20 *m*, taper angle of rotameter is θ = 1°, diameter of spherical bob D_b = 0.025 *m*, drag coefficient for the spherical bob C_D = 0.65.
What is the maximum flow rate that can be measured using this rotameter?

Ex III.39: A rotameter used for measurement of water flow has the following specifications:

- Inlet diameter of Rotameter $d = 0.03\ m$, Length of tube $L = 0.25\ m$
- The taper angle of rotameter is $\theta = 1.2°$, Diameter of spherical bob $D_b = 0.03\ m$, Drag coefficient for the spherical bob $C_D = 0.8$.

What is the maximum flow rate that can be measured using this rotameter?

Ex III.40: A rotameter is to be used for measurement of the flow of air at 2 *bar* and 30°C. The maximum flow rate is 0.01 *kg/s*, the inlet diameter of the meter is 35 *mm*, and the length of the meter is 30 *cm*. The bob is constructed so that its density is 2 times that of air and its volume is 8 *ml* (the shape may be taken as being cylindrical with diameter equal to 35 *mm*.) Calculate the taper for a drag coefficient of 0.65 and determine the meter constant.

Ex III.41: A rotameter is used for an airflow measurement at conditions of 15 °C and 5400 *kPa* gauge pressure. The local barometer reads 740 *mm* mercury and the rotameter is rated at 100 *l/min* (full scale) at standard conditions of 1 *atm* and 20°C. Calculate the mass flow of air for a reading of 50 percent of full scale.

Ex III.42: A rotameter is to be used for measurement of the flow of air at 1.2 *bar* and 15 °C. The maximum flow rate is 0.002*kg/s*, the inlet diameter of the meter is 15 *mm*, and the length of the meter is 200 *mm*. The bob is constructed so that its density is 50 times that of air and its volume is 5 *ml*. Calculate the taper for a drag coefficient of 0.8 and determine the meter constant.

Thermo-physical properties, Radiation properties of surfaces, Gas concentration, Force/Acceleration,torque and power

Module IV considers the measurement of many quantities of interest to mechanical engineers. These occur in important branches of mechanical engineering, viz., thermal engineering, power engineering, design and manufacturing. Chapter 11 deals with the measurement of thermo-physical properties that play important role in thermal as well as manufacturing processes. Chapter 12 deals with radiation properties of surfaces such as the emissivity and reflectivity. The knowledge of these properties is vital in the design of space systems. Chapter 13 deals with the measurement of concentration of gases. With increasing emphasis on pollution control and monitoring in most applications, these measurements assume importance. Mechanical engineers deal with combustion devices and the monitoring of their operation consists largely in the measurement of concentration of gaseous constituents, apart from temperature and pressure that have been considered in earlier Modules. Chapter 14 deals primarily with the measurement of force and the related quantities such as acceleration, torque, power and so on. These occur in such mundane applications like weighing of an object to applications in the study of vibration and noise. Characterization of most power conversion devices (such as turbines and engines) requires the measurement of these quantities. Measurement of elastic properties of materials is not considered here since these form part of specialized courses. Also not considered here will be the measurement of quantities that are traditionally grouped under the head "Metrology". These are important in manufacturing engineering and are dealt with, again, by specialized books.

Chapter *11*

Measurement of thermo-physical properties

Thermo-physical properties are required in all engineering analysis and design. Hence the measurement of these is an important aspect of mechanical measurements. In this chapeter we deal will the measurement of commonly encountered thermo-physical properties such as thermal conductivity, specific heat, calorific value and viscosity.

11.1   Introduction

In mechanical engineering applications material properties are required for accurate prediction of their behavior as well for design of components and systems. The properties we shall be interested in measuring are those that may be referred to generally as thermo-physical properties. However, an important property of fuels, the calorific value is also included here since it is measured by methods that are basically liké the ones that are used for measuring thermo-physical properties.

Thus we shall be discussing the following measurement topics in this chapter:

- Thermal conductivity
- Heat capacity
- Calorific value of fuels
- Viscosity

Thermal conductivity and viscosity are also referred to as transport properties since these are important in heat and fluid flow.

11.2   Thermal conductivity

Thermal conductivity may be measured by either steady state methods or unsteady (transient) methods. This classification is based on the nature of the temperature field within the sample. The steady state methods are generally termed as calorimetric methods since they lead to the formation of a steady temperature field within the sample whose thermal conductivity is intended to be estimated. The heat transfer accompanying the steady temperature field is *estimated* by subtracting all losses from the heat supplied. This is basically what one refers to as calorimetry. In the case of unsteady methods the sample experiences a time varying or transient temperature field. The losses are usually negligible because of short measurement times.

In practice, an one dimensional temperature field - steady or transient - is set up in a sample of known geometric shape and size. In the case of steady state methods the Fourier heat conduction law in one dimension is applied by measuring the temperature gradient in terms of temperature difference and thickness of the specimen. Thermal conductivity is estimated using Fourier law.

In the case of transient techniques, an one dimensional transient temperature field is set up in the specimen and the thermo-physical properties such as thermal conductivity and the thermal diffusivity are estimated by invoking well known solutions to the one dimensional heat equation. The measurement times may be a few milliseconds to several seconds or minutes.

Since a large number of techniques are actually available, here we discuss only a few of them that are more commonly used in practice. The steady state methods that will be discussed are:

- Guarded hot plate method: Solid, Liquid
- Radial heat conduction apparatus: Liquid, Gas
- Thermal conductivity comparator: Solid

As far as unsteady methods are concerned, we discuss only one, viz. the Laser flash apparatus that is used for the determination of thermal properties of a solid material.

11.2.1 Basic ideas

Thermal conductivity is defined through Fourier law of heat conduction. In the case of one-dimensional heat conduction the appropriate relation that defines the thermal conductivity is

$$k = -\frac{q}{\frac{dT}{dx}} = -\frac{\frac{Q}{A}}{\frac{dT}{dx}} \tag{11.1}$$

In Equation 11.1 k is the thermal conductivity in $W/m°C$, q is the conduction heat flux in W/m^2 along the x direction, given by the ratio of total heat transfer by conduction Q and area normal to the heat flow direction A and T represents the temperature. In practice Equation 11.1 is replaced by

$$k = \frac{\frac{Q}{A}}{\frac{|\Delta T|}{\delta}} \tag{11.2}$$

Here $|\Delta T|$ represents the absolute value of the temperature difference across a thickness δ of the sample. Following assumptions are made in writing the above:

Heat conduction is one dimensional: Temperature field is steady and varies with only one space variable. Note that the space variable may be in a planar sample, a cylindrical sample or a spherical sample. Equation 11.2 is appropriate for a sample in the form of a slab.

Temperature variation across the sample is linear: This assumption presupposes that the thermal conductivity is a weak function of temperature, or the temperature difference is very small compared to the mean temperature of the medium

With this background, the following general principles may be enunciated, that are common to all methods of measurement of thermal conductivity:

- Achieve one dimensional temperature field within the medium
- Measure heat flux
- Measure temperature gradient
- Estimate thermal conductivity
- In case of liquids and gases suppress convection - more later
- Parasitic losses are reduced/ eliminated/estimated and are accounted for - in all cases

11.3 Steady state methods

11.3.1 Guarded hot plate apparatus: solid sample

The guarded hot plate apparatus is considered as the primary method of measurement of thermal conductivity of solid materials that are available in the form of a slab (or plate or blanket). The principle of the guard has already been explained in the case of heat flux measurement and heat transfer coefficient measurement in sections 6.1.5 and 6.2. It is a method of reducing or eliminating heat flow in an unwanted direction and making it take place in the desired direction. At once it will be seen that one-dimensional temperature field may be set up using this approach, in a slab of material of a specified area normal to heat flow direction and a sample of uniform thickness. Schematic of a guarded hot plate apparatus is shown in Figure 11.1.

Two samples of identical size are arranged symmetrically on the two sides of an assembly consisting of main and guard heaters. The two heaters are energized by independent power supplies with suitable controllers. The control action is driven by the temperature difference across the gap between the main and guard heaters. The controller may use proportional - integral - derivative control (PID) to adjust the heater powers and drive the temperature difference to zero. Heat transfer from the lateral edges of the sample is prevented by the guard backed by a thick layer of insulation all along the periphery. The two faces of each of the samples are maintained at different temperatures by heaters on one side and the cooling water circulation on the other side. However identical one-dimensional temperature fields are set up in the two samples.

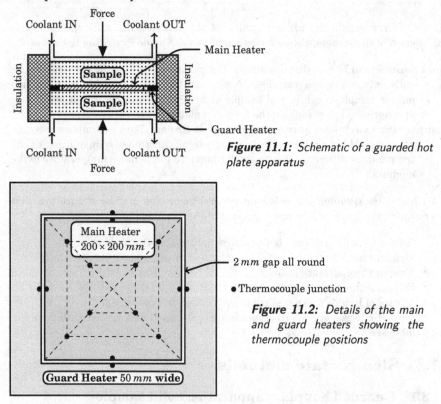

Figure 11.1: *Schematic of a guarded hot plate apparatus*

Figure 11.2: *Details of the main and guard heaters showing the thermocouple positions*

The details of the main and guard heaters along with the various thermocouples that are used for the measurement and control of the temperatures are shown in the plan view shown in Figure 11.2. As usual there is a narrow gap of 1 - 2 mm all round the main heater across which the temperature difference is measured and maintained at zero by controlling the main and guard heater inputs. However the sample is monolithic having a surface area the same as the areas of main, guard and gap all put together. Temperatures are averaged using several thermocouples that are fixed on the heater plate and the water cooled plates on the two sides of the samples. The thermal conductivity is then estimated based on Equation 11.2, where the heat transfer across any one of the samples is half that supplied to the main heater and the area is the face area of one of the samples covering the *main heater* only.

Typically the sample, in the case of low conductivity materials, is 25 *mm* thick and the area occupied by the main heater is 200×200 *mm*. The heat input is adjusted such that the temperature drop across the sample is of the order of 5°C. In order to improve the contact between heater surface and the sample surface a film of high conductivity material may be applied between the two. Invariably an axial force is also applied using a suitable arrangement so that the contact between surfaces is thermally good. However care should be taken to see that the applied force does not change the density of the sample material, especially when the sample is in the form of a blanket.

Example 11.1

A guarded hot plate apparatus is used to measure the thermal conductivity of an insulating material. The specimen thickness is 25 ± 0.5 *mm*. The heat flux is measured within *1%* and is nominally $80W/m^2$. The temperature drop across the specimen under the steady state is $5\pm0.2°C$. Determine the thermal conductivity of the sample along with its uncertainty.

Solution :

The given data is written down as (all are nominal values):

$$q = 80\,W/m^2; \Delta T = 5°C; \delta = 25\,mm = 0.025\,m$$

Using Equation 11.2, the nominal value of the thermal conductivity is

$$k = \frac{q\delta}{\Delta T} = \frac{80 \times 0.025}{5} = 0.4\,W/m°C$$

The uncertainties in the measured quantities specified in the problem are

$$u_q = \pm\frac{80}{100} = \pm0.8\,W/m^2;\ u_{\Delta T} = \pm0.2°C;\ u_\delta = \pm0.5mm = \pm0.0005m$$

Logarithmic differentiation is possible and the error in the thermal conductivity estimate may be written down as

$$u_k = \pm k\sqrt{\left(\frac{u_q}{q}\right)^2 + \left(\frac{u_{\Delta T}}{\Delta T}\right)^2 + \left(\frac{u_\delta}{\delta}\right)^2}$$

$$= \pm0.4 \times \sqrt{\left(\frac{0.8}{80}\right)^2 + \left(\frac{0.2}{5}\right)^2 + \left(\frac{0.0005}{0.025}\right)^2} = \pm0.018\,W/m°C$$

Thus the thermal conductivity is estimated within an error margin of

$$\text{Percent uncertainty in k} = \frac{u_k}{k} \times 100 = \pm\frac{0.018}{0.4} \times 100 \approx \pm4.6\%$$

11.3.2　Guarded hot plate apparatus: liquid sample

Measurement of thermal conductivity of liquids (and gases) is difficult because it is necessary to make sure that the fluid is stationary. In the presence of temperature variations and in the presence of gravity the fluid will start moving around due to natural convection. There are two ways of immobilizing the fluid:

- Use a thin layer of the fluid in the direction of temperature gradient so that the Grashof number is very small and the regime is conduction dominant.
- Set up the temperature field in the fluid such that the hot fluid is above the cold fluid and hence the layer is in the stable configuration.

The guarded hot plate apparatus is suitably modified to achieve these two conditions.

Figure 11.3 shows schematically how the conductivity of a liquid is measured using a guarded hot plate apparatus. The symmetric sample arrangement in the case of a solid is *replaced* by a single layer of liquid sample with a guard heater at the top. Heat flow across the liquid layer is downward (along the direction of gravity) and hence the liquid layer is in a stable configuration. The thickness of the layer is chosen to be very small (of the order of a mm) so that heat transfer is conduction dominant. The guard heat input is so adjusted that there is no temperature difference across the gap between the main and the guard heaters. It is evident that all the heat input to the main heater flows downwards through the liquid layer and is removed by the cooling arrangement. Similarly the heat supplied to the guard heater is removed by the cooling arrangement at the top (not shown in the figure).

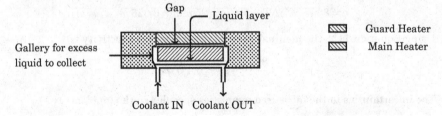

Figure 11.3: *Guarded hot plate apparatus for the measurement of thermal conductivity of liquids*

11.3.3　Radial heat conduction apparatus for liquids and gases

Another apparatus suitable for the measurement of thermal conductivity of fluids (liquids and gases) is one that uses radial flow of heat through a very thin layer of liquid or gas. A cross sectional view of such an apparatus is shown in Figure 11.4. The heater is in the form of a cylinder (referred to as bayonet heater or the plug) and is surrounded by a narrow radial gap that is charged with the liquid or the gas whose thermal conductivity is to be measured. Liquid may be charged in to the gap using a syringe pump. The outer cylinder is actually a jacketed cylinder that is cooled by passing cold water. Heat loss from the bayonet heater except through the annular fluid filled gap is minimized by the use of proper materials. Thermocouples are arranged to measure the temperature across the fluid layer. Since the gap (the thickness of the fluid layer) is very small compared to the diameter of the heater heat conduction across the gap may be very closely approximated by that across a slab. Hence Equation 11.2 may be used in this case also.

Typical specifications of an apparatus of this type are given below:

- Diameter of cartridge heater $D = 37\,mm$
- Radial clearance $= 0.3\,mm$
- Heat flow area $A = 0.0133\,m^2$
- Temperature difference across the gap $\Delta T \approx 5°C$
- Heater input $Q = 20 - 30\,W$

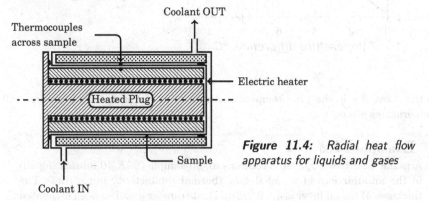

Thermocouples across sample

Coolant OUT

Electric heater

Heated Plug

Sample

Coolant IN

Figure 11.4: *Radial heat flow apparatus for liquids and gases*

The diameter is thus about 75 times the layer thickness. The use of this apparatus requires a calibration experiment using a fluid of known thermal conductivity (usually dry air which may be treated as standard reference material) filling the gap. Thermal conductivity of air is well known (as a function of temperature) and is available as tabulated data in data handbooks. If an experiment is conducted with dry air, heat transferred across the gap may be determined using Fourier law of heat conduction. The heat input to the bayonet heater is measured and the difference between these two should represent the heat loss. The experiment may be conducted with different amounts of heat input (and hence different temperature difference across the air layer) and the heat loss estimated. This may be represented as a function of the temperature difference across the gap. When another fluid is in the annular gap, the heat loss may be *assumed* to be given by the previously measured values. Hence the heat loss may be deducted from the heat input to get the actual heat transfer across the fluid layer. At once Equation 11.2 will give us the thermal conductivity of the fluid.

Heat loss data has been measured in an apparatus of this kind and is shown as a plot in Figure 11.5. The data shows mild nonlinearity. Hence the heat loss Q_L in W may be represented as a polynomial function of the temperature difference across the gap using regression analysis. The heat loss is a function of $\Delta T = T_P - T_J$ and is given by the polynomial

$$Q_L = 0.0511 + 0.206\Delta T + 0.0118\Delta T^2 - 0.000153\Delta T^3 \tag{11.3}$$

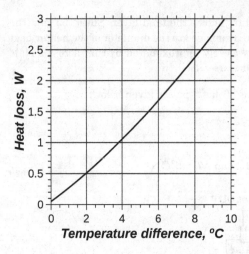

Figure 11.5: *Heat loss calibration data for a radial flow thermal conductivity apparatus*

In the above T_P is the plug temperature and T_J is the jacket temperature. All temperatures are in $°C$.

Example 11.2

A radial heat flow apparatus contains an oil sample (*SAE 40* lubricating oil) in the annular gap of a radial flow thermal conductivity apparatus. The thickness of the oil layer is $\delta = 0.3\,mm$. Heat transfer area has been estimated from the manufacturer data to be $A = 0.0133\,m^2$. The electric heater is specified to have a resistance of $R = 53.5\,\Omega$. The following data was collected in an experiment:

- Heater voltage: $V = 40\,V$
- Plug temperature: $T_P = 32.9°C$
- Jacket temperature: $T_J = 28.2°C$

What is the thermal conductivity of the oil sample? If the measured parameters have the following uncertainties what will be the uncertainty in the estimated value of the thermal conductivity?

$$u_V = \pm 0.5\,V \quad u_T = \pm 0.2°C$$

The heat loss is given by Equation 11.3 and is susceptible to an uncertainty of $\pm 0.5\%$

Solution :

First we determine the nominal value of the thermal conductivity using the nominal values of all the measured quantities.

The electrical heat input to the heaters is given by

$$Q_e = \frac{V^2}{R} = \frac{40^2}{53.5} = 29.91\,W$$

The temperature drop across the sample liquid is

$$\Delta T = T_P - T_J = 32.9 - 28.2 = 4.7°C$$

The heat loss may at once be calculated using Equation 11.3 as

$$Q_L = 0.0511 + 0.206 \times 4.7 + 0.0118 \times 4.7^2 - 0.000153 \times 4.7^3 = 1.26\ W$$

(This agrees with the value shown in Figure 11.5) The heat conducted across the liquid layer is then given by

$$Q = Q_e - Q_L = 29.91 - 1.26 = 28.95\ W$$

Using Equation 11.2 the nominal value of the thermal conductivity of oil sample is

$$k = \frac{Q}{A}\frac{\delta}{\Delta T} = \frac{28.65}{0.0133} \times \frac{0.3 \times 10^{-3}}{4.7} = 0.138\ W/m°C$$

Now we calculate the uncertainty in the nominal value of the thermal conductivity estimated above. Only the heat transferred and the temperatures are susceptible to error. We know that $Q = Q_e - Q_L = \frac{V^2}{R} - Q_L$. Assuming that R is not susceptible to any error, we have

$$\frac{\partial Q}{\partial V} = \frac{2V}{R} = \frac{2 \times 40}{53.5} = 1.495\ W/V;\quad \frac{\partial Q}{\partial Q_L} = -1$$

Hence the error in the measured value of the heat conducted across the liquid layer is given by (using error propagation formula)

$$u_Q = \pm\sqrt{\left(\frac{\partial Q}{\partial V}u_V\right)^2 + \left(\frac{\partial Q}{\partial Q_L}u_L\right)^2} = \pm\sqrt{(1.495 \times 0.5)^2 + (-1 \times 0.05 \times 1.26)^2} = \pm 0.75\ W$$

The errors in the measured temperatures are equal and hence the error in the measured temperature difference is

$$u_{\Delta T} = \pm\sqrt{2}u_T = \pm\sqrt{2} \times 0.2 = 0.283°C$$

The error propagation formula then gives

$$u_k = \pm k\sqrt{\left(\frac{u_Q}{Q}\right)^2 + \left(\frac{u_{\Delta T}}{\Delta T}\right)^2} = \pm 0.138\sqrt{\left(\frac{0.75}{28.65}\right)^2 + \left(\frac{0.283}{4.7}\right)^2} = \pm 0.009\ W/m°C$$

Thus the measured value of thermal conductivity of oil sample is $0.138 \pm 0.009\ W/m°C$.

11.3.4 Thermal conductivity comparator

Thermal conductivity comparator is a method in which the thermal conductivity of a sample is obtained by comparison with another sample of known thermal conductivity. This method is especially useful for the determination of thermal conductivity of good conductors, such as metals and alloys. The principle of the

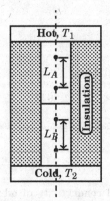

- Thermocouples
- A: Standard Reference Material
- B: Sample of unknown k

Figure 11.6: *Schematic of a thermal conductivity comparator*

method may be explained with reference to Figure 11.6. A sample of standard reference material (SRM - a material whose thermal conductivity is known and certified by the manufacturer) is placed in series with the material whose thermal conductivity needs to be estimated. Both the materials have identical cross section (usually cylindrical) and heat is allowed to flow, under the steady state, as indicated in the figure. Thermocouples are arranged as shown in order to estimate the temperature gradients in each material. Heat loss in the lateral direction is prevented by the provision of insulation as shown. With the nomenclature of Figure 11.6, we have the following:

$$k_A \frac{\Delta T_A}{L_A} = k_B \frac{\Delta T_B}{L_B} \text{ or } k_B = k_A \frac{L_B}{L_A} \frac{\Delta T_A}{\Delta T_B}$$

(11.4)

The above expression is based on the fact that the conduction heat flux through the sample and the SRM is the same. The lengths and temperature differences are the measured quantities, A is the SRM and B is the sample.

Example 11.3

A thermal conductivity comparator uses a standard reference material (SRM) of thermal conductivity $45 \pm 2\%$ $W/m°C$. Two thermocouples placed $22 \pm 0.25\ mm$ apart indicate a temperature difference of $2.5 \pm 0.2°C$. The material of unknown thermal conductivity is in series with the SRM and indicates a temperature difference of $7.3 \pm 0.2°C$ across a length of $20 \pm 0.25\ mm$. Determine the thermal conductivity of the sample and its uncertainty.

Solution :

The given data may be written down using the nomenclature of Figure 11.6.

$$k_A = 45\ W/m°C;\ L_A = 22\ mm;\ \Delta T_A = 2.5°C;\ L_B = 20\ mm;\ \Delta T_B = 7.3°C$$

The nominal value of the thermal conductivity of the sample is obtained using Equation 11.4 as

$$k_B = 45 \times \frac{20}{22} \frac{2.5}{7.3} = 14\ W/m°C$$

The uncertainties specified are all written as percentage uncertainties.

$$u_{k_A} = \pm 2\%;\ u_{L_A} = \pm \frac{0.25}{22} \times 100 = 1.14\%;\ u_{\Delta T_A} = \pm \frac{0.2}{2.5} \times 100 = \pm 8\%$$

$$u_{L_B} = \pm \frac{0.25}{20} \times 100 = 1.25\%; \quad u_{\Delta T_B} = \pm \frac{0.2}{7.3} \times 100 = \pm 2.74\%$$

Since the unknown thermal conductivity depends on the other quantities involving only products of ratios, the percentage error may be directly calculated using percent errors in each of the measured quantities. Thus

$$
\begin{aligned}
u_{k_B} &= \pm \sqrt{u_{k_A}^2 + u_{L_A}^2 + u_{L_B}^2 + u_{\Delta T_A}^2 + u_{\Delta T_B}^2} \\
&= \pm \sqrt{2^2 + 1.14^2 + 1.25^2 + 8^2 + 2.74^2} = \pm 8.85\%
\end{aligned}
$$

The uncertainty is large and the result is of poor quality. Temperatures need to measured with better precision to improve the method.

11.4   Transient method

Though many methods are available under the unsteady category only one of them, viz. the laser flash method, will be considered as a representative one. Also it is the most commonly used method in laboratory practice and laser flash apparatus are available commercially, even though they are very expensive.

11.4.1   Laser flash method

The laser flash method imposes a pulse of heat to a thin sample and monitors the back surface temperature as a function of time. The schematic of the laser flash apparatus is shown in Figure 11.7.

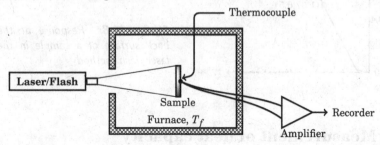

Figure 11.7: *Schematic of the laser flash apparatus*

The sample is in the form of a thin slab and is maintained at the desired temperature by placing it in a furnace. The front face of the slab is heated with a laser or flash pulse and the temperature of the back face is monitored as a function of time. The laser or flash pulse is of such intensity that the temperature will rise by only a few degrees. In other words the temperature rise is very small in comparison with the mean temperature of the sample. If the thermal diffusivity of the material of the sample is α, the non-dimensional time is defined as $Fo = \dfrac{\alpha t}{\delta^2}$ where t is time and δ is the sample thickness. Non-dimensional time is also referred to as the Fourier number. The non-dimensional temperature is defined as $\theta = \dfrac{T}{T_{max}}$ where T_{max} is

the maximum temperature reached by the back surface of the sample. Figure 11.8
shows the shape of the response in terms of the non-dimensional coordinates. Note
that it is not necessary to know the temperature in absolute terms since only the
ratio is involved. It is found that the non-dimensional temperature response is 0.5
at a non-dimensional time of 1.37 as indicated in the figure. The quantity that is
estimated from the measurement is, in fact, the thermal diffusivity of the material.
Thus if $t_{\frac{1}{2}}$ is the time at which the response is 0.5, we have

$$\boxed{\alpha = 1.37\frac{L^2}{t_{\frac{1}{2}}}} \tag{11.5}$$

If the density ρ and specific heat capacity C of the material are known (they may
be measured by other methods, as we shall see later) the thermal conductivity is
obtained as

$$k = \rho C \alpha = 1.37\rho C\frac{L^2}{t_{\frac{1}{2}}} \tag{11.6}$$

In practice the temperature signal is amplified and manipulated by computer
software to directly give the estimate of the thermal diffusivity of the solid sample.

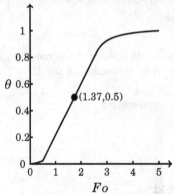

Figure 11.8: *Response at the back surface of a sample in the laser flash method*

11.5   Measurement of heat capacity

Heat capacity is an important thermo-physical property that is routinely measured in the laboratory. In material characterization heat capacity is one of the
important properties whose changes with temperature indicate changes in the
structure of the material itself. The method used for measurement of heat capacity
is usually a calorimetric method, where energy balance in a controlled experiment
gives a measure of the heat capacity.

11.5.1   Heat capacity of a solid

We consider first the measurement of heat capacity of a solid material. The
basic principle of the calorimetric method is the application of the first law of
thermodynamics (heat balance) in a carefully conducted experiment. A weighed
sample of solid material, in granular or powder form, whose heat capacity is to be

measured is heated prior to being transferred quickly into a known mass of water or oil (depending on the temperature level) contained in a jacketed vessel (see Figure 11.9. Let the mass of the solid be m and let it be at a temperature T_3 before it is dropped into the calorimeter. Let T_1 be the initial temperature of the calorimeter and the jacket. Let the mass specific heat product of the calorimeter, stirrer and the liquid in the calorimeter be W. Let T_2 be the maximum temperature reached by the calorimeter with its contents including the mass dropped in it. Let C be the specific heat of the solid sample that is being estimated using the experimental data. We recognize that there is some heat loss that would have taken place over a period of time at the end of which the temperature has reached T_2. Let the heat loss be Q_L.

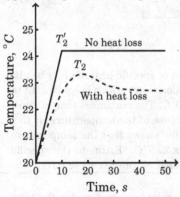

Drop hot sample

Stirrer

② ①

Constant temperature
water jacket

Calorimeter containing
water or oil

1,2: Precision thermometers

Figure 11.9: *Calorimeter for the measurement of specific heat of a solid*

Energy balance requires that

$$W(T_2 - T_1) + Q_L = mC(T_3 - T_2) \tag{11.7}$$

Now let us look at how we estimate the heat loss. For this we show, in Figure 11.10, the typical temperature-time plot that is obtained in such an experiment. If there

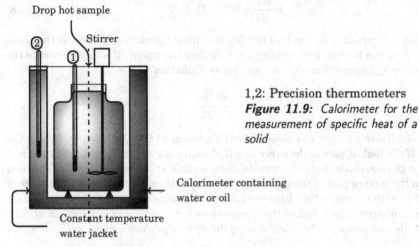

Figure 11.10: *Temperature time plot for the calorimeter*

were to be no loss the temperature-time trace should be like that shown by the solid trace made up of two straight lines. After the temperature reaches T_2' it should remain at that value! However there is heat loss and the temperature-time trace is a curve that follows the broken line. After the temperature reaches a maximum value of T_2 (lower than T_2') it decreases with time. The rate of decrease is, in fact, an indication of how large the heat loss is. If we assume that temperature rise of the calorimeter is very small during the mixing part of the experiment, we may assume the rate of heat loss to be a linear function of temperature difference with respect to the ambient given by $Q_L = K(T - T_\infty)$ where T_∞ is the ambient temperature. Now consider the state of affairs after the maximum temperature has been passed. The calorimeter may be assumed to be a first order system governed by the equation

$$W\frac{dT}{dt} + K(T - T_\infty) = 0 \tag{11.8}$$

Here we assume that K, the loss coefficient during the cooling process is the same as the K in the heating part (mixing part) of the experiment. If we approximate the derivative by finite differences, we may recast Equation 11.8 as

$$K = \frac{W\frac{\Delta T}{\Delta t}}{T - T_\infty} \tag{11.9}$$

The numerator on the right hand side of Equation 11.9 is obtained by taking the slope of the cooling part of the curve at any chosen t after the cooling has started and thus at the corresponding temperature. Now we look at the heat loss term. The heat loss in the mixing part of the experiment may be obtained by integrating the rate of heat loss with respect to time between $t = 0$ and the time t_{max} at which the maximum temperature is reached. In fact the temperature time curve may be *approximated* by a triangle and hence the heat loss during the mixing process is given by

$$Q_L = K\int_0^{t_{max}} (T - T_\infty)dt \approx K\frac{(T_2 - T_\infty)}{2}t_{max} \tag{11.10}$$

This is substituted in Equation 11.7 and rearranged to get the following expression for the estimated value of C:

$$C = \frac{\left(W + \frac{Kt_{max}}{2}\right)}{m(T_3 - T_2)}(T_2 - T_\infty) \tag{11.11}$$

Example 11.4

In a calorimeter experiment for determination of specific heat of glass beads, $0.15\,kg$ of beads at $80°C$ are dropped in to a calorimeter which is equivalent to $0.5\,kg$ of water, at an initial temperature of $20°C$. The maximum temperature of $22.7°C$ of the mixture occurs by linear increase of the temperature in $10\,s$ after the mixing starts. The subsequent cooling shows that the temperature drops at $0.2°C/min$ when the temperature is $22.7°C$. Estimate the specific heat of glass beads. Assume that the specific heat of water is $4.186\,kJ/kg°C$.

Solution :

The given data is written down using the nomenclature introduced in the text:

$$m = 0.15\,kg;\ T_\infty = 20°C;\ T_2 = 22.7°C;\ T_3 = 80°C$$

The thermal capacity of the calorimeter has been specified as being equivalent to $m_w = 0.5\ kg$ of water whose specific heat has been specified as $C_w = 4.186\ kJ/kg°C$. This is known as the *water equivalent* of the calorimeter and hence

$$W = m_w \times C_w = 0.5 \times 4.186 = 2.093\ kJ/°C$$

Cooling rate obtained in the experiment has been specified as

$$\frac{\Delta T}{\Delta t} = 0.2°C/min = 0.2/60 = 0.0033°C/s$$

The corresponding temperature has been given as $T = 22.7°C$. The cooling constant may hence be calculated using Equation 11.9 as

$$K = \frac{2.093 \times 0.0033}{22.7 - 20} = 0.0026\ kW/°C$$

The time of mixing is given as $t_{max} = 10\ s$ and the temperature increase during mixing is linear. Hence the heat loss during the mixing process is approximated as

$$Q_L = \frac{K(T_2 - T_\infty)}{2} t_{max} = \frac{0.0026 \times (22.7 - 20)}{2} \times 10 = 0.035\ kJ$$

The specific heat of glass bead sample may now be estimated using Equation 11.11 as

$$C_{gb} = \frac{2.093 \times (22.7 - 20) + 0.035}{0.15 \times (80 - 22.7)} = 0.661\ kJ/kg°C$$

The thermal mass of the glass beads is

$$W_{gb} = m \times C_{gb} = 0.15 \times 0.661 = 0.099\ kJ/°C$$

Since the thermal mass of the beads is very small compared to the thermal mass of the calorimeter it is alright to ignore it in calculating the heat loss, a point the reader should be able to appreciate.

11.5.2   Heat capacity of liquids

Essentially the same calorimeter that was used in the case of solid samples may be used for the measurement of specific heat of liquids, with the modifications shown in Figure 11.11. The sample liquid is taken in the calorimeter vessel and allowed to equilibrate with the temperature of the constant temperature liquid in the jacket and the rest of the works. The heater is energized for a specific time duration so that a measured amount of thermal energy is added to the sample liquid. Stirrer (not shown) will help the liquid sample to be at a uniform temperature. The temperature increase of the liquid at the end of the heating period is noted down. The liquid specific heat is then estimated using energy balance. Effect of heat loss may be taken into account by essentially the same procedure that was used in the earlier case.

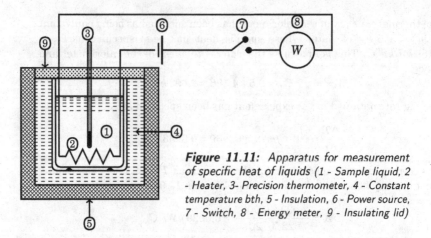

Figure 11.11: Apparatus for measurement of specific heat of liquids (1 - Sample liquid, 2 - Heater, 3- Precision thermometer, 4 - Constant temperature bth, 5 - Insulation, 6 - Power source, 7 - Switch, 8 - Energy meter, 9 - Insulating lid)

11.6   Measurement of calorific value of fuels

Calorific or heating value of a fuel is one of the most important properties of a fuel that needs to be measured before the fuel can be considered for a specific application. The fuel may be in the liquid, gaseous or solid form. The measurement of heating value is normally by calorimetry. While a continuous flow calorimeter is useful for the measurement of heating values of liquid and gaseous fuels, a bomb calorimeter is used for the measurement of heating value of a solid fuel.

11.6.1   Preliminaries

Heating value of a fuel or more precisely, the standard enthalpy of combustion is defined as the enthalpy change ΔH_T^0 that accompanies a process in which the given substance (the fuel) undergoes a reaction with oxygen gas to form combustion products, all reactants and products being in their respective standard states at $T = 298.15\ K = 25°C$, indicated by superscript 0. A bomb calorimeter (to be described later) is used to determine the standard enthalpy of combustion. The process that takes place is at a constant volume. Hence the process that occurs in a calorimeter experiment for the determination of the standard enthalpy of combustion is process I in Figure 11.12. What is desired is the process shown as III in the figure. We shall see how to get the information for process III from that measured using process I.

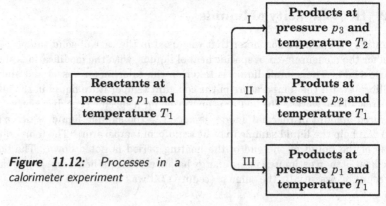

Figure 11.12: Processes in a calorimeter experiment

Let the calorimeter have a heat capacity equal to W. The change in the internal energy of the calorimeter when the temperature changes from T_1 to T_2 is $W(T_2 - T_1)$ assuming that W is independent of pressure. If we neglect work interaction due to stirring, I law applied to a perfectly insulated calorimeter ($Q = 0$) shows that $\Delta U_I = 0$ during process I. In the process shown as II the products and the calorimeter are at temperature T_1. Thus, for process II we should have

$$\Delta U_{II} = \Delta U_I - \int_{T_1}^{T_2} W dT = -\int_{T_1}^{T_2} W dT$$

Thus it is clear that, for process II, since the temperature remains at T_1,

$$\Delta U_{II} = \Delta U_{T_1} = -\int_{T_1}^{T_2} W dT = \Delta U_{T_1}^0$$

The last part results by noting that the internal energy is assumed to be independent of pressure. Note that by definition $\Delta H_{T_1}^0 = \Delta U_{T_1}^0 + \Delta(pV)$. Since the products are in the gaseous state, assuming the products to behave ideally, we have

$$\Delta(pV) = p_2 V_2 - p_1 V_1 = (n_2 - n_1)\Re T_1$$

where the n_2 represents the number of moles of the product and n_1 the number of moles of the reactant and $\Re$ is the universal gas constant. Thus it is possible to get the desired heat of combustion as

$$\Delta H_{T_1}^0 = (n_2 - n_1)\Re T_1 - \int_{T_1}^{T_2} W dT \tag{11.12}$$

In a typical experiment the temperature change is limited to a few degrees and hence the specific heat may be assumed to be a constant. Under this assumption Equation 11.12 may be replaced by

$$\boxed{\Delta H_{T_1}^0 = (n_2 - n_1)\Re T_1 - W(T_2 - T_1)} \tag{11.13}$$

11.6.2 The Bomb calorimeter

Schematic of a bomb calorimeter used for the determination of heating value (standard enthalpy of combustion) of a solid fuel is shown in Figure 11.13. The bomb is a heavy walled pressure vessel within which the combustion reaction will take place at constant volume. At the bottom of the bomb is placed sufficient amount of water such that the atmosphere within the bomb remains saturated with water vapor throughout the experiment. This guarantees that the water that may be formed during the combustion reaction will remain in the liquid state. The bomb is immersed within a can of water fitted with a precision thermometer capable of a resolution of $0.01\,^\circ C$. This assembly is placed within an outer water filled jacket. The jacket water temperature remains the same both before and after the combustion within the bomb. There is no heat gain or loss to the bomb from outside and the process may be considered to be adiabatic. The fuel is taken in the form of a pellet (about $1\,g$) and the combustion is accomplished by initiating it by an electrically heated fuse wire in contact with the pellet. The bomb is filled with oxygen under high pressure (25 *bar*) such that there is more than enough oxygen to guarantee complete combustion. The heating value is estimated after accounting for the heat generated by the fuse wire consumed to initiate combustion.

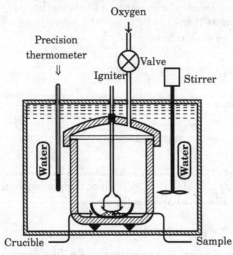

Figure 11.13: *Schematic of a Bomb calorimeter*

The details of construction and the external appearance of a bomb calorimeter are shown in Figure 11.14. The bomb calorimeter has approximately a diameter of 25 *cm* and a height of 30 *cm*.

Figure 11.14: *Photographic details of the "Bomb"*

Benzoic acid ($C_7H_6O_2$ - solid) is used as a standard reference material of known heat of reaction $\Delta H^0 = -3227\,kJ/mol$. Benzoic acid is taken in the form of a pellet and burnt in a bomb calorimeter to provide the data regarding the heat capacity of calorimeter. A typical example is given below.

Example 11.5

A pellet of Benzoic acid weighing 0.103 *g* is burnt in a bomb calorimeter and the temperature of the calorimeter increases by 2.17 °*C*. What is the

heat capacity of the calorimeter? Take the value of universal gas constant as
$\Re = 8.3143\,kJ/mol \cdot K$

Solution :

We indicate the phase of the reactants/products as g: gas; l: liquid; s: solid.
Consider now the combustion reaction.

$$C_7H_6O_2(s) + 7.5O_2(g) \rightarrow 7CO_2(g) + 3H_2O(l)$$

Thus 7.5 moles of oxygen are required for complete combustion of Benzoic acid.
The change in the number of moles during the reaction is given by taking
the gaseous constituents before and after the reaction. We thus have $n_1 =$
number of moles of oxygen = 7.5; $n_2 =$ number of moles of carbon dioxide = 7.
Thus we have

$$\Delta n = n_2 - n_1 = 7.5 - 7 = -0.5\,mol$$

Per mole of Benzoic acid basis, we have

$$\Delta U = \Delta H - \Delta n \Re T = -3227 - (-0.5) \times 8.3143 \times 298 = -3225.8\,kJ/mol$$

We use the following molecular weights:

$$M_C = 12.011;\quad M_H = 1.008;\quad M_O = 15.9995$$

The molecular weight of Benzoic acid M_B is then calculated as

$$M_B = 12.011 \times 7 + 1.008 \times 6 + 15.9995 \times 2 = 122.12\,g/mol$$

Thus 0.103 g of Benzoic acid corresponds to $\dfrac{0.103}{122.12} = 0.000843\,mol$. The heat
released by combustion of this is given by

$$Q = 0.000843 \times 3225.8 = 2.7207\,kJ$$

The temperature increase of the calorimeter has been specified as $\Delta T = 2.17°C$. Hence the heat capacity of the calorimeter is

$$W = \frac{Q}{\Delta T} = \frac{2.7207}{2.17} = 1.2538\,kJ/°C$$

Example 11.6

In a subsequent experiment, the calorimeter of Example 11.5 is used to
determine the heating value of sugar. A pellet of sugar weighing 0.303 g is
burnt and the corresponding temperature rise indicated by the calorimeter is
3.99 $°C$. What is the heat of combustion of sugar?

Solution :

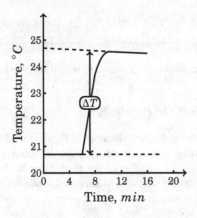

Figure 11.15: Bomb calorimeter data for Example 11.6

Figure 11.15 shows the variation of temperature of the calorimeter during the experiment. Since heat losses are unavoidable the temperature will start reducing as shown in the figure. By using ΔT as shown in the figure the effect of the heat losses may be accounted for. Note that the work due to stirrer may itself compensate for the heat loss to some extent. The slope of the cooling curve takes this into account naturally.

Molecular formula for sugar is $C_{12}H_{22}O_{11}$. The molecular weight of sugar M_S is found as

$$M_S = 12.011 \times 12 + 1.008 \times 22 + 15.9995 \times 11 = 342.2 \, g/mol$$

The mass of sugar pellet of $0.303 \, g$ corresponds to $n_S = \dfrac{0.303}{342.2} = 0.000885 \, mol.$ The heat release in the calorimeter is calculated based on $W = 1.2538 \, kJ/°C$ and $\Delta T = 3.99°C$ as

$$Q = -W\Delta T = -1.2538 \times 3.99 = -5.003 \, kJ$$

The combustion reaction for sugar is given by

$$C_{12}H_{22}O_{11}(s) + 12O_2(g) \rightarrow 12CO_2(g) + 11H_2O(l)$$

Hence there is no change in the number of moles before and after the reaction. Thus the heat of reaction is no different from the heat released and thus

$$\Delta H^0 = -\frac{Q}{n_S} = \frac{5.003}{0.000885} = 5651.9 \, kJ/mol$$

11.6.3 Continuous flow calorimeter

A continuous flow calorimeter is useful for determining the heating value of gaseous fuels. Figure 11.16 shows the schematic of a such a calorimeter. We assume that all the processes that take place on the gas side are at a *mean* pressure equal to the atmospheric pressure. The gas inlet pressure may be just "a few *mm* of water column gauge" and hence this assumption is a good one. As the name indicates, the processes that take place in the calorimeter are in the steady state with continuous flow of the gas air mixture (air provides oxygen for combustion) and the coolant (water) through the cooling coils. As indicated temperatures and flow rates are measured using appropriate devices familiar to us from earlier chapters. The gas air mixture is burnt as it issues through a nozzle that is surrounded by a cooling coil through which a continuous flow of cooling water is maintained. The enthalpy fluxes involved in the apparatus are given below.

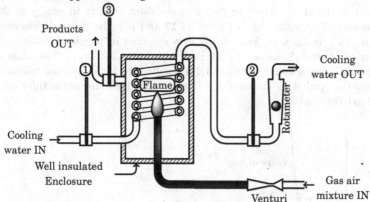

Figure 11.16: *Schematic of a continuous flow calorimeter. 1 - 3 are precision thermometers*

Gas air mixture (m_g) entering in at room temperature ($T_{g,entry}$): Mass flow measured using venturi; Temperature measured using a liquid in glass thermometer

Products of combustion leaving at a higher temperature ($T_{g,exit}$): Temperature measured using a liquid in glass thermometer

Cooling water (m_w)entering at constant temperature ($T_{w,entry}$) Mass flow measured by a rotameter

Cooling water stream leaves at a higher temperature ($T_{w,exit}$) Temperature measured using a liquid in glass thermometer

Energy balance requires that the following hold:

$$\frac{m_g HV}{1+FA} = m_g \left[C_{pp}T_{g,exit} - C_{pm}\right] + m_w C_w \left[T_{w,exit} - T_{w,entry}\right] \qquad (11.14)$$

In the above equation HV is the heating value of the fuel, C_{pp} is the specific heat of products of combustion, C_{pm} is the specific heat of gas air mixture and FA is the fuel to air ratio on mass basis. The determination of C_{pp} will certainly require knowledge of the composition of the products formed during the combustion process. The products of combustion have to be obtained by methods that we shall describe later. In the case of hydrocarbon fuels with complete combustion (possible with enough excess air) the products will be carbon dioxide and water vapor. If the exit temperature of the products is above $100°C$ the water will be in the form of steam or water vapor. The estimated heating value is referred to as the lower heating value (LHV) as opposed to the higher heating value (HHV) that is obtained if the water vapor is made to condense by recovering its latent heat.

11.7   Measurement of viscosity of fluids

Viscosity of a fluid is a thermo-physical property that plays a vital role in fluid flow and heat transfer problems. Viscosity relates the shear stress within a flowing fluid to the spatial derivative of the velocity. A Newtonian fluid follows the relation

$$\tau = \mu \frac{du}{dy} \tag{11.15}$$

In the above τ is the shear stress in Pa, u is the fluid velocity in m/s, y is the coordinate measured as indicated in Figure 11.17 and μ is the dynamic viscosity of the fluid in $Pa\ s$ or $kg/m\ s$. Viscosity is also specified in terms of $Poise$ or P (more usually in $centi-Poise$ or cP) which is equal to 0.1 $kg/m\ s$. The ratio of dynamic viscosity μ to the density of the fluid ρ is called the kinematic viscosity and represented by the symbol ν. The unit of ν is m^2/s. An alternate unit is the $Stoke$ that is equal to 0.0001 m^2/s.

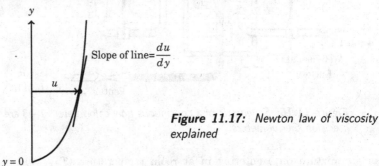

Figure 11.17: *Newton law of viscosity explained*

The shear stress introduced through Equation 11.15 accounts for the slowing down of a higher velocity fluid layer adjacent to a lower velocity fluid layer. It also accounts for the frictional drag experienced by a body past which a fluid is flowing. Measurement of viscosity thus is based on the measurement of any one of these effects.

Three methods of measuring viscosity of a fluid will be discussed here.

Laminar flow in a capillary: The pressure drop in the fully developed part of the
      flow is related to the wall shear that is related in turn to viscosity.
Saybolt viscometer: The time taken to drain a standard volume of liquid through a
      standard orifice is related to the kinematic viscosity.
Rotating cylinder viscometer: The torque required to maintain rotation of a drum
      in contact with the liquid is used as a measure of viscosity.

11.7.1   Laminar flow in a capillary

The nature of flow in a straight circular tube is shown in Figure 11.18. The fluid enters the tube with a uniform velocity. If the Reynolds number is less than about 2000 the flow in the tube is laminar. Because of viscous forces within the fluid and the fact that the fluid satisfies the no slip condition at the wall, the fluid velocity undergoes a change as it moves along the tube, as indicated in the figure. At the end of the so called developing region the fluid velocity profile across the tube has an invariant shape given by a parabolic distribution of u with y. This region is referred to as the fully developed region. The axial pressure gradient is constant in this part of the flow and is related to the fluid viscosity and the flow velocity. The requisite background is available to the reader from a course on Fluid Mechanics. We make use of the results of Hagen Poiseuille flow. The reader may recall that this was used earlier while discussing the transient behavior of pressure sensors (see section 7.2.2).

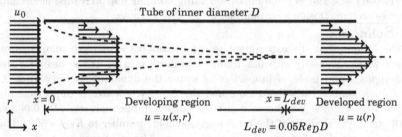

Figure 11.18: *Laminar flow in a circular tube*

A schematic of the laminar flow apparatus used for the measurement of viscosity of a liquid is shown in Figure 11.19. The apparatus consists of a long tube of small diameter (capillary) in which a steady flow is established. By connecting the capillary to a constant head tank on its upstream the pressure of the liquid is maintained steady at entry to the capillary. The pump - overflow arrangement maintains the level of the liquid steady in the constant head tank. The mean flow velocity is measured using a suitable rotameter just upstream of the capillary. The pressure drop across a length L of the capillary in the developed section of the flow is monitored by a manometer as shown. The reason we choose a capillary (a small diameter tube, no more than a few mm in diameter) is to make sure that the pressure drop is sizable (and hence measurable with good accuracy) for a reasonable length of the tube (say a few tens of cm to 1 or 2 m) and the corresponding entry length also is not too long. If necessary the liquid may be maintained at a constant temperature by heating the liquid in the tank and by insulating the capillary. The mean temperature of the liquid as it passes through the capillary is monitored by a suitable thermometer.

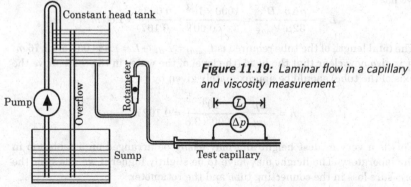

Figure 11.19: *Laminar flow in a capillary and viscosity measurement*

Essentially the same arrangement may be used for the measurement of viscosity of gases by replacing the constant head tank by a gas bottle with a pressure regulator.

A fine needle valve may be used to control the flow rate of the gas. Pressure drop may be measured using a suitable differential pressure gauge. The gas may have to be allowed to escape through a well ventilated hood. In case the gas is toxic it has to be collected by connecting the exit of the capillary to a gas recovery system.

Example 11.7

Viscosity of water is to be measured using laminar flow in a tube of circular cross section. Design a suitable set up for this purpose.

Solution :

We assume that the experiment would be conducted at room temperature, say $20°C$. Viscosity of water from a handbook is used as the basis for the design. From table of properties of water the dynamic viscosity at $20°C$ is approximately $\mu = 0.001\ Pa\ s$ and the density of water is approximately $\rho = 1000\ kg/m^3$. We choose a circular tube of inner diameter $D = 0.003\ m$ ($3\ mm$). We shall limit the maximum Reynolds number to $Re_D = 500$. The maximum value of mean velocity V_{max} may now be calculated as

$$V_{max} = \frac{Re_D \mu}{\rho D} = \frac{500 \times 0.001}{1000 \times 0.003} = 0.167\ m/s$$

The corresponding volume flow rate is the maximum flow rate for the apparatus and is given by

$$Q_{max} = \frac{\pi}{4} D^2 V_{max} = \frac{\pi}{4} \times 0.003^2 \times 0.167 = 1.178 \times 10^{-6}\ m^3/s$$

The volume flow rate is thus about $1.2\ ml/s$. If the experiment is conducted for $100\ s$, the volume of water that is collected (in a beaker, say) is $120\ ml$. The volume may be ascertained with a fair degree of accuracy. The development length for the chosen Reynolds number is

$$L_{dev} = 0.05 D Re_D = 0.05 \times 0.003 \times 500 = 0.075\ m$$

If we decide to have a sizable pressure drop of, say, $1\ kPa$ the length of tube may be decided. Using the results known to us from Hagen Poiseuille theory, we have

$$L = -\frac{\rho \Delta p}{32\mu} \frac{D^2}{V_{max}} = \frac{1000 \times 10^3}{32 \times 0.001} \times \frac{0.003^2}{0.167} = 1.68\ m$$

The total length of the tube required is $L_{total} = L_{dev} + L = 1.68 + 0.075 \approx 1.76\ m$. It is also now clear that the head (the level of the water in the tank above the axis of the tube) should be approximately given by

$$h = -\frac{\Delta p}{\rho g} = \frac{10^3}{1000 \times 9.8} = 0.102\ m$$

This is a very modest height and may easily be arranged on a table top in the laboratory. The height may have to be slightly higher if we allow for the pressure loss in the connecting tube and the rotameter.

11.7.2 Saybolt viscometer

Schematic of a Saybolt viscometer is shown in Figure 11.20. The method consists in measuring the time to drain a fixed volume of the test liquid (60 ml) through a capillary of specified dimensions (as shown in the figure). The test liquid is surrounded by an outer jacket so that the test fluid can be maintained at the desired temperature throughout the experiment. If the drainage time is t s, then the kinematic viscosity v of the test liquid is given by the formula

$$v = \left[0.22018t - \frac{179.3}{t} \right] \times 10^{-6} m^2/s \qquad (11.16)$$

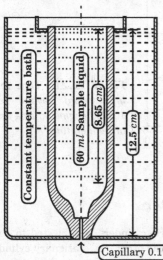

Figure 11.20: *Schematic of Saybolt viscometer*

Capillary 0.1765 cm DIA, 1.225 cm LONG

Example 11.8

A Saybolt viscometer is used to measure the viscosity of engine oil at 40°C. The time recorded for draining 60 ml of engine oil is $275 \pm 1\,s$. Calculate the kinematic viscosity and its uncertainty.

Solution :

We use Equation 11.16 with the nominal value of t as 275 s to get the nominal value of kinematic viscosity of engine oil as

$$v_{oil} = \left[0.22018 \times 275 - \frac{179.3}{275} \right] \times 10^{-6} = 59.9 \times 10^{-6} m^2/s \text{ or } 59.9\, mm^2/s$$

For uncertainty calculation we first calculate the influence coefficient as

$$I_t = \frac{dv_{oil}}{dt} = \left[0.22018 + \frac{179.3}{t^2} \right] \times 10^{-6}$$

$$= \left[0.22018 + \frac{179.3}{275^2} \right] \times 10^{-6} = 2.23 \times 10^{-7} m^2/s^2$$

The uncertainty $u_{v_{oil}}$ in the estimated (nominal) value of the kinematic viscosity with uncertainty in drainage time of $u_t = \pm 1\,s$ is

$$u_{v_{oil}} = \pm I_t \times u_t = \pm 2.23 \times 10^{-7} \times 1 = \pm 2.23 \times 10^{-7} m^2/s$$

The uncertainty is approximately $\pm 0.37\%$.

It is well known that the viscosity of oil varies very significantly with temperature. Hence the drainage time of $60\ ml$ of oil in a Saybolt viscometer varies over a wide range as the temperature of oil is varied. Typically the drainage time varies from around $10000\ s$ at $0°C$ to about $100\ s$ at $100°C$, as shown in Figure 11.21.

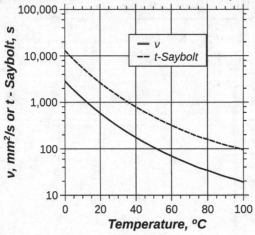

Figure 11.21: *Time to drain in standard Saybolt test of 10W50 engine oil as a function of temperature*

11.7.3　Rotating cylinder viscometer

When a liquid layer between two concentric cylinders is subjected to shear by rotating one of the cylinders with respect to the other, the stationary cylinder will experience a torque due to the viscosity of the liquid. The torque is measured by a suitable technique to estimate the viscosity of the liquid. The schematic of a rotating cylinder viscometer is shown in Figure 11.22. It consists of an inner stationary cylinder and an outer rotating cylinder that is driven by an electric motor. There are narrow gaps a and b where the liquid whose viscosity is to be determined is trapped. The torque experienced by the stationary cylinder is related to the liquid viscosity as the following analysis shows. The nomenclature used in the following derivation is given in Figure 11.22. If the gap a between the two cylinders is very small compared to the radius of any of the cylinders we may approximate the flow between the two cylinders to be "couette" flow familiar to us from fluid mechanics. The velocity distribution across the gap b is linear and the constant velocity gradient is given by $\dfrac{du}{dr} = \dfrac{\omega r_2}{b}$. The uniform shear stress acting on the surface of the inner cylinder all along its length L is then given by $\tau = \mu \dfrac{du}{dr} = \mu \dfrac{\omega r_2}{b}$. The torque experienced by the inner cylinder due to this shear stress is

$$
\begin{aligned}
T_L &= \text{shear stress} \times \text{area} \times \text{moment arm} \\
&= \mu \frac{\omega r_2}{b} \times 2\pi r_1 L \times r_1 = \frac{2\pi \mu \omega r_1^2 r_2 L}{b}
\end{aligned}
\qquad (11.17)
$$

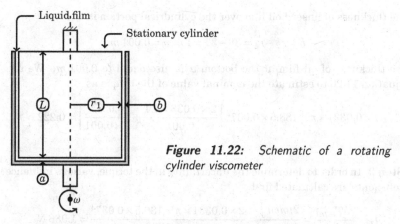

Figure 11.22: Schematic of a rotating cylinder viscometer

There is another contribution to the torque due to the shear stress at the bottom of the cylinder due to the liquid film of thickness a. At a radial distance r from the center of the stationary cylinder the shear stress is given by $\tau(r) = \mu \dfrac{\omega r}{a}$. Consider an elemental area in the form of a circular strip of area $2\pi r dr$. The elemental torque on the stationary cylinder due to this is given by

$$
\begin{aligned}
dT_a &= \text{shear stress} \times \text{area} \times \text{moment arm} \\
&= \mu \frac{\omega r}{a} \times 2\pi r dr \times r = \frac{2\pi \omega \mu}{a} r^3 dr
\end{aligned}
\tag{11.18}
$$

The total torque will be obtained by integrating this expression from $r = 0$ to r_1. Thus we have

$$
T_a = \frac{2\pi \omega \mu}{a} \int_0^{r_1} r^3 dr = \frac{\pi \omega \mu r_1^4}{2a}
\tag{11.19}
$$

The total torque T_{total} experienced by the stationary cylinder is obtained by adding Equations 11.18 and 11.19. Thus

$$
T_{total} = T_L + T_a = \mu \pi \omega r_1^2 \left[\frac{2 r_2 L}{b} + \frac{r_1^2}{2a} \right]
\tag{11.20}
$$

Example 11.9

A rotating cylinder apparatus is run at an angular speed of $1800 \pm 5\, RPM$. The geometric data is specified as: $r_1 = 37 \pm 0.02\, mm$, $r_2 = 38 \pm 0.02\, mm$, $L = 100 \pm 0.5\, mm$ and $a = 1 \pm 0.01\, mm$
What is the torque experienced by the stationary cylinder? What is the power dissipated? Perform an error analysis. The fluid in the viscometer is linseed oil with a viscosity of $\mu = 0.0331\, kg/m\, s$.

Solution :

Step 1 First we use the nominal values to estimate the nominal value of the torque. We convert the angular speed from RPM to rad/s as

$$
\omega = 1800\, RPM = \frac{1800}{60} \times 2\pi = 188.5\, rad/s
$$

The thickness of linseed oil film over the cylindrical portion is

$$b = r_2 - r_1 = 38 - 37 = 1\,mm = 0.001\,m$$

The thickness of oil film at the bottom a is also equal to 0.001 m. We use Equation 11.20 to estimate the nominal value of the torque as

$$T_{total} = 0.0331 \times \pi \times 188.5 \times 0.037^2 \left[\frac{2 \times 0.038 \times 0.1}{0.001} + \frac{0.037^2}{2 \times 0.001} \right] = 0.222\,N\,m$$

Step 2 In order to determine the uncertainty in the torque, various influence coefficients are calculated first.

$$I_{r_1} = \frac{\partial T_{total}}{\partial r_1} = \frac{2\mu\pi\omega r_1^3}{a} = \frac{2 \times 0.0331 \times \pi \times 188.5 \times 0.037^3}{0.001} = 1.986\,N$$

$$I_{r_2} = \frac{\partial T_{total}}{\partial r_2} = \frac{2\mu\pi\omega r_1^2 L}{b} = \frac{2 \times 0.0331 \times \pi \times 188.5 \times 0.037^2 \times 0.1}{0.001} = 5.367\,N$$

$$I_a = \frac{\partial T_{total}}{\partial a} = -\frac{\mu\pi\omega r_1^4}{2a^2} = -\frac{0.0331 \times \pi \times 188.5 \times 0.037^4}{2 \times 0.001^2} = -18.37\,N$$

$$I_b = \frac{\partial T_{total}}{\partial b} = -\frac{2\mu\pi\omega r_1^2 r_2 L}{b^2}$$

$$= -\frac{2 \times 0.0331 \times \pi \times 188.5 \times 0.037^2 \times 0.038 \times 0.1}{0.001^2} = -203.942\,N$$

$$I_L = \frac{\partial T_{total}}{\partial L} = \frac{2\mu\pi\omega r_1^2 r_2}{b} = \frac{2 \times 0.0331 \times \pi \times 188.5 \times 0.037^2 \times 0.038}{0.001} = 2.039\,N$$

$$I_\omega = \frac{\partial T_{total}}{\partial \omega} = \mu\pi r_1^2 \left[\frac{2r_2 L}{b} + \frac{r_1^2}{2a} \right]$$

$$= 0.0331 \times \pi \times 0.037^2 \left[\frac{2 \times 0.035 \times 0.1}{0.001} + \frac{0.037^2}{2 \times 0.001} \right] = 0.00118\,\frac{Nm}{\frac{rad}{s}}$$

Step 3 The errors in various measured quantities are taken from the data supplied in the problem. The expected error in the estimated value for the torque is

$$u_{T_{total}} = \pm \left[(I_{r_1} u_{r_1})^2 + (I_{r_2} u_{r_2})^2 + (I_a u_a)^2 + (I_b u_b)^2 + (I_L u_L)^2 + (I_\omega u_\omega)^2 \right]^{\frac{1}{2}}$$

$$= \pm \left[\left(1.986 \times \frac{0.02}{1000} \right)^2 + \left(5.367 \times \frac{0.02}{1000} \right)^2 \right.$$

$$+ \left(36.736 \times \frac{0.02}{1000} \right)^2 + \left(203.492 \times \frac{0.028}{1000} \right)^2$$

$$\left. + \left(2.039 \times \frac{0.04}{1000} \right)^2 + \left(0.00118 \times \frac{5 \times 2 \times \pi}{60} \right)^2 \right]^{\frac{1}{2}} = \pm 0.0022\,N\,m$$

Note that the factor 1000 in the denominator of some of the terms is to convert the error in mm to error in m. The error in the RPM is converted to appropriate error in ω using the factor $2\pi/60$. We have assumed that the given viscosity of linseed oil has zero error!

Concluding remarks

This chapter has dealt with the measurement of thermo-physical properties of solids, liquids and gases. These are important from the point of view of analysis and design of engineering systems.

Chapter *12*

Radiation properties of
surfaces

*This chapter deals with a special topic that may be of interest to some
readers. Radiation properties of surfaces play a vital role in thermal control
of spacecrafts and hence are of major importance in their design. In recent
times interest has been driven by its application to graphics and animation
where realistic images require a realistic geometric modeling of radiation
properties of surfaces.*

12.1   Introduction

Radiation properties of surfaces are important in many engineering applications. For example, the heat loss from a heated object such as an electronic component fixed on a printed circuit board depends on the surface emissivity of the electronic component. In space and solar energy applications radiation plays a dominant role. Radiation emanates from hot surfaces and the amount of radiation is determined by the radiation properties of the surfaces. We have seen earlier while discussing pyrometry that the surface emissivity (spectral and total) plays an important role (see section 4.5) in that application.

In recent times radiation properties of surfaces have also become important in computer graphics where shadowing and shading patterns are based on the realistic simulation of surface properties such as the reflectance pattern of the surface, due to directional illumination of the surface.

In practice both spectral and total properties are required, depending on the application. Surface properties are also angle dependent (incident angle for incident radiation and viewing angle for emitted or reflected radiation) and hence detailed angular distributions are required in many applications. The methods of measurement that are discussed in this chapter will deal with all these aspects.

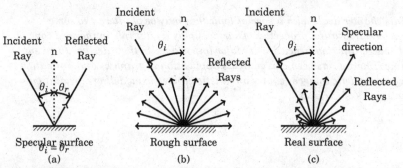

Figure 12.1: *Schematic representation of reflection at an opaque surface: (a) Specular surface (b) Diffuse surface (c) Real surface*

We restrict our discussion here to a surface that is opaque and hence radiation is either reflected or absorbed at the surface. Consider first an optically smooth flat surface. The surface roughness (for example, the arithmetic average of the surface height distribution indicated by R_a) as a fraction of the wavelength (λ) of radiation should be small, say $\dfrac{R_a}{\lambda} < 0.1$, for a surface to be considered optically smooth. Such a surface reflects radiation specularly following the laws of geometric optics. The incident ray, the reflected ray and the normal to the surface, at the point of incidence are all in the same plane. Also the angle made by the incident ray with the surface normal is equal to the angle made by the reflected ray with the surface normal (see Figure 12.1(a)). Engineering surfaces are seldom smooth, and, in the extreme, may be considered perfectly rough (Figure 12.1(b)). Many reflected rays are produced going into all the directions of a hemisphere that may be constructed at the point of incidence (the figure is restricted to two dimensions and is only for illustration).

The reflection is direction *independent* and we refer to the surface as being perfectly diffuse or simply diffuse. Real surfaces, however, may have reflectance behavior in between the specular and diffuse extremes as shown schematically by Figure 12.1(c). This case exhibits the most complex measurement situation. The reflectance may be some what larger along the specular direction compared to all other directions represented by the hemisphere centered at the point of incidence.

12.1.1 Definitions

Consider a surface that is maintained at a constant temperature of T_s. Let the spectral emissive power of this surface be $E_\lambda(T_s)$. A black body at the same temperature would have a spectral emissive power given by $E_{b\lambda}(T_s)$. This is described by the familiar Planck distribution function. The spectral hemispherical emissivity (or simply the spectral emissivity) $\varepsilon_{\lambda h}$ of the surface is defined as

$$\varepsilon_{\lambda h} = \frac{E_\lambda(T_s)}{E_{b\lambda}(T_s)} \tag{12.1}$$

The corresponding total quantity is obtained by integrating over wavelength from 0 to ∞. Equation 12.1 will then be replaced by

$$\varepsilon_h = \frac{E(T_s)}{E_b(T_s)} = \frac{\int_0^\infty E_\lambda(T_s)d\lambda}{\int_0^\infty E_{b\lambda}(T_s)d\lambda} = \frac{1}{\sigma T_s^4} \int_0^\infty E_\lambda(T_s)d\lambda \tag{12.2}$$

Angular quantities (angle dependent surface properties) are defined now. Consider radiation incident within a cone defined by the incident direction (θ, ϕ) as shown in Figure 12.2. Radiation reflected in a direction defined by (θ', ϕ') is indicated by the reflected cone in the figure. The incident radiation is characterized by the incident intensity $I_\lambda(\theta, \phi)$ while the reflected intensity is characterized by $I_\lambda(\theta', \phi')$. Ratio of the reflected intensity to the incident intensity is thus a function of (θ, ϕ) and (θ', ϕ') and represents the so called bidirectional (or bi-angular) reflectivity $\rho_\lambda(\theta, \phi; \theta', \phi')$ of the surface.

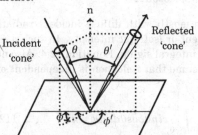

Figure 12.2: *Bi-angular reflection at a surface*

Our interest here is only with the hemispherical quantities. For this we consider two cases.

Directional hemispherical reflectivity

Consider radiation incident along (θ, ϕ) direction and that reflected along all the directions of the hemisphere as shown in Figure 12.3. The incident radiant flux is given by $I_\lambda(\theta,\phi)cos\theta d\Omega$ where $d\Omega$ is the elemental solid angle represented by the incident cone. The hemispherical reflected flux is given by

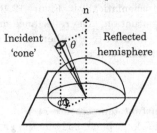

Figure 12.3: *Directional hemispherical reflectivity*

$$\text{Hemispherical reflected flux} = \int_{\Omega'=2\pi} I_\lambda(\theta',\phi')cos\theta' d\Omega'$$

The integration is over the hemisphere represented by the total solid angle of $\Omega' = 2\pi$. We may rewrite the reflected flux as

$$\text{Hemispherical reflected flux} = \int_{\Omega'=2\pi} \rho_\lambda(\theta,\phi;\theta',\phi')I_\lambda(\theta,\phi)cos\theta' d\Omega'$$

where $\rho_\lambda(\theta,\phi;\theta',\phi')$ is the bidirectional reflectivity. Noting that $I_\lambda(\theta,\phi)$ is constant since the incident direction is fixed, this may be taken outside the integral sign. The directional hemispherical reflectivity is then given by

$$\rho_{\lambda h}(\theta,\phi) = \frac{1}{cos\theta d\Omega} \int_{\Omega'=2\pi} \rho_\lambda(\theta,\phi;\theta',\phi')cos\theta' d\Omega' \qquad (12.3)$$

If the incident radiation is coming in through the entire hemisphere we may define the hemispherical reflectivity of the surface as

$$\rho_{\lambda h} = \frac{\int_{\Omega'=2\pi} \rho_\lambda(\theta,\phi;\theta',\phi')I_\lambda(\theta,\phi)cos\theta' d\Omega'}{\int_{\Omega=2\pi} I_\lambda(\theta,\phi)cos\theta d\Omega} \qquad (12.4)$$

In engineering applications our interest is mostly with diffuse incident radiation that is characterized by an angle independent or isotropic intensity. In such a case $I_\lambda(\theta,\phi) = I_\lambda$ and may be removed out of the integral sign in Equation 12.4. Also note that $d\Omega = sin\theta d\theta d\phi$. If in addition, we assume that reflection is independent of ϕ the integral in the denominator becomes

$$\int_{\Omega=2\pi} I_\lambda(\theta,\phi)cos\theta d\Omega = I_\lambda \int_{\theta=0}^{\frac{\pi}{2}} \int_{\phi=0}^{2\pi} sin\theta cos\theta d\theta d\phi = \pi I_\lambda \qquad (12.5)$$

Equation 12.4 then takes the form

$$\rho_{\lambda h} = \frac{1}{\pi} \int_{\Omega=2\pi} \rho_{\lambda h}(\theta,\phi)cos\theta d\Omega \qquad (12.6)$$

If the surface is opaque (most engineering applications deal with opaque surfaces) the hemispherical reflectivity is related to the hemispherical absorptivity $\alpha_{\lambda h}$ as

$$\alpha_{\lambda h} = 1 - \rho_{\lambda h} \qquad (12.7)$$

Hemispherical directional reflectivity

Let the incident radiation come from all the directions of the hemisphere as shown in Figure 12.4. The incident flux is then given by

$$\text{Incident flux} = \int_{\Omega=2\pi} I_\lambda(\theta,\phi)\cos\theta d\Omega$$

The hemispherical directional reflectivity is then given by

$$\rho_\lambda(\theta',\phi') = \frac{\int_{\Omega=2\pi} \rho_\lambda(\theta,\phi;\theta',\phi')I_\lambda(\theta,\phi)\cos\theta d\Omega}{\int_{\Omega=2\pi} I_\lambda(\theta,\phi)\cos\theta d\Omega} \tag{12.8}$$

The reflected direction represented by (θ',ϕ') remains fixed in the above expression.

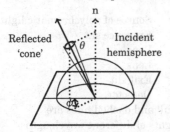

n

Reflected
'cone'

θ

Incident
hemisphere

$d\Omega$

Figure 12.4: *Hemispherical directional reflectivity*

12.2 Features of radiation measuring instruments

Radiation measurements are, in general, expensive because of high quality instruments that are needed. The measurement of spectral properties (pertaining to different incident light frequencies or wavelengths) requires expensive monochromator. Suitable sources of radiation and detectors are required (for covering the ultraviolet, visible and infrared regions of electromagnetic radiation). Total reflectance (integrated over all frequencies of incident radiation) is normally easier to measure and less expensive. If a surface is perfectly rough there is no angular variation of reflectivity and hence easy to measure. In what follows we shall look into the details of each of these.

12.2.1 Components of a reflectivity measuring instrument

The components of a radiation measuring instrument are shown schematically in Figure 12.5. The instrument may either be a single or double beam instrument depending on the requirement. The source of radiation is appropriately chosen with the requirement of the instrument in mind. Special lamps are used for measurement in the ultraviolet part of the spectrum. The visible part of the spectrum is covered by the use of a filament type source. To cover the infrared part of the spectrum the source is usually a "glower" that consists of an electrically conducting element in the form of a long cylinder. A DC current is directly passed through the element to run it at a temperature of around 1000 K. The glower is cooled by circulating water at the two ends. The light from the source is collimated and is passed through a slit before it is passed on to the monochromator.

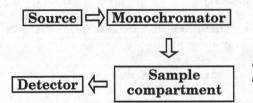

Figure 12.5: *Schematic of a radiation measuring instrument*

Monochromator is a device that selects a single wavelength to emerge out of it before it is incident on the test surface. It may use a prism (Figure 12.6), grating (Figure 12.7) or an interferometer (Figure 12.8). In the last case the spectrum is obtained by suitable processing of the signal by converting the time domain signal to the frequency domain.

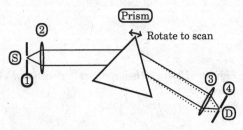

S: Source of polychromatic light
1:    Entrance slit
2:    Lens
3:    Lens
4:    Exit slit
D: Detector
Full and dashed lines are
lights of different wavelengths

Figure 12.6: *Schematic of a Prism Monochromator*

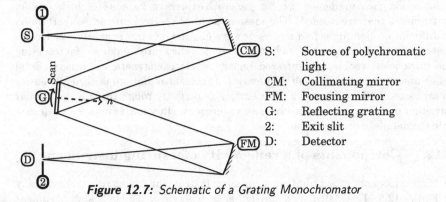

S: Source of polychromatic light
CM: Collimating mirror
FM: Focusing mirror
G: Reflecting grating
2:      Exit slit
D: Detector

Figure 12.7: *Schematic of a Grating Monochromator*

The sample compartment allows the incident light to be incident on the surface of the specimen at the desired incident angle. It also has a suitable arrangement by which the reflected radiation in a particular direction or in all the directions of the hemisphere is collected and passed on to the detector. The last action is possible by the use of an integrating sphere that will be discussed next.

12.3   Integrating sphere

An integrating sphere is a useful accessory for the measurement of hemispherical properties of a surface. An integrating sphere consists of a spherical shell, usually

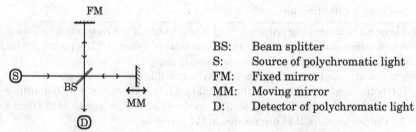

BS: Beam splitter
S: Source of polychromatic light
FM: Fixed mirror
MM: Moving mirror
D: Detector of polychromatic light

Figure 12.8: *Schematic of a Fourier Transform Infrared Spectrometer (FTIR)*

made of aluminum, with the inside surface coated with a highly reflecting diffuse coating (typically magnesium oxide or proprietary paints). Suitable ports are provided for allowing light to enter and leave the integrating sphere. Arrangement is also provided for placing a test surface within the sphere and orient it in a desired orientation. The integrating sphere may have a diameter of a few centimeters to a few meters. Coating used in the integrating sphere is chosen such that it has a constant reflectivity value in the useful wavelength range of interest.

The working principle of the integrating sphere is explained by taking the case of measurement of hemispherical properties, as examples.

12.3.1 Hemispherical emissivity

In order to measure the hemispherical emissivity of a surface it is necessary that we gather radiation leaving the test surface in to all the directionns in a hemisphere i.e. in to 2π solid angle. An integrating sphere is used for this purpose. Typically the experiment may be arranged as shown schematically in Figure 12.9.

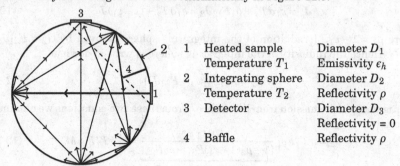

1	Heated sample	Diameter D_1
	Temperature T_1	Emissivity ϵ_h
2	Integrating sphere	Diameter D_2
	Temperature T_2	Reflectivity ρ
3	Detector	Diameter D_3
		Reflectivity = 0
4	Baffle	Reflectivity ρ

Figure 12.9: *Integrating sphere arranged for measurement of hemispherical emissivity*

It is possible to analyze an integrating sphere by different methods. These are:

- Method of detailed balance coupled with geometric concepts[1]

---

[1]Victor Sandgren, Characterization of an integrating sphere radiation reference source, Report No. EX084/2011, Chalmers University of Technology, SE-412 96 Göteborg, Sweden

• Radiosity irradiation method[2]

Analysis of an integrating sphere: Here we make use of the radiosity irradiation method to analyze radiation transfer in an integrating sphere with a port to which a heated sample is attached while a detector is mounted on another port. The two ports are shielded from each another by a suitable baffle as shown in Figure 12.9. The baffle is attached to the sphere wall and has both surfaces with a diffuse reflecting coating which is the same as that on the sphere. The legend in the figure shows all the parameters that are needed in the analysis.

Typical paths for radiation with one or more reflections at the sphere surface are indicated in the figure. Since the radiation is reflected diffusely at each reflection, radiation entering the detector does so along all possible directions of the hemisphere. The dashed line indicates that the detector is in the shadow region and hence does not have direct radiant exchange with the source. Thus the view factor for direct heat exchange between the source and the detector $F_{13} = 0$. By reciprocity view factor between detector and source $F_{13} = 0$. Since all radiation leaving the detector or the source reaches the effective area of integrating sphere of $A'_2 = A_2 - A_1 - A_3$, (A represents area) we have $F_{12} = F_{32} = 1$. By reciprocity we than have $F_{21} = \dfrac{A_1}{A'_2} = f_1$ and $F_{23} = \dfrac{A_3}{A'_2} = f_3$. By sum rule we than have the self view factor for the integrating sphere given by $F_{22} = 1 - F_{21} - F_{23} = 1 - f_1 - f_3 = 1 - f$ where $f = f_1 + f_3$.

The radiosity irradiation formulation follows. Radiosity of the detector is zero i.e. $J_3 = 0$ since it absorbs all radiation incident on it and it does not reflect any radiation. Irradiation of the source is given by $G_1 = F_{12}J_2$. Hence the radiosity of the source J_1 is written down as

$$J_1 = \varepsilon_1 \sigma T_1^4 + \rho_1 F_{12} J_2 = \varepsilon_1 \sigma T_1^4 + (1 - \varepsilon_1) J_2 \qquad (12.9)$$

where $\rho_1 = 1 - \varepsilon_1$. Irradiation of the integrating sphere is given by $G_2 = F_{21}J_1 + F_{22}J_2$. Hence the radiosity of the integrating sphere is written down as

$$J_2 = \rho_2 [F_{21}J_1 + F_{22}J_2] \qquad (12.10)$$

after ignoring the emission from its surface. From these two equations we eliminate J_1 to get

$$J_2 = \sigma T_1^4 F_{21} \underbrace{\frac{\rho_2 \varepsilon_1}{[1 - \rho_2(1 - \varepsilon_1)F_{21} - \rho_2 F_{22}]}}_{\text{Sphere multiplier, } M} = \varepsilon_1 \sigma T_1^4 F_{21} M \qquad (12.11)$$

In the special case when surface 1 is black i.e. $\varepsilon_1 = 1$ the sphere multiplier simplifies to

$$M_{\varepsilon_1 = 1} = \frac{\rho_2}{[1 - \rho_2 F_{22}]} \qquad (12.12)$$

Radiosity of surface 2 then becomes

$$J_{2, \varepsilon_1 = 1} = \sigma T_1^4 F_{21} M_{\varepsilon_1 = 1} \qquad (12.13)$$

---

[2]S.P. Venkateshan, Heat Transfer, Ane Books, 2011

We may rewrite the expression for M in the general case, after some algebraic manipulations, as

$$M = M_{\varepsilon_1=1} \frac{\varepsilon_1}{(1 - M_{\varepsilon_1=1}F_{21}) + \varepsilon_1 M_{\varepsilon_1=1}F_{21}} \qquad (12.14)$$

Irradiation on the detector is given by

$$G_3 = F_{32}J_2 = J_2 \qquad (12.15)$$

Hence the radiation flux entering the detector is given by

$$q_3 = J_3 - G_3 = 0 - J_2 \qquad (12.16)$$

Hence the power received by the detector is

$$Q_3 = q_3 A_3 = -J_2 A_3 \qquad (12.17)$$

If the detector response is proportional to the power received by the detector with sensitivity of $S\ V/W$ the detector output isgiven by

$$V_3 = SQ_3 \qquad (12.18)$$

Example 12.1

A heated surface at $T = 1000\ K$ and diameter of $D_1 = 0.01\ m$ is mounted on the wall of an integrating sphere of diameter $D_2 = 0.3\ m$ somewhere on its surface. A detector of diameter $(D_3 = 0.005\ m)$ is placed at a different location on the wall of the sphere. Diffuse reflectivity of the surface of the integrating sphere is $\rho = 0.99$. In one experiment the heated surface was replaced by a black body while in the subsequent experiment it was a sample surface of unknown emissivity. Suitable baffling is used to prevent direct radiation from the sample reaching the detector. Detector has a sensitivity of $S = 1\ V/W$. What will be the detector output in the first case? What is the hemispherical emissivity of the sample if the detector response is $1\ mV$?

Solution :

Step 1 Areas of the three surfaces are

$$A_1 = \pi \frac{0.01^2}{4} = 7.854 \times 10^{-5}\ m^2$$

$$A_2 = \pi \frac{0.3^2}{4} = 0.28274\ m^2$$

$$A_3 = \pi \frac{0.005^2}{4} = 1.964 \times 10^{-5}\ m^2$$

$$A_2' = A_2 - A_1 - A_3 = 0.28274 - 7.854 \times 10^{-5} - 1.964 \times 10^{-5} = 0.2827\ m^2$$

Step 2 We now calculate all the view factors required in the problem based.

$$F_{21} = \frac{7.854 \times 10^{-5}}{0.2827} = 2.779 \times 10^{-4}$$

$$F_{23} = \frac{1.964 \times 10^{-5}}{0.2827} = 6.947 \times 10^{-5}$$

$$F_{22} = 1 - 6.947 \times 10^{-5} - 2.779 \times 10^{-4} = 0.9997$$

Step 3 When the heated surface is a black body source $\epsilon_1 = 1$ and the expression for the sphere multiplier (Equation 12.11) reduces to

$$M = M_a = \frac{\rho_2}{[1 - \rho_2 F_{22}]} = \frac{0.99}{[1 - 0.99 \times 0.9997]} = 95.71$$

Using this we calculate the radiosity of the sphere surface as

$$J_2 = \sigma T_1^4 F_{21} M = 5.67 \times 10^{-8} \times 2.779 \times 10^{-4} \times 95.71 = 1507.94 \, W/m^2$$

Step 4 Hence the power received by the detector, based on Equation 12.17 is

$$Q_3 = -J_2 A_3 = -1507.94 \times 1.964 \times 10^{-5} = -0.0296 \, W = -29.6 \, mW$$

The negative sign indicates that the heat transfer is to the detector. With detector sensitivity of $S = 0.1 \, V/W$ the detector output $V_{3,\epsilon_1=1}$ is given by

$$V_{3,\epsilon_1=1} = SQ_3 = 0.1 \times 0.0296 = 0.00296 \, V = 2.96 \, mV$$

Step 5 When surface 1 is a gray surface with unknown emissivity the detector output has been given as $V_{3,\epsilon_1} = 1 \, mV$. It may be seen that the output of the detector in this case is given by $V_{3,\epsilon_1} = S\sigma T_1^4 F_{21} M$. Represent the ratio of detector outputs in the two cases as $R = \dfrac{V_{3,\epsilon_1}}{V_{3,\epsilon_1=1}} = \dfrac{1}{2.96} = 0.338$. We use Equation 12.14 and get, after some minor manipulations, the following for the emissivity of the surface.

$$\epsilon_1 = \frac{R(1 - M_{\epsilon_1=1}F_{21})}{(1 - M_{\epsilon_1=1}F_{21}R)} = \frac{0.338(1 - 95.71 \times 2.779 \times 10^{-4})}{(1 - 95.71 \times 2.779 \times 10^{-4} \times 0.338)} = 0.332$$

12.3.2 Hemispherical directional reflectivity

Hemispherical directional reflectivity means (see section 12.1.1) the reflectivity along a particular direction when the radiation is incident on the test surface from all directions represented by a hemisphere. The integrating sphere makes it possible to illuminate the test surface from all directions represented by a hemisphere.

The arrangement shown in Figure 12.10 is suitable for measuring the hemispherical directional reflectivity of a test surface. A narrow pencil of light is incident on to some location on the integrating sphere through an entrance port. Since the integrating sphere is highly reflecting and diffuse this light is sent in to all the directions of a hemisphere. Any ray among these reaches the test surface placed at a location elsewhere on the sphere either directly after just a single reflection or after a number of reflections off the surface of the integrating sphere. Hence it is clear that the test surface is illuminated by isotropic radiation intensity. The light

reflected off the test surface may be sampled along the desired direction as shown in
the figure.

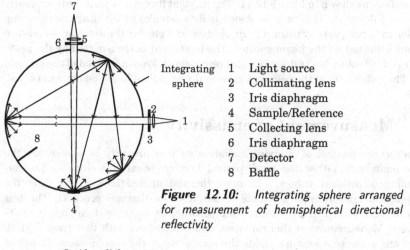

Integrating   1   Light source
sphere   2   Collimating lens
          3   Iris diaphragm
          4   Sample/Reference
          5   Collecting lens
          6   Iris diaphragm
          7   Detector
          8   Baffle

Figure 12.10: *Integrating sphere arranged
for measurement of hemispherical directional
reflectivity*

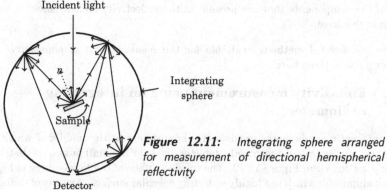

Figure 12.11: *Integrating sphere arranged
for measurement of directional hemispherical
reflectivity*

The light reflected off the surface is measured using a detector to yield a signal
proportional to the reflected intensity. A light detector placed anywhere on the
sphere will indicate a signal proportional to the flux incident on the test surface.
Ratio of these two will yield the hemispherical directional reflectivity of the test
surface. If the incident radiation is taken from a monochromator the measurement
yields spectral reflectivity. If the incident radiation is from a black body the
reflectivity will be the total reflectivity.

12.3.3 Directional hemispherical reflectivity

Directional hemispherical reflectivity means the reflectivity of the test surface in
to all directions represented by a hemisphere, of radiation incident along a given
direction (see section 12.1.1). The integrating sphere makes it possible to collect the
radiation leaving the test surface along all directions represented by a hemisphere.

In this case the test surface is mounted on a suitable platform that may be oriented
such that it is illuminated by a pencil of radiation along a desired direction. The

reflected radiation leaving the test surface along all the directions of the hemisphere is collected by the integrating sphere and is incident on a detector placed elsewhere on its surface, as shown in Figure 12.11. The incident flux is measured independently by a second detector. This may be done, as for example, by dividing the incoming radiation in to two parts by using a beam splitter, and placing the detector to measure the beam reflected off the beam splitter. The beam that is transmitted by the beam splitter is incident on the test surface and represents a known fraction of the original beam. The reflectivity of the test surface is again estimated based on the ratio of two signals.

12.4 Measurement of emissivity

An alternate method of studying radiation surface property is to measure the surface emissivity, either the total/spectral hemispherical emissivity or the corresponding directional values. In most thermal applications of radiation the hemispherical emissivity is adequate to describe the thermal process. The test surface is maintained at an elevated temperature by heating it so that it emits radiation. Measurement of this radiation and comparison with that from a black body at the same temperature yields the emissivity of the test surface. However, many of the components that are present in the reflectivity measurement are also present in this case.

There are several methods available for the measurement of emissivity. We consider a few of them here.

12.4.1 Emissivity measurement using an integrating radiometer

The integrating radiometer consists of a hemisphere at the middle of which the heated test surface and a thermopile detector are kept at small but equal distances off center as shown in Figure 12.12. The inside surface of the hemisphere is highly polished aluminum which is a highly reflecting specular surface. Any ray of radiation that leaves the test surface (one such is shown in the figure) is reflected specularly and is incident on the thermopile. The integrating sphere (actually the hemisphere) thus brings all the rays that leave the test surface to incident on the thermopile. Hence the hemispherical emissivity of the surface is obtained by comparing the signal registered by the thermopile with the test surface against a black body at the same temperature. The sphere itself needs to be cooled such that the sphere is at a low temperature while the test surface is maintained at an elevated temperature. Otherwise emission from the sphere will contribute to error in the measurement.

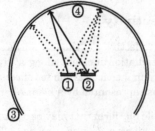

1: Heated test surface
2: Thermopile detector
3: Water cooled aluminum sphere
4: Mirror surface (specular)

Figure 12.12: *Schematic of the integrating radiometer*

12.4.2 Emissivity by transient cooling in vacuum

A second method of emissivity of a surface is by performing a cooling experiment. The test surface is taken in the form of a thin plate with a substrate of high thermal conductivity. The test plate is suspended by thin supporting wires inside a large vacuum environment. The walls of the vacuum chamber are cooled by a suitable arrangement and maintained at a low temperature. The test plate is heated to an elevated temperature and allowed to cool starting at $t = 0$. If all other modes of heat transfer are negligible in comparison with surface radiation, the cooling of the plate is governed by the equation

$$MC\frac{dT}{dt} = -S\varepsilon_h\sigma(T^4 - T_{bkg}^4) \qquad (12.19)$$

It is assumed that the heated plate behaves as a lumped system. In Equation 12.19 M represents the mass of the plate, C the specific heat of the plate material, S the surfsce area of plate, T_{bkg} the constant background temperature provided by the walls of the vacuum chamber and σ is the Stefan-Boltzmann constant.

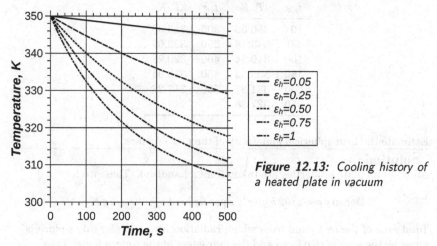

Figure 12.13: Cooling history of a heated plate in vacuum

The cooling rate and hence the cooling history of the plate is a function of hemispherical emissivity ε_h. Typically the temperature history as a function of ε_h is as shown in Figure 12.13. This figure is based on a simulation where ε_h is assumed known, along with all the other parameters that occur in Equation 12.19. In practice the temperature time history is recorded by a data logger. The background temperature is measured independently. The mass of the plate is determined by using a precision balance and the specific heat may be either measured by an independent experiment or obtained from the manufacturer.

The unknown emissivity may be estimated by standard parameter estimation techniques, using, for example the least square method. If the measured temperature as function of time is represented by T_m and the simulated temperature obtained by solving Equation 12.19 as an initial value problem by assuming an emissivity value is T_s, then we require that

$$S = \sum_i (T_{s,i} - T_{m,i})^2 \qquad (12.20)$$

be minimized. The minimization may be done, in general, based on non-linear least squares using a suitable optimization method.

Example 12.2

A square aluminum plate of size $0.05 \times 0.05 \times 0.0015\ m$ is heated to a temperature of around $350\ K$ and is allowed to cool in a high vacuum environment in a specially designed vacuum chamber. The walls of the chamber are mainteined at $300\ K$ by passing cold water through embedded pipes. In a certain experiment the following temperature time data was recorded.

t, s	T, K	t, s	T, K
0	350.60	300	325.38
50	343.78	350	322.67
100	340.24	400	320.26
150	335.52	450	318.04
200	331.87	500	317.22
250	327.38		

Estimate the hemispherical emissivity of the plate surface.

Solution :

Aluminaum plate properties are taken from a handbook. These are:

$$\text{Density } \rho = 2702\ kg/m^3; \quad \text{Specific heat } c = 903\ J/kg\,K$$

Total area of the test plate from which radiation heat transfer takes place is given by the area of two faces and the four edges of the square plate. Thus

$$S = 2 \times 0.05^2 + 4 \times 0.05 \times 0.0015 = 0.0053\ m^2$$

The mass of plate is given by the product of plate volume and the density.

$$M = 0.05^2 \times 0.0015 \times 2702 = 10.1325 \times 10^{-3}\ kg$$

The governing equation 12.19 may be recast as

$$\frac{d\theta}{d\zeta} = \theta^4 - 1$$

where $\theta = \dfrac{T}{T_{bkg}}, \zeta = \dfrac{t}{\tau}$ where τ is a characteristic time given by $\tau = \dfrac{MC}{S\sigma\epsilon_h T_{bkg}^3}$.

The governing equation is subject to the initial condition $T(0) = T_0$ or $\theta(0) =$

$\dfrac{T_0}{T_{bkg}}$. Fortunately the governing equation is amenable to exact solution as shown in the reference quoted below.[3] The solution is given by

$$\zeta = \frac{t}{\tau} = \frac{1}{4}\ln\left[\frac{(\theta_0-1)(\theta+1)}{(\theta_0+1)(\theta-1)}\right] - \frac{1}{2}\left[\tan^{-1}\theta_0 - \tan^{-1}\theta\right]$$

This equation may be rearranged as

$$\tau = \frac{t}{\zeta} = \frac{t}{\frac{1}{4}\ln\left[\frac{(\theta_0-1)(\theta+1)}{(\theta_0+1)(\theta-1)}\right] - \frac{1}{2}\left[\tan^{-1}\theta_0 - \tan^{-1}\theta\right]}$$

Making use of the temperature time history given in the problem we make a table as shown below.

t, s	$T, K(Data)$	$\theta, Data$	ζ	$\tau = \dfrac{t}{\zeta}$
0	349.90	1.166	0.0000	-
50	348.47	1.162	0.0057	8762.60
100	345.66	1.152	0.0175	5700.09
150	344.76	1.150	0.0216	6956.64
200	342.28	1.141	0.0330	6054.97
250	341.48	1.138	0.0369	6771.89
300	339.28	1.131	0.0481	6235.63
350	337.82	1.126	0.0559	6259.32
400	336.38	1.121	0.0640	6253.38
450	335.10	1.117	0.0715	6295.62
500	333.78	1.113	0.0796	6282.54

If the data were to be perfect with no errors we should have the same value for the ratio $\dfrac{t}{\zeta}$ in the last column. Ignoring the first two entries that seem to be grossly in error, the best value of τ is obtained by averaging all the other entries. The best value of τ is then obtained as $\tau = 6388.75$. We use the expression for τ given earlier to estimate ε_h as

$$\varepsilon_h = \frac{MC}{S\sigma T_{bkg}^3 \tau} = \frac{10.1325 \times 10^{-3} \times 903}{0.0053 \times 5.67 \times 10^{-8} \times 300^3 \times 6388.75} = 0.177$$

The data compares well with the solution obtained using the cooling model with $\varepsilon_h = 0.177$ as indicated by Figure 12.14(a). The quality of the fit is also brought out by the small magnitude of the rms error between the data and the fit of $0.334\,K$.

In order to find out how close we are to the optimum value of the emissivity a simulation was made where the rms error was obtained with various assumed emissivity values. A plot of rms error vs assumed value of emissivity is made as shown in Figure 12.14(b). It appears that $\varepsilon_h = 0.180$ is the best estimate based on the simulation (rms error equal to $0.306\,K$). The value obtained earlier is pretty close to this.

---

[3] Chapter 17 in Heat Transfer by S.P. Venkateshan, ANE Books 2008

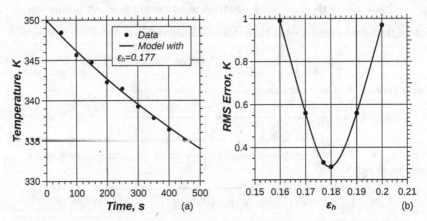

Figure 12.14: *(a) Comparison of data with fit in Example 12.2 (b) Variation of rms error with chosen value of ε_h*

12.4.3 Calorimetric method of emissivity measurement

Instead of the unsteady method described above one may use a steady sate calorimetric technique for the measurement of emissivity. The test surface is bonded on to a heater plate which is provided with temperature sensors. The plate is suspended in a vacuum chamber using very fine wires. The walls of the vacuum chamber are maintained at a low temperature by cooling them with liquid nitrogen ($T \sim 77K$). The test plate is heated and brought to the desired steady temperature. The heat input to the plate (Q) is measured as also the temperature of the plate (T_p) and the walls of the vacuum chamber (T_w). The emissivity is then estimated using the relation

$$\varepsilon_h = \frac{Q}{S(T_p - T_w)^4} \qquad (12.21)$$

where S is the surface area of the plate assembly. Any losses by conduction and residual convection may be estimated by performing the experiment with a surface of known emissivity. The heat loss may be estimated as a function of plate temperature and used as a correction on Q obtained in the experiment, with a surface of unknown emissivity.

Example 12.3

A square aluminum plate (surface treated to get a high emissivity) of size $0.05 \times 0.05 \times 0.0015\ m$ is maintained at a temperature of $350 \pm 0.2\ K$ when placed in a high vacuum environment in a specially designed vacuum chamber. The walls of the chamber are maintained at $265\ K$ by passing chilled coolant through embedded pipes. Heat is input to the plate using an embedded electric heater. Heat supplied has been measured to be $2.000 \pm 0.002\ W$. Estimate the hemispherical emissivity of the plate and its uncertainty. Assume that conduction losses through attached wires is 0.5% of the heat input to the heater.

Solution :

Step 1 The given data is written down as follows:

Plate length, L	$0.05\,m$		Plate temperature, T_p	$350\,K$
Plate thickness, t	$0.0015\,m$	Background temperature, T_{bkg}		$265\,K$
Heat input to plate, Q	$1.200\,W$		Conduction loss Q_c	$0.006\,W$

Uncertainties specified in the problem are written down as shown below:

δL	$0.0005\,m$	δt	$0.00002\,m$
δT	$0.2\,K$	δQ	$0.002\,W$

Step 2 Surface area of plate losing heat by radiation is given by

$$S = 2L^2 + 4Lt = 2 \times 0.05^2 + 4 \times 0.05 \times 0.0015 = 0.0053\,m^2$$

We take the value of Stefan Boltzmann constant as $\sigma = 5.67 \times 10^{-8}\,W/m^2K^4$ and obtain the nominal value of hemispherical emissivity ϵ_h as

$$\epsilon_h = \frac{Q - Q_c}{\sigma S(T_p^4 - T_{bkg}^4)} = \frac{1.200 - 0.006}{5.67 \times 10^{-8} \times 0.0053(350^4 - 265^4)} = 0.394$$

Step 3 Uncertainty in area is calculated now using the familiar error propagation formula. The influence coefficients are given by

$$I_L = \frac{\partial S}{\partial L} = 4L + 4t = 4(0.05 + 0.0015) = 0.206$$

$$I_t = \frac{\partial S}{\partial t} = 4L = 4 \times 0.05 = 0.2$$

Uncertainty in area may then be estimated as

$$\delta S = \sqrt{(I_L \delta L)^2 + (I_t \delta t)^2} = \sqrt{(0.206 \times 0.0005)^2 + (0.2 \times 0.00002)^2} = 1.031 \times 10^{-4}$$

Hence $\dfrac{\delta S}{S} = \dfrac{1.031 \times 10^{-4}}{0.0053} = 1.945 \times 10^{-2}$.

Step 4 Uncertainty in the difference in fourth power of temperatures are calculated now. The influence coefficients are given by

$$I_{T_p} = \frac{\partial(T_p^4 - T_{bkg}^4)}{\partial T_p} = 4T_p^3 = 4 \times 350^3 = 1.715 \times 10^8$$

$$I_{T_{bkg}} = \frac{\partial(T_p^4 - T_{bkg}^4)}{\partial T_p} = 4T_{bkg}^3 = 4 \times 265^3 = 7.4 \times 10^7$$

With these we get

$$\delta(T_p^4 - T_{bkg}^4) = \sqrt{\left(I_{T_p}\delta T_p\right)^2 + \left(I_{T_{bkg}}\delta T_{bkg}\right)^2}$$

$$= \sqrt{(1.715 \times 10^8 \times 0.2)^2 + (7.4 \times 10^7 \times 0.2)^2} = 3.7392 \times 10^7$$

Hence $\dfrac{\delta(T_p^4 - T_{bkg}^4)}{T_p^4 - T_{bkg}^4} = \dfrac{3.7392 \times 10^7}{350^4 - 265^4} = 3.7114 \times 10^{-3}$.

Step 5 We assume that there is no error in the conduction loss. Hence the fractional uncertainty in Q is given by $\dfrac{\delta(Q-Q_c)}{Q-Q_c} = \dfrac{0.002}{1.194} = 1.675 \times 10^{-3}$.

Step 6 Finally we obtain the uncertainty in the estimated hemispherical emissivity as

$$
\delta\epsilon_h = \epsilon_h \sqrt{\left(\frac{\delta(Q-Q_c)}{Q-Q_c}\right)^2 + \left(\frac{\delta S}{S}\right)^2 + \left(\frac{\delta\left(T_p^4 - T_{bkg}^4\right)}{T_p^4 - T_{bkg}^4}\right)^2}
$$

$$
= 0.394\sqrt{\left(1.675\times10^{-3}\right)^2 + \left(1.945\times10^{-2}\right)^2 + \left(3.7114\times10^{-3}\right)^2} = 0.008
$$

Thus the hemispherical emissivity of the plate surface is $\epsilon_h = 0.394 \pm 0.008$.

12.4.4   Commercial portable ambient temperature emissometer

This portable instrument for the measurement of emissivity near room temperature is supplied by the Devices & Services Company, Texas, U.S.A. If two surfaces are placed close to each other the heat transfer is proportional to the difference in fourth power of temperature and the emissivity of the two surfaces. If the gap is small compared to the size of the surface we have

$$
q_r(1-2) = \frac{\sigma(T_1^4 - T_2^4)}{\frac{1}{\varepsilon_1} + \frac{1}{\varepsilon_2} - 1} \tag{12.22}
$$

Here the subscripts identify the two surfaces that face each other. Consider two surfaces with emissivities of ε_h (high emissivity surface) and ε_l (low emissivity surface) and at very nearly equal temperatures of $T(T_h \sim T_l \sim T)$, and a third surface of emissivity ε and at ambient temperature T_∞. If these are arranged such that the first two surfaces exchange radiation with the third surface, we have

$$
q_r(1-3) = \frac{\sigma(T_h^4 - T_\infty^4)}{\frac{1}{\varepsilon_h} + \frac{1}{\varepsilon} - 1} \tag{12.23}
$$

$$
q_r(2-3) = \frac{\sigma(T_l^4 - T_\infty^4)}{\frac{1}{\varepsilon_l} + \frac{1}{\varepsilon} - 1} \tag{12.24}
$$

Since the temperatures of the high and low emissivity surfaces are close to each other, the numerators of the two expressions in Equations 12.23 and 12.24 are practically equal. Because $\varepsilon_h > \varepsilon_l$ the denominator of $q_r(1-3)$ is smaller than the denominator of $q_r(2-3)$. Hence $q_r(1-3)$ is greater than $q_r(2-3)$. Because of this a small temperature difference exists between the surfaces 1 and 2. In the D&S emissometer the design is such that this temperature difference is proportional to the emissivity of the third surface, which is the test surface. Schematic of the D&S emissometer is shown in Figure 12.15. It consists of a detector head which has four sectored radiation detectors as shown. The detectors are thermopiles with

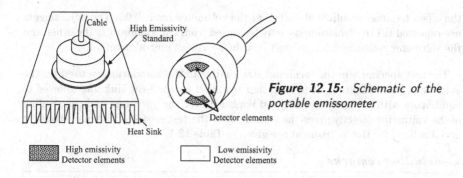

Figure 12.15: *Schematic of the portable emissometer*

two sectors having high emissivity and the other two having low emissivity. The test specimen and the standard samples that are supplied with the emissometer are placed on a heat sink as shown during the measurement process. The heat sink helps to maintain the samples at room temperature. The detector head has cable connection for both heating the detector head and for taking out the differential output across the thermopiles.

In the emissometer the detector head is heated to about 355 K. The standard surfaces supplied with the instrument have diameters of 66.7 mm and thickness of 4 mm. The detector head views a 50 mm diameter area of specimen from a distance of 4.3 mm. The instrument output is connected to a digital voltmeter that directly reads the emissivity of the specimen with two significant digits. The instrument requires

Table 12.1: *Specifications of Emissometer Model AE*

Output	2.4 mV nominal with a sample emissivity of 0.9 at 25°C
Readout	D&S Scaling Digital Voltmeter with resolution of 0.01 mV
Output impedance	150Ω, nominal
Linearity	Detector output linear with emissivity and to within 0.01 unit
Sample Temperature	Maximum 55°C or (328 K)
Drift	Negligible during measurement
Time constant	10 s, nominal
Standards	Four calibration standards are provided with the emissomete

around 30 minutes of warm up time. The two standards are placed on the heat sink so that both of them attain the ambient temperature. In order to have good thermal contact between the standards and the heat sink the air gaps are filled with distilled water. The detector head is then placed over the high emissivity standard and the gain of the voltmeter is adjusted so that it reads 0.89, after allowing about 90 s for equilibration. The detector head is then placed over the low emissivity standard and the offset trimmer is adjusted such that the voltmeter reads 0.06. The adjustments are repeated till the emissometer may be moved from one standard to the other and the voltmeter readings indicate the two values without any adjustment.

The test specimen in the form and size similar to the standards are used for the emissivity determination. The specimen is placed on the heat sink and allowed to equilibrate with it. The detector head is placed over the specimen and the reading of the voltmeter directly gives the emissivity of the test surface. The manufacturer specifications for the instrument are given in Table 12.1.

Concluding remarks

> This chapter has described the basic principles involved in the measurement of radiation properties of surfaces. Application of an integrating sphere has been elaborated. An introduction has been made for determining hemispherical emissivity by a transient technique. A commercial emissometer has also been described.

Chapter *13*

Gas concentration

Monitoring of poluution in the atmosphere has become important with increased awareness of the environment. With tight pollution control laws it is essential to measure and monitor various pollutants. Gas concentration measurement discussed in this chapter is directed towards this aspect.

13.1   Introduction

In recent times there is a lot of concern regarding the presence and the concentration levels of harmful gases in the earth's atmosphere. Quantitative analysis of these is important from these as well as other considerations. For example, monitoring the health of a power plant requires the monitoring the levels of various gases that leave through the stack. In the case of internal combustion engines the analysis of the exhaust leaving through the tail pipe of a car provides important information about the performance of the engine. Laws require that the concentration of pollutant gases be within specified limits. In view of these we look at the following in what follows:

- Flue gas monitoring
- Atmospheric pollution monitoring
- Exhaust gas analysis in IC Engines

However, before we embark on these we shall look at ways of specifying the concentration of gases. The concentration of a candidate gas in a mixture of gases may either be specified on volume basis or on mass basis. On volume basis the units are either ppm_V (parts per million by volume) or ppb_V (parts per billion by volume). For example, a ppm_V of gas on volume basis is defined as

$$1\,ppm_V = \frac{\text{Volume of candidate gas}}{\text{Total volume of gas mixture} \times 10^6} \tag{13.1}$$

Note that $1\,ppm_V$ is also equal to 0.0001% by volume. On mass basis the units normally employed are mg/m^3 or $\mu g/m^3$. These mean respectively that $1\,mg$ or $1\,\mu g$ of the candidate gas is present in a cubic meter of the gas mixture. The unit also represents the density of the candidate gas per unit volume of the gas mixture. In many applications the volume may simply be taken as the volume of all the other gases put together since the concentration of the candidate gas is very small or required to be very small. It is possible to convert concentration from volume basis to mass basis by the use of gas laws applied to mixture of gases. We assume that the gas mixture is mostly standard dry air with a small concentration of the candidate gas. Let the density of the candidate gas in the mixture be ρ_c given in mg/m^3. Let the total pressure be one standard atmosphere equal to $P = 101325\,Pa$. Let the temperature of the gas mixture be $T\,K$. The universal gas constant is given by $\Re = 8314.47\,J/mol\,K$. The partial pressure of the candidate gas is then given by

$$p_c = \rho_c \frac{\Re T}{M_c}$$

where M_c is the molecular weight of the candidate gas. The total gas pressure is given by

$$P = \rho \frac{\Re T}{M}$$

where M is the molecular weight of air and ρ is air density. The ratio of pressures is the same as the ratio of volumes and is given by division of these two expressions as

$$\frac{p_c}{P} = \frac{V_c}{V} = \frac{\rho_c M}{\rho M_c} = \frac{\rho_c}{M_c} \frac{\Re T}{P} \tag{13.2}$$

If the candidate gas density is specified in mg/m^3, the above may be simplified to

$$\frac{V_c}{V} = \frac{\rho_c}{M_c} \times \frac{8314.47 \times T}{101325} = \frac{\rho_c \times T}{12.185 \times M_c} \, ppm_V \qquad (13.3)$$

Note that the above is strictly valid for small amount of the candidate gas being present in the mixture. Table 13.1 shows the mass concentration corresponding 1 ppm_V of some candidate gases present in trace amounts.

Table 13.1: *Mass concentration in mg/m^3 of 1 ppm_V at $25°C$*

Carbon monoxide	1.145
Nitric oxide	1230
Nitrogen dioxide	1.850
Ozone	1.963
Sulfur dioxide	2.617

Example 13.1

Gaseous pollutant NO_2 has a mass concentration of *20 mg/m^3* in an air sample at $30°C$ and $1\,atm$. Express this in ppm_V by volume basis.

Solution :

The temperature of the gas sample is given as $T = 30°C = 30 + 273 = 303\,K$. Molecular weight of NO_2 is $M_c = 46.01$. The density of NO_2 in the gas sample is specified to be $\rho_c = 20\,mg/m^3$. Using Equation 13.3 we may represent the concentration in volume basis as

$$NO_2 \text{ concentration on volume basis} = \frac{20 \times 303}{12.185 \times 46.01} = 10.81\,ppm_V$$

Example 13.2

Air sample at $25°C$ has $2.5\,ppm_V$ of SO_2. What is the concentration in mass basis if the total pressure of the gas mixture is $1\,atm$?

Solution :

The temperature of the gas sample is given as $T = 25°C = 25 + 273 = 298\,K$. Molecular weight of SO_2 is $M_c = 64$. The concentration of the pollutant is specified as $2.5\,ppm_V$. The mass concentration in mg/m^3 is then obtained by the use of Equation 13.3 as

$$\rho_c = 2.5 \times \frac{12.185 \times 64}{298} = 6.543\,mg/m^3$$

Alternately we may use the information given in Table 13.1 to obtain the pollutant density as

$$\rho_c = 2.5 \times 2.617 = 6.543\,mg/m^3$$

Sometimes the pollutant concentrations are given in more appropriate units with certain applications in mind. For example pollutant levels in the case of automobile exhaust are specified in g/km. Depending on the fuel used by the automobile, this way of specifying the level of pollutant, sets limits on fuel economy, as shown in Example 13.3.

Combustion products commonly met with are:

- Oxides of Carbon - CO and CO_2
- Oxides of Nitrogen - NO, N_2O, NO_2 etc., referred to collectively as NO_x
- Un burnt Fuel (HC) - HC referring to hydrocarbon
- Particulate matter - Soot

Table 13.2 gives emission limits of combustion products commonly met with in exhaust from internal combustion engines.

Table 13.2: *Examples of Emission limits*

	CO (g/km)	$HC + NO_x$ g/km	
BS I	2.72/3.16	0.97/1.30	
BS II	2.20	0.50	
		HC	NO_x
BS III	2.30	0.20	0.15
BS IV	1.00	0.10	0.08

Applicable limits in India

Example 13.3

An car manufacturer claims that his car will consume $1\,l$ of petrol in covering a distance of $18\,km$. The petrol used is known to produce $2.12\,kg$ of CO_2 per l. What is the emission level of this car? Will it meet the European Union (EU) requirement of $95\,g/km$ in 2020? What should be the performance value that will satisfy this?

Solution :

The manufacturer specified performance of the car is $18\,km/l$ of petrol. With a CO_2 emission of $2.12\,kg/l$ of petrol, we have

$$CO_2\text{ emission per }km = \frac{2.12 \times 1000}{18} = 117.8\,g/km$$

Obviously the car does not meet the EU requirement by 2020. In order to meet the EU requirement the car should have a minimum fuel economy of $\frac{117.2 \times 18}{95} = 22.3\,km/l$.

13.1.1 Methods of gas concentration measurement

Broadly the gas concentration measurement methods may be classified as Non separation methods and Separation methods. In the former there is no effort made to isolate the candidate gas from the gas mixture. In the latter the candidate gas is physically separated before being measured. We shall describe a few of the methods available in these two broad categories.

Non separation methods:

- Non Dispersive Infrared Analyzer (NDIR)
- Differential Absorption LIDAR (DIAL)
- Chemiluminescence NO_x detection

Separation methods:

- Gas Chromatography
- Orsat gas analyzer

These are discussed in detail in what follows.

13.2   Non separation methods

13.2.1   Non Dispersive Infrared Analyzer (NDIR)

Most pollutant gases that are of interest to us absorb in the infrared part of the electromagnetic spectrum. The absorption by a gas depends on the product of concentration level and the path length over which the electromagnetic radiation travels through the gas. For example, CO absorbs roughly 3% of radiation in the wavelength range $4.4 - 4.6$ μm for a path length concentration product of 100 ppm_V m. On the other hand CO_2 absorbs about 10% of radiation in the wavelength range $4.2 - 4.3$ μm for the same path length concentration product. HC absorbs approximately 8% of radiation in the $3.3 - 3.5$ μm band under the same condition.

It is thus seen that small concentrations of the pollutant gases are measurable based on absorption of radiation of suitable wavelength even when the gas sample contains a mixture of these gases. In principle there is thus no need to separate the candidate gas from the mixture before making the measurement of concentration. One of the most popular methods is the non dispersive infrared detection where the radiation used is broad band radiation. No monochromator (see section 12.2) is required in this method. Just how a particular gas is detected will become clear from the discussion on the acousto-optic detector that follows.

Figure 13.1 shows the constructional details of an acousto-optic cell. The cell consists of a rigid vessel that contains the gas that is to be detected. Collimated infrared radiation is allowed in to the cell through a suitable window. The infrared radiation is chopped using a wheel with a set of holes arranged along the periphery of the wheel. The wheel is rotated at a constant speed using a suitable motor. A pressure transducer (usually a condenser microphone, see section 7.4.4) is placed within the acousto-optic cell as shown. When the infrared radiation passes into the cell a part of it which is in the absorption band of the gas is absorbed by the candidate gas. This heats the gas and since the gas is confined within a rigid vessel, the volume is held fixed and hence the pressure goes up. When the incoming radiation is chopped (it enters the cell intermittently) the pressure within the cell varies as shown schematically in Figure 13.1. The pressure transducer picks up this and generates a signal proportional to the pressure change. The pressure change is a function of the candidate gas concentration within the cell. Since the cell is initially filled with a certain concentration of the candidate gas and sealed the pressure change is proportional to the amount of infrared radiation that enters it.

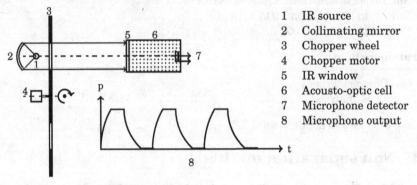

1	IR source
2	Collimating mirror
3	Chopper wheel
4	Chopper motor
5	IR window
6	Acousto-optic cell
7	Microphone detector
8	Microphone output

Figure 13.1: *Schematic of an acousto-optic cell*

Now consider the situation shown in Figure 13.2 where a sample cell is placed in the path of infrared radiation in front of the acousto-optic cell. The sample cell is provided with two windows that allow the infrared radiation to pass through with negligible absorption. If the sample cell contains a certain concentration of the candidate gas that is also contained in the acousto-optic cell the amount of radiation in the absorption band of the candidate gas passed on in to the acousto-optic cell is less than when the sample cell is absent or the sample gas does not contain the candidate gas. It is thus clear that the pressure change in the acousto-optic cell is reduced in direct proportion to the concentration of the candidate gas in the sample cell.

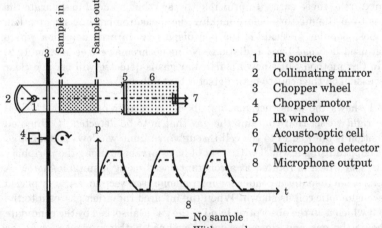

1	IR source
2	Collimating mirror
3	Chopper wheel
4	Chopper motor
5	IR window
6	Acousto-optic cell
7	Microphone detector
8	Microphone output

— No sample
-- With sample.

Figure 13.2: *Sketch for explaining the principle of NDIR*

We are now ready to describe a full fledged NDIR system. Figure 13.3 shows all the components of NDIR.

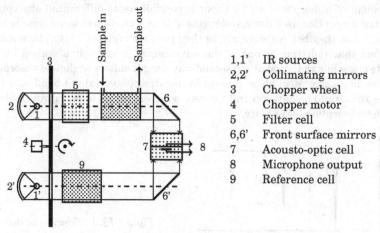

1,1'	IR sources
2,2'	Collimating mirrors
3	Chopper wheel
4	Chopper motor
5	Filter cell
6,6'	Front surface mirrors
7	Acousto-optic cell
8	Microphone output
9	Reference cell

Figure 13.3: Details of a typical NDIR system

The instrument is a double beam instrument with two IR sources with collimating optics for each beam. Beam 2 is a reference beam that sends the infrared radiation through a reference cell that contains all other gases (including gases that have absorption overlap with the candidate gas) other than the candidate gas. There is thus *no absorption* of the band that is absorbed by the candidate gas. Beam 1 passes through the sample gas cell as well as a filter cell that contains large concentration of those gases that have significant absorption in a region that overlaps absorption by the candidate gas. This way only the radiation that is absorbed by the candidate gas is available for absorption by the sample cell. The acousto-optic detector is now in the form of a two chambered device with a diaphragm between the two chambers. Beam 1 and beam 2 are steered in to the two chambers that are formed by the diaphragm. The diaphragm also forms a part of a condenser microphone as shown in the figure. The microphone thus responds to the difference in pressure across it. The signal thus developed is proportional to the concentration of the candidate gas in the sample cell. Calibration of the instrument is possible with special gas mixtures having known concentrations of candidate gas in the mixture, taken in the sample cell. The NDIR needs to use different filter cells and acousto-optic detectors for different gases.

For example NDIR is used for monitoring CO in ambient air. The concentration may be between 0 and 100 ppm_V. CO in flue gases may vary between $0-250$ ppm_V to as high as 1% by volume. The IR radiation used is in the $4-4.5$ μm range. NDIR is also useful for measuring SO_2 and NO_x in flue gases. In the case of NO_x removal of interference by CO_2, SO_2, water and Hydrocarbons is necessary while in the measurement of SO_2 interference by NO_2 is to be removed.

13.2.2 Differential Absorption LIDAR (DIAL)

Remote measurement of gas concentration in the atmosphere is necessitated because of difficulties in measurement in situ and also because of large path

lengths involved. Either airborne or satellite based or ground based instruments
are required to make gas concentration measurements. For example, airborne
measurement of atmospheric water vapor is possible using differential absorption.
Differential absorption uses two wavelengths of laser radiation that are close to each
other such that they behave similarly as they pass through the atmosphere except
for the fact that radiation at one of the wavelengths is strongly absorbed by the
candidate gas while radiation at the second wavelength suffers negligible absorption
by the candidate gas. Most gases present in the atmosphere exhibit band structure
with fine structure i.e. have several discrete wavelengths or lines within the band
over which absorption takes place.

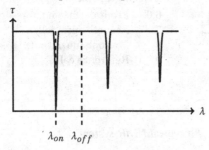

Figure 13.4: *Principle of differential absorption*

Figure 13.4 shows the choice of the two wavelengths for differential absorption.
The ordinate shows the transmittance τ . Smaller the τ larger is the absorption. The
wavelength corresponding to that close to the center of the 'line' is referred to as λ_{on}
while that in the space between two 'lines' is referred to as λ_{off}. Radiation at λ_{on}
suffers significant absorption while radiation at λ_{off} suffers very little absorption.
However all other phenomena like scattering by particulate matter and reflection
from clouds etc. are virtually the same for the two wavelengths since they are chosen
very close to each other. For example, in a specific case the two wavelengths are
chosen between $0.813 - 0.818\,\mu m$.

Schematic of the DIAL system is shown in Figure 13.5. It consists of pulsed lasers
working at two wavelengths needed for differential absorption, optical arrangement
for sending these two laser beams through the atmosphere, collection optics to collect
the backscattered radiation and then detectors to take a ratio of the two signals
to get the concentration of the candidate gas. Backscatter may take place from
smoke or cloud or a target placed on a balloon or some such arrangement. Apart
from absorption the beams will undergo attenuation of the intensity due to beam
divergence, scattering by particulates and transmission losses in the optics. As
mentioned earlier all these are the *same* for both wavelengths. The path lengths
for both the laser beams are also the same. For the beam at wavelength λ_{on} the total
or extinction cross section σ_t is the sum of absorption cross section σ_a and all other
processes represented by σ_s. For the beam at wavelength λ_{off} the cross section is
just σ_s.

By Beer's law we know that the attenuation of radiation follows the relation

$$I_\lambda = I_{\lambda,0}e^{-c\sigma_t L}$$

where I_λ the intensity of radiation transmitted across a path length of L, $I_{\lambda,0}$ is
the intensity of entering radiation, c is the concentration of the candidate gas and

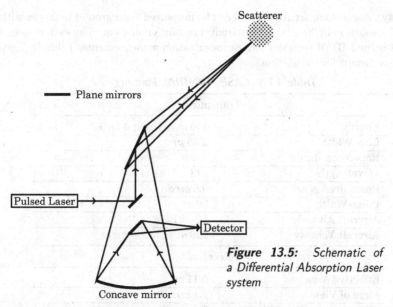

Figure 13.5: Schematic of a Differential Absorption Laser system

σ_t is the total cross section. We apply Beer's law to the laser radiation at the two wavelengths to get

$$I_{\lambda_{on}} = I_{\lambda_{on},0} e^{-c(\sigma_a + \sigma_s)L} \tag{13.4}$$

$$I_{\lambda_{off}} = I_{\lambda_{off},0} e^{-c\sigma_s L} \tag{13.5}$$

It will of course be good to choose the intensities at the two wavelengths the same even though it is not really necessary. The ratios $R_1 = \dfrac{I_{\lambda_{on}}}{I_{\lambda_{on},0}}$ an $R_2 = \dfrac{I_{\lambda_{off}}}{I_{\lambda_{off},0}}$ are estimated from the measured intensity values. Using Expressions 13.4 and 13.5 we see that

$$\frac{R_1}{R_2} = \frac{e^{-c(\sigma_a + \sigma_s)L}}{e^{-c\sigma_s L}} = e^{-c\sigma_a L} \tag{13.6}$$

If the absorption cross section of the candidate gas is known the concentration of the candidate gas is given by

$$c = -\frac{1}{\sigma_a L} \ln\left(\frac{R_1}{R_2}\right) \tag{13.7}$$

In practice the laser source is pulsed with a known pulse width as well as repetition rate. The return signal may be gated such that the signal is collected after a known time delay thus making it possible to vary or choose the required path length (this is similar to the pulse echo method using ultrasound discussed earlier in section 8.5.1). The ratio in Equation 13.7 is averaged over a large number of shots of the laser and averaged to improve signal to noise ratio. The detector used is a high sensitivity Avalanche Photo Diode (APD).

An example of this is the "LIDAR Atmospheric Sensing Experiment or LASE of NASA" for airborne water vapor measurement. This is capable of measuring water concentration in the range of 0.01 g/kg to 20 g/kg (this is also referred to as specific humidity). Aerosol backscatter ratios can be measured from ground to $20\,km$ with a vertical resolution of $30\,m$ and a horizontal resolution of $40\,m$. The system uses an aircraft borne LIDAR to make water concentration measurements. Table 13.3 gives the specifications for the system.

Table 13.3: *LASE H_2O DIAL Parameters*

Transmitter	
Energy	$150\,mJ$ (On and Off)
Line Width	$0.25\,pm$
Repetition Rate	$5\,Hz$
Wavelength	$813-818\,nm$
Beam divergence	$0.6\,mrad$
Pulse Width	$50\,ns$
Aircraft Altitude	$16-21\,km$
Aircraft Velocity	$200\,m/s$ ($720\,km/h$)
Receiver	
Effective Area	$0.11\,m^2$
Field of View	$1.1\,mrad$
Filter Bandwidth (FWHM)	$0.4\,nm$ (Day), $1\,nm$ (Night)
Optical Transmittance (Total)	23% (Day), 49% (Night)
Detector Efficiency	80% Silicon APD
Noise Equivalent Power	$2.5 \times 10^{-14}\,W/\sqrt{Hz}$ (at $1.6\,MHz$)
Excess Noise Factor (APD)	2.5

Refer to the web site: asd-www.larc.nasa.gov/lase

13.2.3   Chemiluminescence NO_x detection

As mentioned earlier NO_x refers to a combination of both NO and NO_2. If NO_x is subject to high temperature of around $1000°C$ NO_2 breaks down to NO. The sample will thus contain only NO. If NO is reacted with ozone the following exothermic reaction takes place.

$$NO + O_3 \rightarrow NO_2 + O_2$$

Since the reaction is exothermic the NO_2 formed is in the excited state and hence emits radiation.   This is called chemiluminescence.   The emitted radiation is proportional to the concentration of NO_2 and hence provides a method of measuring NO_x concentration. Schematic of the chemiluminescent NO_x analyzer is shown in Figure 13.6. The required flow rate of the sample gas is maintained by adjusting the flow controller. The sample gas is heated to convert NO_x to NO. In the reaction chamber, NO reacts with O_3 to generate NO_2 in the excited sate. The emitted radiation from the stream of excited NO_2 is collected by suitable optics and measured by the photo detector. The signal is related to the concentration of NO_x in the sample. Figure 13.7 shows the photograph of a typical portable rack mounted exhaust gas analyzer station that is placed near an internal combustion engine under test. The installation shows NO_x analyzer along with the pressure, temperature and flow meters that are required in the test.

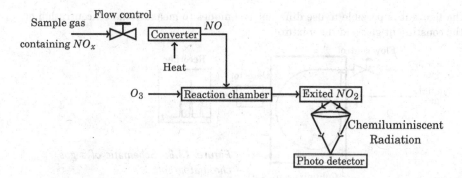

Figure 13.6: Schematic of a chemiluminescent NO_x analyzer

Figure 13.7: Photograph of a chemiluminescent NO_x analyzer

13.3 Separation methods

13.3.1 Gas Chromatography

Gas chromatography is a technique by which a pulse of gas mixture is physically separated in to its constituents that arrive at a suitable detector as temporally separated pulses. The separation takes place in a gas chromatograph column or GC column that contains a suitable adsorber. As the gas mixture which is carried along with a carrier gas (like Helium) moves through the column different constituents in the gas mixture pass through at different speeds (they are adsorbed and eluted or released at different rates) and hence get separated as the mixture covers a long path through the column. The first to arrive out of the column is the carrier gas. Other gases arrive as temporally separated pulses. The concentration of each gas is ascertained from these pulses.

Schematic of a gas chromatograph is shown in Figure 13.8. A controlled carrier gas flow is maintained using the flow controller. A small metered amount of sample gas is added to it at a point downstream of the flow controller and before the gas chromatograph column. The column is in the form of coil of a capillary whose inside surface is coated with a suitable adsorber. A suitable detector at the end of the column completes the chromatograph instrument. The column is maintained at a constant temperature by placing the column inside an oven as shown. As shown in the figure it is possible to use different techniques to measure the concentration of the constituent gases in the mixture.

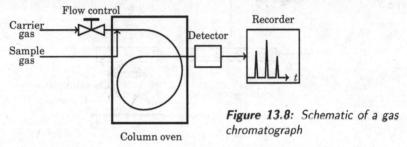

Figure 13.8: *Schematic of a gas chromatograph*

Some of the detection techniques are:

- Thermal conductivity detector
- Infrared spectrometer - The method is referred to as GC IR
- Mass spectrometer - The method is referred to as GC MS
- Flame ionization detector (or FID)

Thermal conductivity detector

Thermal conductivity detector is akin to the Pirani gauge that was discussed in the measurement of vacuum (see section 7.6.2). At the exit of the GC column a steady stream of carrier gas will emerge. A candidate gas in the sample gas will emerge some time after the sample gas is injected in to the carrier gas stream. The thermal conductivity of the candidate gas is, in general, different from that of the carrier gas. Examples of thermal conductivities of different gases of interest to us are given in Table 13.4. The thermal conductivity values are normalized with respect to thermal conductivity of dry air. Either argon or helium may be used as carrier gas. Gases whose relative thermal conductivities are close to one another are not resolved well. However, with prior knowledge it is possible to resolve these. Note that the relative thermal conductivity values change to some extent with temperature. The values quoted in the table are at room temperature.

Thermal conductivity of a mixture of Hydrogen and carbon dioxide (is taken as an example) varies with percentage of Hydrogen in the mixture as shown in Figure 13.9. We see that if Hydrogen is present in small quantities the thermal conductivity varies more or less linearly with H_2 concentration up to about 20%. Thermal conductivity detector is shown schematically in Figure 13.10. It essentially consists of a full bridge circuit with four cells containing heated wires. Two of these are bathed by a stream of carrier gas while the other two are bathed by carrier gas plus the candidate gas whenever it emerges out of the GC column. When all the heated wires are bathed by carrier gas alone, the bridge is under balance. Whenever the candidate gas is also present the thermal conductivity changes and the bridge goes out of balance. This is because of the slight change in the temperature of the wire in contact with the candidate gas. The imbalance potential difference V is recorded and the output of the instrument looks schematically like that shown

Table 13.4: Relative thermal conductivities of various gases

Gas	Relative thermal conductivity	Gas	Relative thermal conductivity
Air	1.00	Hydrogen	6.39
Argon (Carrier gas)	0.63	Nitric oxide	0.90
Carbon dioxide	0.55	Nitrogen	0.92
Carbon monoxide	0.88	Nitrous oxide	0.57
Chlorine	0.30	Oxygen	0.92
Ethane	0.69	Sulfur dioxide	0.31
Ethylene	0.65	Water vapor	0.63
Helium (Carrier gas)	5.42		

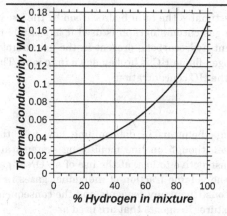

Figure 13.9: Thermal conductivity of a mixture of Hydrogen and Carbon dioxide

in Figure 13.8. Calibration of the instrument is possible using sample mixtures of known composition. Limit of detection is around $0.4 \, mg/m^3$.

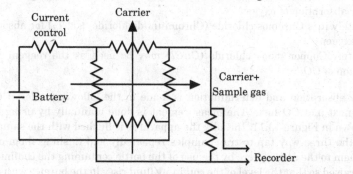

Figure 13.10: Thermal conductivity detector schematic

GC with Infrared spectrometer - GC IR

The GC column is interfaced with an infrared spectrometer as detector for measuring the gas concentration. The stream coming out of the GC column is sent through a cell which is placed in the path of the IR beam in an IR spectrophotometer. The absorption in a suitable absorption band of the candidate gas may be used to determine the candidate gas composition using Beer's law.

GC with Mass spectrometer - GC MS

The GC column is interfaced with a mass spectrometer as detector for measuring the gas concentration. A mass spectrum of the gas stream emerging from the GC column is recorded. The peak heights of different mass fragments may be calibrated with gas mixtures of known composition.

Flame ionization detector (or FID)

Total hydrocarbon analyzer or THC estimates the total hydrocarbon by the use of a flame ionization detector or FID. If a gas chromatograph is used it is possible to determine the concentrations of different hydrocarbons present in the gas sample. When a hydrocarbon is burnt in a hydrogen flame HC is broken down in to ions. The ion current is measured and related to the HC concentration.

13.3.2   Orsat gas analyzer

Flue gas composition in power plants helps us in determining how well the combustion system is functioning. Even though on line monitoring systems are common in modern power plants it is instructive to look at the use of an Orsat gas analyzer. Orsat Analyzer determines the volume fraction of individual gases in a mixture by selective absorption of the constituents by reagents and the consequent reduction in volume of a sample gas mixture. Reagents that are used are:

- For absorption of CO_2 it is 40% solution of potassium hydroxide
- Mixture of Pyrogallic acid (1,2,3 - Trihydroxybenzene) and solution of KOH is used for absorption of oxygen
- Alternately it is Chromos chloride (Chromium dichloride, $CrCl_2$) for absorption of oxygen
- Alternately Copper mono chloride ($CuCl$) may be used as the reagent for absorption of CO

Analysis is by absorption and determination is made in the following order: CO_2 first, Oxygen next and CO last. The measurement is done manually by using an apparatus shown in Figure 13.11. Initially the apparatus is flushed with the sample gas by using the three way tap to trap samples repeatedly and flushing them out by rejecting them to the atmosphere by the use of the bottle containing the confining liquid. If it is raised so that the level of the confining fluid rises in the burette with all the two way taps in the closed position the gas within the apparatus may be released to atmosphere. If now the three way valve is opened to the inlet hose the gas sample may be allowed to enter the burette by lowering the bottle. As the level of the liquid goes down the sample gas will fill the burette. This operation is repeated such that

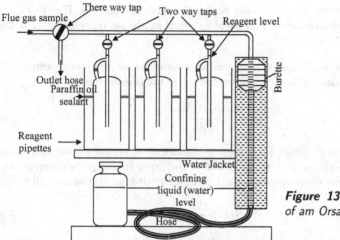

Wooden block

Figure 13.11: *Schematic of am Orsat analyzer*

any gas that was present in the burette to start with is completely flushed by the sample gas. Once the sample gas has been taken in the three way tap is closed. The sample gas trapped in the burette may now be sent in to the absorbing liquid by opening the appropriate two way tap and rising the bottle. The sample gas comes in contact with the reagent and sufficient time is allowed for the absorption of the appropriate constituent gas to take place. The bottle is now lowered and the level of the liquid in the burette will indicate the change in the volume of the sample gas and hence the volume of the gas constituent that is absorbed by the reagent. The two way tap is closed and the operation repeated with the second reagent, third reagent and so on. For low CO_2 (<4% by volume) or high O_2 (>1.5%) the burette must have 0.1 ml subdivisions.

13.3.3 Particulate matter - Soot (or smoke)

Most engines and furnaces used in power plants put out particulate matter in the form of soot or smoke. Soot/smoke refers to particles of unburnt carbon remaining after combustion of a hydrocarbon fuel. These particulates have very small size and cause pollution of the atmosphere. For example, the mean diameter of smoke particles in a diesel engine exhaust is about 30 nm. These may be measured either by collection and weighing or by measuring opacity.

There are several issues that are to be considered in the measurement of smoke. These are basically related to sampling.

Particle losses in sampling tube
Inertial losses: Especially in curved tubes due to centrifugal force field
Gravitational losses: In horizontal sampling tube due to settling of the particles under the action of gravity. These losses may be avoided by vertical orientation of the sampling tube
Iso-kinetic sampling : Iso-kinetic sampling is taking out or collecting a sample under dynamic conditions that prevail in the process. Velocity, temperature

and pressure should not change during the collection process. If any of these change the measurement may yield values that are not representative of the process.

Let us examine the requirements that need to be satisfied for iso-kinetic sampling to hold. For this purpose we define a parameter called the Stokes number as

$$Stk = \frac{\rho_p D_p^2 V_0}{18\mu L_{ch}} \tag{13.8}$$

where D_p = Particle diameter, ρ_p = particle density, V_0 = sample air velocity in the free stream, L_{ch} = characteristic dimension of the probe (possibly its diameter D_s) and μ is the dynamic viscosity of the fluid. Alternately it may also be defined by the relation

$$Stk = \frac{\tau V_0}{D_s} \tag{13.9}$$

where τ is the particle relaxation time. The particle relaxation time may thus be identified as

$$\tau = \frac{\rho_p D_p^2}{18\mu} \tag{13.10}$$

If $Stk < 0.01$ sampling is iso-kinetic. In addition, the sample probe should be aligned parallel to the gas stream. Gas velocity entering the probe should be the same as the free gas stream velocity.

Particles in a gas move with the same velocity as the gas if the diameter of the particle is small. Larger particles tend to move, for example, in a straight line when the gas encounters a change in direction. If the sampling is not iso-kinetic the velocity of the sample within the probe is different from the free stream velocity. There is a change in the velocity and the streamlines bend suitably to adjust to the flow field within the probe. It is thus possible that the larger or the smaller particles may not be sampled properly. Thus the measurement of particulates will under sample or over sample particles of different diameter particles in the sample.

Collection or capture efficiency of a sampling tube is dependent on the diameter and velocity. For non iso-kinetic conditions the collection efficiency is described by the formula

$$A(V, Stk) = 1 + \left[\frac{V_0}{V} - 1\right]\left[1 - \frac{1}{Stk(2 + 0.63\frac{V}{V_0})}\right] \tag{13.11}$$

In Equation 13.11 V is the flow velocity in the sampling tube. The collection efficiency is defined as the ratio of the concentration in the sampled flow to that in the free stream, for particles of a specified diameter.

Example 13.4

Determine the diameter of a sampling tube if the largest particle diameter is expected to be $2\ \mu m$. Particle density is specified to be $2500\ kg/m^3$. Other pertinent data: $V_0 2\ m/s$ and flow rate through sample tube is not to be less than $100\ ml/s$. The sample gas is air at 1 atmosphere and $25°C$.

Solution :

We shall design the sample tube for a flow rate of $Q = 100\ ml/s = 10^{-4}\ m^3/s$. Assuming iso-kinetic sampling to prevail, the velocity inside the sampling tube is set to $V = V_0 = 2m/s$. The volume flow rate is given by $Q = \pi \dfrac{D_s^2}{4} V_0$ and hence the sampling tube diameter is given by

$$D_s = \sqrt{\frac{4Q}{\pi V_0}} = \sqrt{\frac{4 \times 10^{-4}}{\pi \times 2}} = 0.00798m \approx 0.008\ m$$

We now calculate the Stokes number to determine whether the above diameter is satisfactory. We are given that the air sample is at 1 standard atmosphere and at a temperature of $25°C$. Dynamic viscosity of air is read off table of properties as $\mu = 18.5 \times 10^{-6} kg/m\ s$. Diameter of largest particles is given to be $D_p = 2\ \mu m$. The Stokes number is then calculated using Equation 13.8.

$$Stk = \frac{(2 \times 10^{-6})^2 \times 2500 \times 2}{18 \times 18.5 \times 10^{-6} \times 0.008} = 0.00752 \approx 0.008$$

Since this is less than 0.01, the sampling tube of 0.008 m diameter is satisfactory and guarantees iso-kinetic sampling.

Collection of particulate matter may be accomplished by drawing or aspirating the particulate laden gas through a filter paper or a membrane. Filter paper is useful for particles between 0.5 and 1 μm while a membrane may be useful down to 0.1 μm. The filter is operated for a metered duration during which the particles collect on the filter element. If the flow rate of the sample gas is maintained constant during the collection process we can calculate the volume of gas that has passed through the filter element. The particulates collected may be determined either by weighing the filter element before and after the collection process or by measuring the optical reflectance of the filter element in comparison with a virgin filter element. The mass of particulates collected divided by the volume of gas that has passed through the filter paper will give the mass of particulates per unit volume of the gas.

The collection of smoke particles is to be made by passing 300 ml of smoke containing stream at 1 atm and 298 K across the paper using an electronically controlled diaphragm pump. The diffuse reflectance of the smoke blackened (R_S) as well as virgin paper (R_V) are measured using a lamp and detector arrangement. The relative reflectance R_R is then defined by the relation

$$R_R = \frac{R_S}{R_V} \times 100\% \tag{13.12}$$

The paper blackening index or the Bosch Smoke Number (BSN) is then defined as

$$BSN = \frac{100 - R_R}{10} \tag{13.13}$$

Thus we see that the smoke number is zero with a clean smokeless stream and has a maximum possible value of 10 for the dirtiest stream! In the latter case the filter paper will be completely covered with carbon having zero reflectance.

The measurement of opacity of flue gases may be made in situ in the stack without separating the particulates. Since no sample collection is involved many of the problems associated with sampling are obviated. Codes specify the way such a measurement is to be made. Telonic Berkeley TBM300[1] portable opacity meter is a suitable measuring device for this purpose.

Opacity meter

Schematic of how an opacity meter may be used for in situ measurement of smoke in a stack is shown in Figure 13.12.

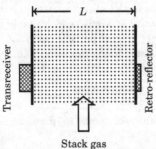

Figure 13.12: In situ opacity measurements in a stack

The instrument is single ended and uses a light source detector combination as the transceiver. At the other end we have a retro reflector to return the light. The light source is normally a green LED (light emitting diode) with a spectral peak between 550 and 570 nm (0.55 and 0.57 μm). The receiver is a photo detector. An instrument available from Preferred Utilities Manufacturing Corporation (JC-30D Smoke Opacity Monitor) has the specifications given in Table 13.5. Opacity is defined

Table 13.5: Specifications of an opacity meter

Spectral peak and mean width	500 - 600 nm
Angle of projection	$< 5°$
Response time	$< 10s$ for 95% change in opacity
Sighting distance	1 to 8 ft (0.3 to 2.5 m) smoke duct
Path length	3 to 10 ft (1 to 3 m)
Accuracy	1%
Calibration drift	$< 0.5\%$

Instrument is double ended with light source and detector facing each other

as 1 minus the transmittance τ, a quantity that was introduced earlier during the discussion on differential absorption technique (see section 13.2.2). We define, as we did earlier an extinction coefficient β to account for loss of intensity due to absorption and scattering by the particulates present in the stack gas. This factor may be related to the particle density if we know the extinction cross section of the particles. Example 13.5 demonstrates this, by considering a typical case.

Example 13.5

Smoke opacity has been measured by a double ended system to be 40%. The

---

[1]visit http://www.telonicberkeley.com/TBM300.html

path length has been noted to be 10 cm. The temperature of the gas is $°C$. It has been ascertained that the soot density is $\rho_s = 600$ $\mu g/m^3$. What is the extinction cross section of soot particles? If the soot may be considered as carbon particles with a mean diameter of $D_p = 0.2$ μm, and of density $\rho_c = 2300$ kg/m^3 determine the number density of soot particles.

Solution :

Opacity of the gas is given to be 40%. Thus the transmittance of the gas is 100 - 40 = 60%. By Beer's law we have

$$\tau = e^{-\beta L} = 0.6$$

Noting that the path length is $L = 10cm = 0.1$ m, taking natural logarithms of the above, we get

$$\beta = -\frac{1}{L}\ln\tau = -\frac{1}{0.1}\ln 0.6 = 5.1083 \ m^{-1}$$

The extinction coefficient β may also be written as the product of soot density ρ_s and the extinction cross section σ_e. Thus the extinction cross section is obtained as

$$\sigma_e = \frac{\beta}{\rho_s} = \frac{5.1083}{600 \times 10^{-6}} = 8513.8 \ m^2/kg$$

The mass of each particle may be obtained from the given data, $D_p = 0.2$ μm and particle density $\rho_c = 2300$ kg/m^3 as

$$m_p = \rho_c\frac{\pi D_p^3}{6} = 2300 \times \frac{\pi \times (0.2 \times 10^{-6})^3}{6} = 9.6343 \times 10^{-12} \ kg$$

We see that the soot density is the product of number density n and the mass of each particle. Hence the number density is determined as

$$n = \frac{\rho_s}{m_p} = \frac{600 \times 10^{-6}}{9.6343 \times 10^{-12}}/m^3 = 6.228 \times 10^7 \ m^{-3}$$

Concluding remarks

Principles of techniques of measuring low concentration of gases has been the theme of this chapter. Several methods, non-dispersive and spectroscopic methods have been discussed. We have also dealt with the measurement of soot that is commonly present in exhaust gases from combustion devices.

Chapter *14*

Force/Acceleration, torque and power

Engineering laboratory applications routinely involve the measurement of mechanical quantities such as force, acceleration, torque and power. Power producing devices such as engines and motors are characterized by torque at the shaft, rotational speed, power output etc. These are measured using suitable instruments as described in this chapter. Accelerometers - instrument used for measuring acceleration - find applications in monitoring loads on machinery or structures that are exposed to dynamic loads. Earthquake forces are also measured using sensitive accelerometers.

14.1   Introduction

In many mechanical engineering applications force/acceleration, torque and power need to be measured. Some of these applications are listed below:

Force/Stress measurement: Weighing of an object, Dynamics of vehicles, Control applications such as deployment of air bag in a vehicle, Study of behavior of materials under different types of loads, Vibration studies, Seismology or monitoring of earthquakes

Torque measurement: Measurement of brake power of an engine, Measurement of torque produced by an electric motor, Studies on a structural member under torsion

Power measurement: Measurement of brake horse power of an engine, Measurement of power consumed by an electric motor

As is apparent from the above the measurement of force, torque and power are involved in dynamic systems and hence cover a very wide range of mechanical engineering applications such as power plants, engines, transport vehicles and so on. Other areas where these quantities are involved are in biological applications, sports medicine, ergonomics and mechanical property measurements of engineering materials. Since the list is very long we cover only some of the important applications in what follows.

14.2   Force Measurement

There are many methods of measurement of a force. Some of these are given below:

Force measurement by mechanical balancing: Simple elements such as the lever may be used for this purpose. A platform balance is an example - of course mass is the measured quantity since acceleration is equal to the local acceleration due to gravity

Transform force to displacement: Using a spring element - it may be an actual spring or an elastic member that undergoes a strain

Convert force to hydraulic pressure: Conversion is using a piston cylinder device. The pressure itself is measured using a pressure transducer (see section 7.4)

Force measurement using a piezoelectric transducer: Piezoelectric transducer directly converts the force to an electrical signal

14.2.1   Platform balance

The platform balance is basically a weighing machine that uses the acceleration due to gravity to provide forces and uses levers to convert these in to moments that are balanced to ascertain the weight of an unknown sample of material. The working principle of a platform balance may be understood by looking at the cross sectional skeletal view of the balance shown in Figure 14.1. The weight W to be measured may be placed anywhere on the platform. The knife edges on which the platform rests share this load as W_1 and W_2. Let T be the force transmitted by the vertical link as shown in the figure. There are essentially four levers and the appropriate Moment equations are given below.

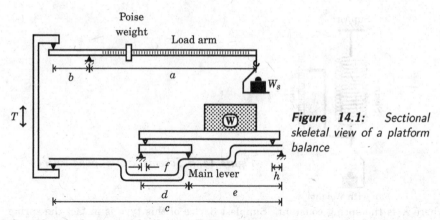

Figure 14.1: Sectional skeletal view of a platform balance

1) Consider the horizontal load arm. Taking moments about the fixed fulcrum, we have

$$Tb = W_s a \qquad (14.1)$$

2) For the main lever we balance the moments at the fixed fulcrum to get

$$Tc = W_2 h + W_1 \frac{f}{d} e \quad \text{or} \quad T = \frac{e}{c}\left[W_2 \frac{h}{e} + W_1 \frac{f}{d}\right] \qquad (14.2)$$

3) The ratio $\frac{h}{e}$ is chosen equal to $\frac{f}{d}$ so that the above becomes

$$T = \frac{h}{c}(W_1 + W_2) = \frac{h}{c}W \qquad (14.3)$$

4) Thus T the force transmitted through the vertical link is independent of where the load is placed on the platform. From Equations in 14.1 and 14.3 we get

$$\boxed{T = \frac{a}{b}W_s = \frac{h}{c}W \quad \text{or} \quad W = \frac{ac}{bh}W_s} \qquad (14.4)$$

Thus the gauge factor for the platform balance is $\frac{ac}{bh}$ such that $W = GW_s$. In practice the load arm floats between two stops and the weighing is done by making the arm stay between the two stops opposite a mark. The main weights are added into the pan (usually hooked on to the end of the arm) and small poise weight is moved along the arm to make fine adjustments. Obviously the poise weight W_p, if moved by a unit distance along the arm is equivalent to an extra weight of $\frac{W_p}{a}$ added in the pan. The unit of distance on the arm on which the poise weight slides is marked in this unit!

14.2.2 Force to displacement conversion

A spring balance (or a spring scale) is an example where a force may be converted to a displacement based on the spring constant. For a spring element (it need not actually be a spring in the form of a coil of wire) the relationship between force F and displacement x is linear and given by

$$F = Kx \qquad (14.5)$$

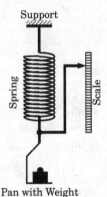

Support

Spring

Scale

Figure 14.2: *Schematic of a spring balance*

Pan with Weight

where K is the spring constant. Simplest device of this type is in fact the spring balance whose schematic is shown in Figure 14.2. The spring is fixed at one end and at the other end hangs a pan. The object to be weighed is placed in the pan and the position of the needle along the graduated scale gives the weight of the object. For a coiled spring like the one shown in the illustration, the spring constant is given by

$$K = \frac{E_s D_w^4}{8 D_m^3 N} \qquad (14.6)$$

In this equation E_s is the shear modulus of the material of the spring, D_w is the diameter of the wire from which the spring is wound, D_m is the mean diameter of the coil and N is the number of coils in the spring.

Example 14.1

A spring balance has been designed using a helical spring made of spring steel whose shear modulus is $E_s = 8 \times 10^9 \ Pa$. The spring is made from a wire of diameter $D_w = 0.57 \ mm$ and the mean diameter of the coil is $D_m = 7.5 \ mm$. The spring has 25 turns in the coil i.e. $N = 25$. If the maximum weight to be measured by this balance is $9.8 \ N$ (i.e. a mass of $1 \ kg$ under standard $g = 9.8 \ m/s^2$) what is the maximum deflection of the spring? If the range on the scale is divided in to 100 intervals what is the least count of the spring balance?

Solution :

Based on Equation 14.6 the spring constant may be evaluated as

$$K = \frac{8 \times 10^9 \times (0.57 \times 10^{-3})^4}{8 \times (7.5 \times 10^{-3})^3 \times 25} = 100.09 \ N/m$$

The maximum deflection z_{max} is to occur when the load applied on the spring is $W = 9.8 \ N$. Hence the maximum spring deflection is given by

$$z_{max} = \frac{W}{K} = \frac{9.8}{100.09} \times 1000 = 97.92 \ mm$$

When the range is divided in to 100 divisions, each division is equal to $0.98 \ mm$ and represents a least count of $0.098 \ N$ or weight of a mass equal to $10 \ g$.

It is interesting to note that Progen Scientific, UK, supplies spring balance 'Kern 281-601' that has a range of $1000\,g$ and least count of $10\,g$.

An elastic element may be used to convert a force to a displacement. Any elastic material follows Hooke's law within its elastic limit and hence is a potential spring element. Several examples are given in Figure 14.3 along with appropriate expressions for the applicable spring constants. Spring constants involve E, the Young's modulus of the material of the element and the geometric parameters indicated in the figure. In case of an element that undergoes bending the moment of inertia of the cross section is the appropriate geometric parameter. The expressions for spring constant are easily derived and are available in any book on strength of materials.

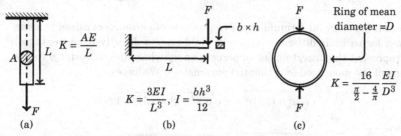

Figure 14.3: *Several configurations for measuring force with the appropriate spring constants: (a) Rod in tension or compression (b) Cantilever beam (c) Thin ring*

Example 14.2 is presented to get an idea about typical numbers that characterize force transducers.

Example 14.2

A cantilever beam made of spring steel (Young's modulus $200\ GPa$) $25\ mm$ long has a width of $2\ mm$ and thickness of $0.8\ mm$. Determine the spring constant. If all lengths are subject to measurement uncertainties of 0.5% determine the percent uncertainty in the estimated spring constant. What is the force if the deflection of the free end of the cantilever beam under a force acting there is $3\ mm$? What is the uncertainty in the estimated force if the deflection itself is measured with an uncertainty of 0.5%?

Solution :

Figure 14.4 describes the situation. The cantilever beam will bend as shown in the figure.

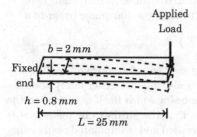

—— Beam shape before loading
- - - - Beam shape after loading

Figure 14.4: *Explanatory figure for Example 14.2*

The given data is written down as:

$$\text{Young's Modulus } E = 200 \times 10^9 \, Pa \qquad \text{Width of beam } b = 0.002 \, m$$
$$\text{Thickness of beam } t = 0.0008 \, m \qquad \text{Length of beam } L = 0.025 \, m$$

The moment of inertia is calculated using the well known formula as

$$I = \frac{bt^3}{12} = \frac{0.002 \times 0.0008^3}{12} = 8.5333 \times 10^{-14} \, m^4$$

Using the formula for the spring constant given in Figure 14.3, we have

$$K = \frac{3EI}{L^3} = \frac{3 \times 200 \times 10^9 \times 8.5333 \times 10^{-14}}{0.025^3} = 3277 \, N/m$$

Since all the relevant formulae involve products of parameters raised to some power, logarithmic differentiation will yield results directly in percentages. We represent the uncertainties in percent as u_p where the subscript p refers to any of the measured or estimated parameters. We have:

$$u_I = \pm\sqrt{u_b^2 + (3u_t)^2} = \pm\sqrt{0.5^2 + (3 \times 0.5^2)} = 1.58\%$$

Hence

$$u_K = \pm\sqrt{u_I^2 + (3u_L)^2} = \pm\sqrt{1.58^2 + (3 \times 0.5)^2} = 2.18\%$$

Thus the spring constant may be specified as

$$K = 3277 \pm \frac{2.18}{100} \times 3277 = 3277 \pm 71.4 \, N/m$$

For the given deflection under the load of $y = 3mm = 0.003 \, m$ the nominal value of the force may be calculated as

$$F = Ky = 3277 \times 0.003 = 9.83 \, N$$

Further the error may be calculated as

$$u_F = \pm\sqrt{u_K^2 + u_y^2} = \pm\sqrt{2.18^2 + 0.5^2} = 2.24\%$$

The estimated force may then be specified as

$$F = 9.83 \pm \frac{2.24}{100} \times 9.83 = 9.83 \pm 0.22 \, N$$

Note that the deflection may easily be measured with a vernier scale held against the free end of the beam element. Alternately a dial gauge (refer to a book on Metrology) may be made use of.

In the case considered in Example 14.2 the force may be inferred from the displacement measured at the free end where the force is also applied. Alternately one may measure the strain at a suitable location on the beam which itself is related to the applied force. The advantage of this method is that the strain to be measured is converted to an electrical signal which may be recorded and manipulated easily using suitable electronic circuits.

14.2.3 Proving ring

Proving ring is a device that is useful in measuring force or load. It was developed at the National Institute of Standards and Technology (NIST), USA. For a description of the proving ring design refer to NIST web site.[1] Proving rings are available for for measuring forces in the $2 - 2000\,kN$ range.

The proving ring consists of a metal ring made of alloy steel with external bosses to facilitate application of load or force to be measured (Figure 14.5) across a diameter. It has a rectangular section as shown by the sectional view across AA. Under a compressive load the ring changes its shape with negative strain of the inner periphery of the ring and positive strain of the outer surface of the ring. Thus the strain gages 1 and 2 suffer reduction in the resistance while strain gages 3 and 4 suffer increase in resistance. If the four gages are connected in the form of a full bridge shown in Figure 14.5 the bridge output is proportional to the applied load. Thus the imbalance of the bridge is amplified by a factor of two as compared to a "half bridge" arrangement.

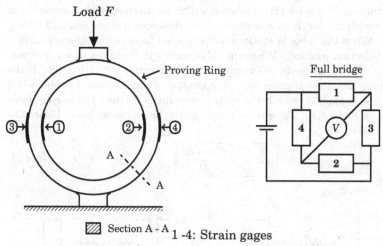

Figure 14.5: *Schematic of a proving ring*

14.2.4 Conversion of force to hydraulic pressure

While discussing pressure measurement we have discussed a dead weight tester (see section 7.3) that essentially consisted of a piston cylinder arrangement in which the pressure was converted to hydraulic pressure. It is immediately apparent that this arrangement may be used for measuring a force. If the piston area is accurately known, the pressure in the hydraulic liquid developed by the force acting on the piston may be measured by a pressure transducer. This pressure when multiplied by the piston area gives the force. The pressure may be measured by using several transducers that were discussed earlier in Chapter 4.

---

[1]"http://www.nist.gov/pml/div684/grp07/provingringdesign.cfm"

14.2.5 Piezoelectric force transducer

A piezoelectric material develops an electrical output when it is compressed by the application of a force. This signal is proportional to the force acting on the material. We shall discuss more fully this in section 14.3.3.

14.3 Measurement of acceleration

Acceleration measurement is closely related to the measurement of force. Effect of acceleration on a mass is to give rise to a force. This force is directly proportional to the mass, which if known, will give the acceleration when the force is divided by it. Force itself may be measured by various methods that have been discussed previously in section 14.2.

14.3.1 Preliminary ideas

Consider a spring mass system as shown in Figure 14.6. We shall assume that there is no damping. The spring is attached to a table (or an object whose acceleration is to be measured) as shown with a mass M attached to the other end and sitting on the table. When the table is stationary there is no force in the spring and it is in the zero deflection position. When the table moves to the right under a steady acceleration a, the mass tends to extend the spring because of its inertia. If the initial location of the mass is given by x_i, the location of the mass on the table when the acceleration is applied is x. If the spring constant is K, then the tension force in the spring is $K(x - x_i)$. The same force is acting on the mass also. Hence the acceleration is given by

$$a = \frac{F}{M} = \frac{K(x - x_i)}{M} \qquad (14.7)$$

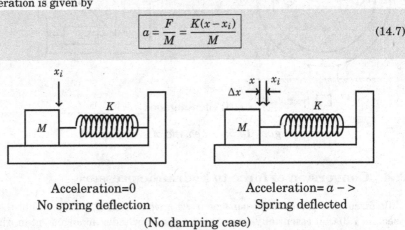

Acceleration=0
No spring deflection

Acceleration= a – >
Spring deflected

(No damping case)

Figure 14.6: Spring mass system under the influence of an external force

This is, of course, a simplistic approach since it is difficult to achieve the zero damping condition. In practice the acceleration may not also be constant. In deed we may want to measure acceleration during periodic oscillations or vibrations of a system. We consider these later on.

Example 14.3

An accelerometer has a seismic mass of $M = 0.02/;kg$ and a spring of spring constant equal to $K = 2000\ N/m$. Maximum displacement allowed is $1\ cm$. What is the maximum acceleration that may be measured? What is the natural frequency of the accelerometer?

Solution :

Figure 14.7 explains the concept in this case. Stops are provided so as not to damage the spring during operation by excessive displacement.

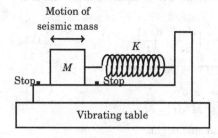

Motion of seismic mass

Figure **14.7:** Accelerometer scheme specified in Example 14.3

Thus the maximum acceleration that may be measured corresponds to the maximum allowed displacement of $\pm 1\ cm$ or $\pm 0.01\ m$. Thus, using Equation 14.7, the maximum acceleration that may be measured is

$$a_{max} = \frac{K\Delta x}{M} = \pm \frac{2000 \times 0.01}{0.02} = \pm 1000\ m/s^2$$

The corresponding acceleration in terms of standard "g" is given by

$$a_{max} = \frac{\pm 1000}{9.8} g = \pm 102\,g$$

The natural frequency of the accelerometer is given by the familiar result

$$f_n = \frac{1}{2\pi}\sqrt{\frac{K}{M}} = \frac{1}{2 \times \pi}\sqrt{\frac{2000}{0.02}} = 50.33\,Hz$$

14.3.2   Characteristics of a spring - mass - damper system

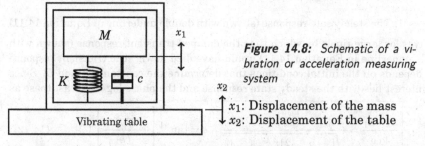

Figure **14.8:** Schematic of a vibration or acceleration measuring system

x_1: Displacement of the mass
x_2: Displacement of the table

Consider the dynamics of the system shown in Figure 14.8. In vibration measurement the vibrating table executes vibrations in the vertical direction and may be represented by a complex wave form. It is however possible to represent it as a Fourier series involving vibrations at a series of frequencies. Let one such component be represented by the input $x_1 = x_{1,0}\cos(\omega_1 t)$. Here x stands for the displacement, t stands for time and ω represents the circular frequency. Force experienced by the mass due to its acceleration (as a response to the input) is $M\dfrac{d^2 x_2}{dt^2}$. Due to the displacement of the spring the mass experiences a force given by $K(x_1 - x_2)$. The force of damping (linear case - damping is proportional to velocity) is $c\left(\dfrac{dx_1}{dt} - \dfrac{dx_2}{dt}\right)$ where c is the damping coefficient. For dynamic equilibrium of the system, we have

$$M\frac{d^2 x_2}{dt^2} = K(x_1 - x_2) + c\left(\frac{dx_1}{dt} - \frac{dx_2}{dt}\right) \tag{14.8}$$

Dividing through Equation 14.8 by M and rearranging we get

$$\frac{d^2 x_2}{dt^2} + \frac{c}{M}\frac{dx_2}{dt} + \frac{K}{M}x_2 = \frac{c}{M}\frac{dx_1}{dt} + \frac{K}{M}x_1 \tag{14.9}$$

We now substitute the input on the right hand side to get

$$\frac{d^2 x_2}{dt^2} + \frac{c}{M}\frac{dx_2}{dt} + \frac{K}{M}x_2 = x_{1,0}\left[\frac{K}{M}\cos(\omega_1 t) - \frac{c}{M}\omega_1\sin(\omega_1 t)\right] \tag{14.10}$$

This is a second order ordinary differential equation that is reminiscent of the equation we encountered while discussing the transient response of a U tube manometer in section 7.2.2. The natural frequency of the system is $\omega_n = \sqrt{\dfrac{K}{M}}$ and the critical damping coefficient is $c_c = 2\sqrt{MK}$. The solution to this equation may be worked out easily to get

$$x_2 - x_1 = e^{-\frac{c}{2M}t}[A\cos(\omega t) + B\sin(\omega t)] + \underline{\underline{\frac{M\omega_{1,0}^2\cos(\omega_1 t - \phi)}{\sqrt{(K - M\omega_1^2)^2 + (c\omega_1)^2}}}} \tag{14.11}$$

In the above equation

$$\omega = \sqrt{\frac{K}{M} - \left(\frac{c}{2M}\right)^2}, \quad \phi = \tan^{-1}\left[\frac{c\omega_1}{K - M\omega_1^2}\right]$$

for $\dfrac{c}{c_c} < 1$. The steady sate response (shown with double underline in Equation 14.11) survives for large times by which time the damped transient response (shown with single underline in Equation 14.11) would have died down. The transient response also depends on the initial conditions that determine the constants A and B. Since our interest lies with the steady state response and the phase lag we recast these as

$$\frac{x_2 - x_1}{x_{1,0}} = \frac{\left(\frac{\omega_1}{\omega_n}\right)^2}{\sqrt{\left[1 - \left(\frac{\omega_1}{\omega_n}\right)^2\right]^2 + \left[\frac{2c\omega_1}{c_c\omega_n}\right]^2}}, \quad \phi = \tan^{-1}\left[\frac{\frac{2c\omega_1}{c_c\omega_n}}{1 - \left(\frac{\omega_1}{\omega_n}\right)^2}\right] \tag{14.12}$$

The solution presented above is the basic theoretical framework on which vibration measuring devices are built. In case we want to measure the acceleration, the input to be followed is the second derivative with respect to time of the displacement given by

$$a(t) = \frac{d^2x_1}{dt^2} = -x_{1,0}\omega_1^2\cos(\omega_1 t)$$

The amplitude of acceleration is hence equal to $x_{1,0}\omega_1^2$ and hence we may rearrange the steady state response as

$$\frac{x_2 - x_1}{x_{1,0}\left(\frac{\omega_1^2}{\omega_n^2}\right)} = \frac{1}{\sqrt{\left[1 - \left(\frac{\omega_1}{\omega_n}\right)^2\right]^2 + \left[\frac{2c\omega_1}{c_c\omega_n}\right]^2}} \tag{14.13}$$

The above is nothing but the acceleration response of the system. Plots help us make suitable conclusions.

Figure 14.9 shows the frequency response of a second order system. Note that the ordinate is non-dimensional response normalized with the input amplitude as the normalizing factor. The frequency ratio is used along the abscissa. Note that the

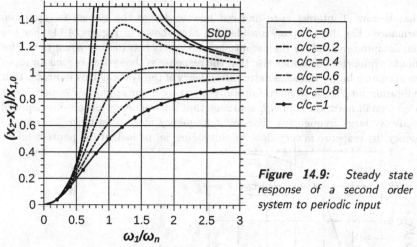

Figure 14.9: *Steady state response of a second order system to periodic input*

amplitude response of the spring mass damper system will take very large values close to resonance where $\omega_1 = \omega_n$. The second order system may not survive such a situation and one way of protecting the system is to provide amplitude limiting stops as was shown in Figure 14.7. In Figure 14.9 the stops are provided such that the amplitude ratio is limited to 25% above the input value.

We now look at the phase relation shown plotted in Figure 14.10 for a second order system. The output always lags the input and varies from 0° for very low frequencies to 180° for infinite frequency. However, the phase angle varies very slowly for high frequencies.

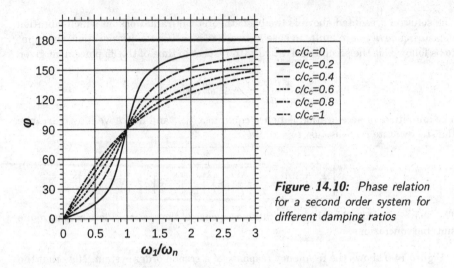

Figure 14.10: Phase relation for a second order system for different damping ratios

Design of second order system for optimum response

What is now of interest is to find out how to design the system for optimum performance. For this purpose consider the case shown in Figure 14.11. For the chosen damping ratio of 0.7 the amplitude response is less than or equal to one for *all* input frequencies. We notice also that the response of the system is good for input frequency much larger than the natural frequency of the system. This indicates that the vibration amplitude is more faithfully given by a spring mass damper device with very small natural frequency, assuming that the frequency response is required at relatively large frequencies. For input frequency more than twice the natural frequency the response is very close to unity. Now let us look at the phase relation

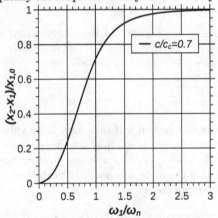

Figure 14.11: Amplitude response of a system with damping ratio of 0.7

for a second order system with the same damping ratio shown in Figure 14.12. It is noted that for input frequency in excess of about 4 times the natural frequency the phase angle varies linearly with the input frequency. This is a very useful property of a second order system as will become clear from the following discussion. An

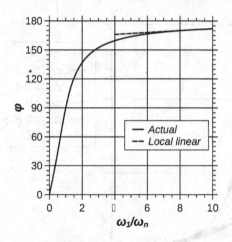

Figure 14.12: *Phase response of a system with damping ratio of 0.7*

arbitrary periodic function with a fundamental frequency of ω_1 may be written in the form of a Fourier series given by

$$x_1 = x_{1,0}\cos(\omega_1 t) + x_{1,1}\cos(2\omega_1 t) + x_{1,2}\cos(3\omega_1 t) + \ldots \qquad (14.14)$$

If ω_1 corresponds to a frequency in the linear phase lag region of the second order system and ϕ is the corresponding phase lag, the phase lag for the higher harmonics are multiples of this phase lag. Thus the fundamental will have a phase lag of ϕ, the next harmonic a phase lag of 2ϕ and so on. Thus the steady state output response will be of the form

$$\text{Output} \sim x_{1,0}\cos(\omega_1 t - \phi) + x_{1,1}\cos(2\omega_1 t - 2\phi) + x_{1,2}\cos(3\omega_1 t - 3\phi) + \ldots \qquad (14.15)$$

This simply means that the output is of the form

$$\text{Output} \sim x_{1,0}\cos(\omega_1 t') + x_{1,1}\cos(2\omega_1 t') + x_{1,2}\cos(3\omega_1 t') + \ldots \qquad (14.16)$$

Thus the output retains the shape of the input.

Now we shall look at the desired characteristics of an accelerometer or an acceleration measuring instrument. We make a plot of the acceleration response of a second order system as shown in Figure 14.13. It is clear from this figure that for a faithful acceleration response the measured frequency must be very small compared to the natural frequency of the second order system. Damping ratio does not play a significant role.

In summary we may make the following statements:

- Displacement measurement of a vibrating system is best done with a transducer that has a very small natural frequency coupled with damping ratio of about 0.7. The transducer has to be made with a large mass with a soft spring.
- Accelerometer is best designed with a large natural frequency. The transducer should use a small mass with a stiff spring. Damping ratio does not have significant effect on the response.

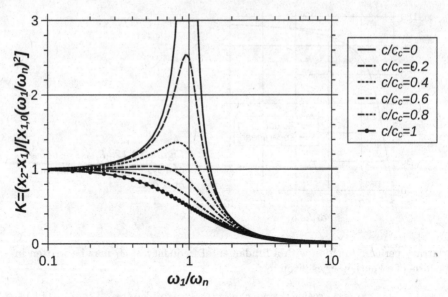

Figure 14.13: *Acceleration response of a second order system*

Example 14.4

A big seismic instrument is constructed with $M = 100\ kg$, $\dfrac{c}{c_c} = 0.7$ and a spring of spring constant $K = 5000\ N/m$. Calculate the value of linear acceleration that would produce a displacement of 5 mm on the instrument. What is the frequency ratio $\dfrac{\omega}{\omega_n}$ such that the displacement ratio is 0.99? What is the useful frequency of operation of this system as an accelerometer?

Solution :

Displacement is $x = 5\ mm = 0.005\ m$ and the spring constant is $K = 5000\ N/m$. Hence the spring force corresponding to the given displacement is

$$F = Kx = 5000 \times 0.005 = 25\ N$$

The seismic mass is $M = 100\ kg$. Hence the linear acceleration is given by

$$a = \frac{F}{M} = \frac{25}{100} = 0.25\ m/s^2$$

Represent $\dfrac{\omega}{\omega_n}$ by the symbol y. The damping ratio has been specified as 0.7. From the response of a second order system given earlier the condition that needs to be satisfied is

$$\text{Amplitude ratio} = 0.99 = \frac{y^2}{\sqrt{(1-y^2)^2 + (1.4y)^2}}$$

We shall substitute $z = y^2$. The above equation then becomes

$$\frac{z}{\sqrt{(1-z)^2 + 1.96z}} = 0.99$$

This equation may be simplified to get the following quadratic equation for z.

$$0.0203z^2 + 0.04z - 1 = 0$$

The quadratic equation has a meaningful solution given by $z = 6.1022$. The positive square root of this, in fact, gives the desired result $y = \sqrt{6.1022} = 2.47$. The natural frequency of the system is given by

$$\omega_n = \sqrt{\frac{K}{M}} = \sqrt{\frac{5000}{100}} = 7.07 \, rad/s$$

The frequency at which the amplitude ratio is equal to 0.99 is thus given by

$$\omega = y \times \omega_n = 2.47 \times 7.07 = 17.5 \, rad/s$$

Again we shall assume that the useful frequency is such that the acceleration response is 0.99 at this cut off frequency. We have

$$\text{Acceleration ratio} = 0.99 = \frac{1}{\sqrt{(1-y^2)^2 + (1.4y)^2}}$$

We shall substitute $z = y^2$. The equation to be solved then becomes

$$\frac{1}{\sqrt{(1-z)^2 + 1.96z}} = 0.99$$

This equation may be simplified to get the following quadratic equation for z.

$$z^2 - 0.04z - 1 = 0$$

The quadratic equation has a meaningful solution $z = 0.1639$. The positive square root of this gives the desired result $y = \sqrt{0.1639} = 0.405$. The frequency at which the amplitude ratio is equal to 0.99 is thus given by

$$\omega = y \times \omega_n = 0.405 \times 7.07 = 2.86 \, rad/s$$

Earthquake waves tend to have most of their energy at periods (the time from one wave crest to the next) of ten seconds to a few minutes. These correspond to frequencies of $\omega = 2\pi/T = 2 \times \pi/10 = 0.63 \, rad/s$ and $\omega = 2\pi/T = 2 \times \pi/60 = 0.105 \, rad/s$. The accelerometer being considered in this example is eminently suited for this application.

Example 14.5

A vibration measuring instrument is used to measure the vibration of a machine vibrating according to the relation $x = 0.007 \cos(2\pi t) + 0.0015 \cos(7\pi t)$ where the amplitude x is in m and t is in s. The vibration measuring instrument has an undamped natural frequency of $0.4 \, Hz$ and a damping ratio of 0.7. Will the output be faithful to the input? Explain.

Solution :

The natural frequency of the vibration measuring instrument specified is $f_n = 0.4\,Hz$. The corresponding circular frequency is given by

$$\omega_n = 2\pi \times 0.4 = 0.8\pi\ rad/s$$

For the first part of the input the impressed frequency is $\omega_1 = 2\pi\ rad/s$. Hence the frequency ratio for this part is

$$\frac{\omega_1}{\omega_n} = \frac{2\pi}{0.8\pi} = 2.5$$

The corresponding amplitude ratio is obtained using first of Equation 14.12 as

$$\text{Amplitude ratio} = \frac{2.5^2}{\sqrt{(1-2.5^2)^2 + (1.4 \times 2.5)^2}} = 0.9905$$

The phase angle is obtained using second of Equation 14.12 as

$$\phi = \tan^{-1}\left(\frac{2 \times 0.7 \times 2.5}{1 - 2.5^2}\right) = 2.554\ rad$$

For the second part of the input the impressed frequency is $\omega_2 = 7\pi$. Hence the frequency ratio for this part is

$$\frac{\omega_2}{\omega_n} = \frac{7\pi}{0.8\pi} = 8.75$$

The corresponding amplitude ratio is

$$\text{Amplitude ratio} = \frac{8.75^2}{\sqrt{(1-8.75^2)^2 + (1.4 \times 8.75)^2}} = 1.000$$

The corresponding phase angle is given by

$$\phi = \tan^{-1}\left(\frac{2 \times 0.7 \times 8.75}{1 - 8.75^2}\right) = 2.981\ rad$$

The output of the accelerometer is thus given by

$$\text{Output} = 0.9905 \times 0.007\cos(2\pi t - 2.554) + 0.0015 \times 1.000\cos(7\pi t - 2.981)$$

or

$$\text{Output} = 0.0069\cos(2\pi t - 2.554) + 0.0015\cos(7\pi t - 2.981)$$

Introduce the notation $\theta = 2\pi t - 2.554$. The quantity 3.5θ may be written as follows:

$$3.5\theta = 3.5(2\pi t - 2.554) = 7\pi t - 8.939$$

The above may be rewritten as

$$3.5\theta = (7\pi t - 2.981) + (2.981 - 8.939)$$

The term shown with underline may be rewritten as $-2\pi + 0.3252$. Thus the output response may be written as

$$\text{Output} = 0.0069\cos(\theta) + 0.0015\cos(3.5\theta + 0.3252)$$

The term -2π does not play a role and hence the second term has an effective lead with respect to the first term of $0.3252\ rad$. Hence the output does not follow the input faithfully. However, the amplitude part is followed very closely for both the terms.

14.3.3   Piezoelectric accelerometer

The principles that were dealt with above help us in the design of piezoelectric accelerometers. These are devices that use a mass mounted on a piezo ceramic as shown in Figure 14.14. The entire assembly is mounted on the device whose acceleration is to be measured. The ENDEVCO MODEL 7703A-50-100 series of piezo-electric accelerometers are typical of such accelerometers[2]. The diameter of the accelerometer is about 16 mm. When the mass is subject to acceleration it applies a force on the piezoelectric material. A charge is developed that may be converted to a potential difference by suitable electronics.

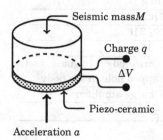

Seismic mass M

Charge q

ΔV

Piezo-ceramic

Acceleration a

Figure 14.14: Piezoelectric accelerometer

The following relations describe the behavior of a piezoelectric transducer.

$$q = Fd, \quad q = C\Delta V \quad \text{and} \quad C = \kappa \frac{A}{\delta} \qquad (14.17)$$

Combining all these we get

$$\Delta V = \frac{d}{\kappa}\frac{\delta}{A}F = \frac{d}{\kappa}\frac{\delta}{A}Ma \qquad (14.18)$$

In Equation 14.18 the various symbols have the following meanings:
C = Capacitance of the piezoelectric element, κ = Dielectric constant, d = Piezoelectric constant, A = Area of Piezo ceramic, δ = Thickness of Piezo ceramic, M = Seismic mass, q = Charge, ΔV = Potential difference and a = Acceleration.

Charge sensitivity is defined as

$$S_q = \frac{q}{a} \qquad (14.19)$$

which has a typical value of $50 \times 10^{-8} Coul/g$. The acceleration is in units of g, the acceleration due to gravity. The voltage sensitivity is defined as

$$S_V = \frac{\Delta V}{a} = \frac{d}{\kappa}\frac{\delta}{A}M \qquad (14.20)$$

Piezo transducers are supplied by many manufacturers. They have specifications typically shown in Table 14.1.

---

[2]https://www.endevco.com

Table 14.1: *Typical product data Brüel & Kjaer 8200*

Range	1000 to 5000 N
Charge sensitivity	$4 \times 10^{-12}\ Coul/N$
Capacitance	$25 \times 10^{-12}\ F$
Stiffness	$5 \times 10^8\ N/m$
Resonance frequency with 5 g mounted on top load	35 kHz
Effective seismic mass:	
Above Piezo-electric element	3 g
Below Piezo-electric element	18 g
Piezoelectric material	Quartz
Transducer housing material	SS 316
Diameter	~ 18mm
Transducer mounting	Threaded spigot and tapped hole in the body
Signal conditioning	Charge amplifier
Useful frequency range	~ 10 kHz

14.3.4   Laser Doppler Vibrometer

We have earlier discussed the use of laser Doppler for measurement of fluid velocity. It is possible to use the method also for the remote measurement of vibration. Consider the optical arrangement shown in Figure 14.15. A laser beam is directed towards the object that is vibrating. The reflected laser radiation is split in to two beams that travel different path lengths by the arrangement shown in the figure. The two beams are then combined at the detector. The detector signal contains information about the vibrating object and this may be elucidated using suitable electronics.

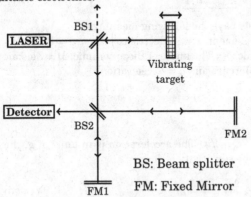

BS: Beam splitter

FM: Fixed Mirror

Figure 14.15: *Schematic to explain the working principle of a Laser Doppler Vibrometer*

Consider the target to vibrate in a direction coinciding with the laser incidence direction. Let the velocity of the target due to its vibratory motion be $V(t)$. The laser beam that is reflected by the vibrating target is split in to two beams by the beam splitter 2. One of these travels to the fixed mirror 1 and reaches the photo detector after reflection off the beam splitter 2. The other beam is reflected by fixed mirror 2 and returns directly to the photo detector. The path lengths covered by the two beams are *different*. Let Δl be the extra distance covered by the second beam with respect to the first. The second beam reaches the detector after a time delay of $t_d = \dfrac{\Delta l}{c}$ where c is the speed of light. Since the speed of light is very large (i.e. $3 \times 10^8 m/s$) the time delay is rather a small quantity, i.e. $\dfrac{\Delta l}{c} \ll 1$.

We know that the reflected radiation will have a slightly different frequency than the incident laser frequency and this is due to the Doppler shift we have discussed earlier. If we look carefully we see that the two beams that reach the detector have in fact undergone slightly different Doppler shifts because of the time delay referred to above. When we combine these two beams at the detector we will get a beat frequency given by

$$f_{beat} = \frac{2}{\lambda}\left[V(t) - V\left(t - \frac{\Delta l}{c}\right)\right] \tag{14.21}$$

If the velocity of the target contains several frequencies one may represent it in terms of a Fourier integral. The velocity is thus given by the relation

$$V(t) = \int_0^\infty A(\omega)\sin[\omega t - \phi(\omega)]d\omega \tag{14.22}$$

Substitute Expression 14.22 in Expression 14.21 to get

$$f_{beat} = \frac{2}{\lambda}\int_0^\infty A(\omega)\left\{\sin[\omega t - \phi(\omega)] - \sin\left[\omega\left(t - \frac{\Delta l}{c}\right) - \phi(\omega)\right]\right\} \tag{14.23}$$

Noting now that $\dfrac{\Delta l}{c} \ll 1$ the integrand may be approximated, using well known trigonometric identities as

$$\sin[\omega t - \phi(\omega)] - \sin\left[\omega\left(t - \frac{\Delta l}{c}\right) - \phi(\omega)\right] \approx 2\frac{\omega\Delta l}{c}\cos[\omega t - \phi(\omega)] \tag{14.24}$$

Thus the beat frequency is given by

$$f_{beat} = \frac{4}{\lambda}\frac{\Delta l}{c}\int_0^\infty A(\omega)\cos[\omega t - \phi(\omega)]d\omega \tag{14.25}$$

We notice that the integral in Equation 14.25 represents the instantaneous acceleration of the vibrating object. Thus the beat frequency is proportional to the instantaneous acceleration of the vibrating target.

In practice the physical path length needs to be very large within the requirement $\dfrac{\Delta l}{c} \ll 1$. A method of achieving this is to use a long fiber optic cable in the second path between the beam splitter 2 and the fixed mirror 2. Recently Rothberg et. al. have described the development of a Laser Doppler Accelerometer[3].

---

[3] S Rothberg, A Hocknell and J Coupland, Developments in laser Doppler accelerometry (LDAc) and comparison with laser Doppler velocimetry, Optics and Lasers in Engineering, V-32, No. 6, 2000, pp. 549-564.

14.3.5   Fiber Optic Accelerometer

Another interesting recent development is a fiber optic accelerometer described by Lopez - Higuera et. al[4]. The schematic of the instrument is shown in Figure 14.16.

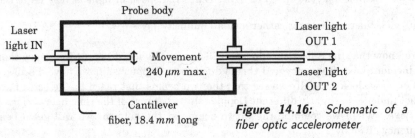

Figure 14.16: Schematic of a fiber optic accelerometer

In this transducer a laser beam is fed in through a cantilever fiber that vibrates with the probe body that is mounted on a vibrating object. The laser beam emerging out of the cantilever is incident on a fiber optic pair that is rigidly fixed and does not vibrate. The vibrating cantilever modulates the light communicated to the two collecting fibers. The behavior of the fiber optic cantilever is given by the following response function which represents the ratio of the relative displacement and the input acceleration.

$$H(\omega) = \frac{1}{\omega^2} \cdot \frac{\cos(\alpha\sqrt{\omega}) + \cosh(\alpha\sqrt{\omega})}{1 + \cos(\alpha\sqrt{\omega})\cosh(\alpha\sqrt{\omega})} \tag{14.26}$$

In the above equation $\alpha = \left[\frac{\rho L^4}{EI}\right]^{\frac{1}{4}}$ with ρ = linear mass density of fiber, E = Young's modulus of the fiber material, L = length of cantilever and I = transverse moment of inertia. For frequencies significantly lower than about 20% of the natural frequency of the cantilever beam given by $f_n = \dfrac{3.516}{2\pi\sqrt{\alpha}}$ the response function is a constant to within 2%. Thus the lateral displacement is proportional to the acceleration. In the design presented in the paper the maximum transverse displacement is about 5 μm with reference to the collection optics. The performance figures for this are: Range: 0.2 – 140 Hz, Sensitivity: 6.943 V/g at a frequency of 30 Hz which converts to 700 $\dfrac{mV}{m/s^2}$. For more details the reader may refer to the paper.

14.4   Measurement of torque and power

Torque and power are important quantities involved in power transmission in rotating machines like engines, turbines, compressors, motors and so on. Torque and power measurements are made by the use of a dynamometer. In a dynamometer the torque and rotational speed are independently measured and the product of these gives the power.

We describe here representative methods of measurement of torque and rotational speed or angular velocity. These are given below:

---

[4]Lopez - Higuera et. al. Journal of Light Wave Technology, Vol.15, No.7, 1997, pp.1120-30.

Torque Measurement: Mechanical brake arrangement - Electrical loading, Engine drives a generator - Measure shear stress on the shaft

Measurement of rotational speed: Tachometer, a mechanical device - Non contact optical rpm meter

14.4.1 Mechanical brake arrangement - Prony brake

The brake drum dynamometer is a device by which a known torque can be applied on a rotating shaft that may belong to any of the devices that were mentioned earlier. Schematic of the brake drum dynamometer is shown in Figure 14.17.

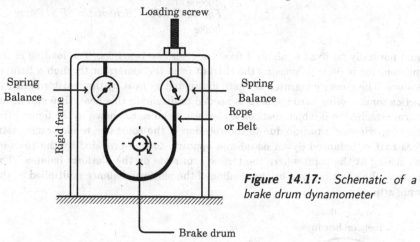

Figure 14.17: *Schematic of a brake drum dynamometer*

A rope or belt is wrapped around the brake drum attached to the shaft. The two ends of the rope or belt are attached to rigid supports with two spring balances as shown. The loading screw may be tightened to increase or loosened to decrease the frictional torque applied on the drum. When the shaft rotates the tension on the two sides will be different. The difference is just the frictional force applied at the periphery of the brake drum. The product of this difference multiplied by the radius of the drum gives the torque. Alternate way of measuring the torque using essentially the brake drum dynamometer is shown in Figure 14.18. This arrangement is named after Prony, a french engineer[5]. The torque arm is adjusted to take on the horizontal position by the addition of suitable weights in the pan after adjusting the loading screw to a suitable level of tightness. The torque is given by the product of the torque arm and the weight in the pan. It is to be noted that the power is absorbed by the brake drum and dissipated as heat. In practice it is necessary to cool the brake drum by passing cold water through tubes embedded in the brake blocks or the brake drum.

14.4.2 Electric generator as a dynamometer

An electric generator is mounted on the shaft that is driven by the power device whose output power is to be measured as shown in Figure 14.19. The stator that

---

[5]Gaspard Clair François Marie Riche de Prony 1755-1839,a French mathematician and engineer

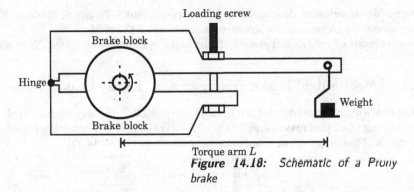

Figure 14.18: *Schematic of a Prony brake*

would normally be fixed is allowed float between two bearings. The loading of the dynamometer is done by passing the current from the generator through a bank of resistors. The power generated is again dissipated as heat by the resistor bank. In practice some cooling arrangement is needed to dissipate this heat. The stator has an arm attached to it which rests on a platform balance as shown in the figure. The stator experiences a torque due to the rotation of the rotor due to electromagnetic forces that is balanced by an equal and opposite torque provided by the reaction force acting at the point where the torque arm rests on the platform balance. The torque on the shaft is given by the reading of the platform balance multiplied by the torque arm.

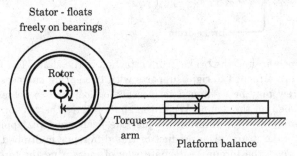

Figure 14.19: *Electric generator used as a dynamometer*

14.4.3   Measure shear stress on the shaft

Torque may also be estimated by measuring the shear stress experienced by the shaft that is being driven. If the shaft is subject to pure torsion the principal stresses and the shear stress are identical as shown by Figure 14.20. The first part of the figure shows the stresses on the surface of the shaft while the second part of the figure shows the Mohr's circle familiar to the reader from Strength of Materials. In order to avoid any bending stresses developing at the location of the load cells, they are located close to a bearing or support. We notice that the principal stresses are tensile and compressive along the two diagonals while the shear stresses are along the longitudinal and circumferential directions. If load cells are mounted in

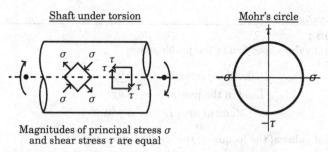

Shaft under torsion   Mohr's circle

Magnitudes of principal stress σ
and shear stress τ are equal

Figure 14.20: *Stresses on the surface of a shaft in torsion*

the form of a bridge with the arms of the bridge along the edges of a square oriented at $45°$ to the shaft axis, the imbalance voltage produced by the bridge is a measure of the principal stress that is also equal to the shear stress. The arrangement of the load cells is as shown in Figure 14.21. This arrangement improves the gage sensitivity by a factor of 2. The bridge circuit diagram shows the arrangement of the load cells and indicates how the gauge factor is improved. Load cells 1 and 3 experience elongation due to tensile stress and hence the resistances of these two load cells increase with respect to the no load value. In the case of load cells 2 and 4 the opposite happens, i.e. the load cells experience compression due to compressive stress and hence the resistances of these two load cells decrease with respect to the no load value. We know that $\tau = \dfrac{TR}{J}$. Here T is the torque, R is the radius of the shaft and $J = \dfrac{\pi R^4}{2}$ is the polar moment of inertia. Using the load cell measured value of τ, we may calculate the torque experienced by the shaft. In this method the dynamometer is used only for applying the load and the torque is measured using the load cell readings.

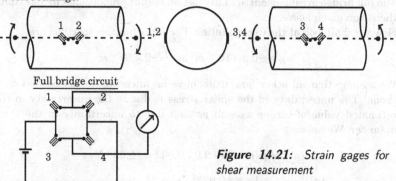

Full bridge circuit

Figure 14.21: *Strain gages for shear measurement*

Example 14.6

A shaft transmitting power from an engine is made of steel and has a diameter of $D = 25 \pm 0.2\,mm$. The torque on the shaft is applied by a Prony brake (see Figure 14.18) in which the load in the pan has a mass of $5\,kg$ and the moment arm is measured to be $L = 350 \pm 2\,mm$. What will be the shear stress on the surface of the shaft? What will be the strain experienced by a load cell bonded on the shaft surface that makes an angle of $45°$ with the shaft axis? Determine the uncertainties also.

Solution :

The nominal values specified in the problem are:

$$\text{Shaft diameter } D = 0.025\, m$$
$$\text{Load in the pan } M = 5\, kg$$
$$\text{Moment arm } L = 0.350\, m$$

The nominal value of the torque on the shaft is given by

$$T = MgL = 5 \times 9.8 \times 0.350 = 17.15\, N\, m$$

The polar moment of inertia is obtained as

$$J = \frac{\pi R^4}{2} = \frac{\pi D^4}{32} = \frac{\pi \times 0.025^4}{32} = 3.835 \times 10^{-8}\, m^4$$

The shear stress on the surface of the shaft may now be calculated as

$$\tau = \frac{TR}{J} = \frac{TD}{2J} = \frac{17.15 \times 0.025}{2 \times 3.835 \times 10^{-8}} = 5.59 \times 10^6\, Pa$$

The Young's modulus of steel is taken from data handbook and is $E = 200 GPa$. The strain in the load cell is equal in magnitude to the longitudinal strain induced by the principal stress (magnitude equal to the shear stress) that is either tensile or compressive. The magnitude of the strain is given by

$$\text{Strain} = s = \frac{\tau}{E} = \frac{5.59 \times 10^6}{200 \times 10^9} = 27.95 \times 10^{-6} \text{ or } 27.95\, \mu - \text{strain}$$

The full bridge arrangement actually has an output proportional to four times the strain given above.

Now we shall look at the uncertainties. The uncertainties specified are

$$u_D = \pm 0.2 \times 10^{-3}\, m, \ u_L = \pm 2 \times 10^{-3}\, m$$

We assume that all other quantities have no uncertainties associated with them. The uncertainty in the shear stress is due to the uncertainty in the estimated value of torque as well as that due to uncertainty in the shaft diameter. We have:

$$u_T = Mgu_L = \pm 5 \times 9.8 \times 0.002 = \pm 0.098\, N\, m$$

$$u_J = \frac{4\pi D^3}{32} u_D = \pm \frac{4 \times \pi \times 0.025^3}{32} \times 0.2 \times 10^{-3} = 1.227 \times 10^{-9}\, m^4$$

With these the uncertainty in the shear stress is calculated (using logarithmic differentiation) as

$$u_\tau = \pm \tau \sqrt{\left(\frac{u_T}{T}\right)^2 + \left(\frac{u_D}{D}\right)^2 + \left(\frac{u_J}{J}\right)^2}$$

$$= \pm 5.59 \times 10^6 \sqrt{\left(\frac{0.098}{17.15}\right)^2 + \left(\frac{0.0002}{0.025}\right)^2 + \left(\frac{1.227 \times 10^{-9}}{3.835 \times 10^{-8}}\right)^2}$$

$$= 0.182 \times 10^6 \, Pa$$

The uncertainty in the strain is easily obtained as

$$u_s = \frac{u_\tau}{E} = \pm \frac{0.182 \times 10^6}{200 \times 10^9} = 0.91 \, \mu - \text{strain}$$

14.4.4 Tachometer - Mechanical Device

This is a mechanical method of measuring the rotational speed of a rotating shaft. This is very commonly used as automobile speed indicator. The tachometer is mechanically driven by being coupled to the rotating shaft. The rotary motion is either transmitted by friction or by a gear arrangement as shown in Figure 14.22. The device consists of a magnet which is rotated by the drive shaft. A speed cup made of aluminum is mounted close to the rotating magnet with an air gap as indicated.

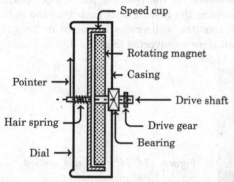

Figure 14.22: Sectional view of a speedometer or a tachometer

The speed cup is restrained by a hair spring and has a pointer attached to its own shaft that moves over a dial. When the magnet rotates due to the rotation of its own shaft the speed cup tends to be dragged along by the moving magnet and hence experiences a torque and tends to rotate along with it. The speed cup moves and takes up a position in which the rotating magnet induced torque is balanced by the restraining torque provided by the spring. Knowing the drive gear speed ratio one may calibrate the angular position on the dial in terms of the rotational speed in *RPM*.

14.4.5 Non contact optical *RPM* meter

This is a non contact method of rotational speed of a shaft. However it requires a wheel with openings to be mounted on the rotating shaft. The optical arrangement is shown in Figure 14.23. The arrangement is essentially like that was used for chopping a light beam in applications that were considered earlier. The frequency of interruption of the beam is directly proportional to the *RPM* of the wheel, that is usually the same as the *RPM* of the shaft to be measured and the number of holes provided along the periphery of the wheel. If there is only one hole the beam is interrupted once every revolution. If there are n holes the beam is interrupted n times per revolution. The rotational speed of the shaft is thus equal to the frequency of interruptions divided by the number of holes in the wheel.

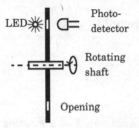

LED — Photo-detector

Rotating shaft

Opening

Figure 14.23: *Optical RPM measurement*

The LED photo detector wheel assembly is available from suppliers as a unit readily useable for *rpm* measurement.

A variant of the above is a single ended measurement that may be made using light reflected from a white reflector attached on the rotating part. The optical speedometer is held close to but not touching the moving part such that the light is reflected back to the instrument each time the white reflector passes in front of the meter. The number of pulses in a unit time is converted to give a digital display of the *RPM* directly as shown in Figure 14.24

Figure 14.24: *Single ended optical RPM measurement*

Example 14.7

An engine is expected to develop $5\,kW$ of mechanical output while running at an angular speed of $1200\,RPM$. A brake drum of $250\,mm$ diameter is available. It is proposed to design a Prony brake dynamometer using a spring balance as the force measuring instrument. The spring balance can measure a maximum force of $100\,N$. Choose the proper torque arm for the dynamometer.

Solution :

The schematic of the Prony brake is shown in Figure 14.25. Even though the spring balance is well known in its traditional linear form (see Figure 14.2) a dial type spring balance is commonly employed in dynamometer applications. The spring is in the form of a planar coil which will rotate a needle against a dial to indicate the force.

The given data is written down: Expected power is

$$P = 5\,kW = 5000\,W$$

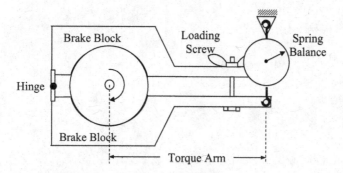

Figure 14.25: *Prony brake arrangement for Example 14.7*

Rotational speed of the engine is given as $N = 1200\ RPM$. This may be converted to angular velocity ω as

$$\omega = \frac{2\pi N}{60} = \frac{2 \times \pi \times 1200}{60} = 125.66\ rad/s$$

We know that the power developed is the product of torque and the angular speed. Hence we get

$$P = T\omega \text{ or } T = \frac{P}{\omega} = \frac{5000}{125.66} = 39.79\ N\,m$$

We shall assume that the force registered by the spring balance is limited to 90% of the maximum, i.e. $F = 0.9 \times 100 = 90\ N$. The required torque arm L may hence be obtained as

$$T = FL \text{ or } L = \frac{T}{F} = \frac{39.79}{90} = 0.442\ m$$

Concluding remarks

This chapter has considered measurement of force, acceleration etc. that are routine in any laboratory. Force measurement has been considered in an earlier chapter while discussing measurement of pressure. Measurement of rotational speed of a shaft, torque experienced by the shaft or power at the shaft all are important in engineering applications. Several techniques of measuring these have been discussed in detail.

Exercise IV

IV.1 Thermo-physical properties

Ex IV.1: Thermal conductivity of water is measured by using a radial flow conductivity apparatus. The apparatus has a heater that is 125 mm long and 37 mm diameter. The annular gap is 0.625 mm. The temperature drop across the gap has been measured to be 2.6°C when the heat supplied is 36.3 W. Determine the thermal conductivity of water. If all length measurements are susceptible to measurement error of ±0.5%, power input is measured with a possible error of ±0.1% and the temperature difference is susceptible to a measurement error of ±0.1°C, what is the uncertainty in the estimated thermal conductivity of water? Which measurement has contributed most to the error in the estimated value of thermal conductivity? Suggest how you would improve the measurement.

Ex IV.2: A thermal conductivity comparator uses a standard reference material (SRM) of thermal conductivity 15 ± 2% $W/m°C$. Two thermocouples placed 12 ± 0.2 mm apart indicate a temperature difference of 3.8 ± 0.2°C. The material of unknown thermal conductivity is in series with the SRM and indicates a temperature difference of 14.0 ± 0.2°C across a length of 25.0 ± 0.2 mm. Determine the thermal conductivity of the sample and its uncertainty. Mention all the assumptions you make. What is the heat flux through the sample? Specify an uncertainty on the heat flux.

Ex IV.3: A thermal conductivity comparator uses a standard reference material (SRM) of thermal conductivity 45±2% $W/m°C$. Two thermocouples placed 22 ± 0.5 mm apart indicate a temperature difference of 2.5 ± 0.2°C. The material of unknown thermal conductivity is in series with the SRM and indicates a temperature difference of 7.3±0.2°C across a length of 20±0.5 mm. Determine the thermal conductivity of the sample and its uncertainty. Mention all the assumptions you make.

Ex IV.4: The specific heat of a certain material is to be estimated by a calorimetric experiment. The quantity of heat supplied to 0.050 kg of material has been estimated as 0.150 kJ. The temperature rise was observed to be 3.55°C. The mass is measured within an uncertainty of $\pm 0.002\,kg$. The heat supplied has a $\pm 1\%$ uncertainty. The temperature measurement is fairly accurate and has an uncertainty of $\pm 0.05°C$. What is the specific heat of the material and its uncertainty? Express the uncertainty also as a percentage of the estimated value.

Ex IV.5: The specific heat of a certain material is to be estimated by a calorimetric experiment. The quantity of heat supplied to $100\,g$ of the material has been estimated to be $200\,J$. The temperature raise was $2°C$. The mass of the material has an error of $\pm 0.2\,g$ while the heat supplied is measured with a $\pm 1\%$ error. The temperature raise is measured very accurately and the error is no more than $\pm 0.01°C$. What is the specific heat and its uncertainty?

Ex IV.6: ASTM D240-92 describes the standard test method of hydrocarbon fuels by bomb calorimeter.
Parr Instrument Company (http://parrinst.com) supply "6300 - Automatic Isoperibol Calorimeter".
Access both these and make a short report on the measurement of calorific value of a coal sample.

Ex IV.7: A Saybolt viscometer is used to measure the viscosity of a liquid. The time recorded is $415 \pm 0.5\,s$. The density of the liquid has been measured separately to be $750 \pm 0.25\,kg/m^3$. Calculate the kinematic viscosity along with its uncertainty. Calculate also the dynamic viscosity along with its uncertainty.

Ex IV.8: A Saybolt viscometer is used to measure the viscosity of engine oil. The time recorded is $205 \pm 0.5\,s$. Calculate the kinematic viscosity along with its uncertainty. The specific gravity of the engine oil has been estimated to be 0.8 ± 0.02. Determine the dynamic viscosity of engine oil along with its uncertainty.

Ex IV.9: A rotating cylinder apparatus has the following dimensions:
$r_1 = 30 \pm 0.02\,mm$; $r_2 = 33 \pm 0.02\,mm$; $L = 75 \pm 0.04\,mm$; $a = 1 \pm 0.01\,mm$; $b = 1.5 \pm 0.02\,mm$ and $\omega = 800 \pm 2\,RPM$.
What is the torque experienced by the rotor if the fluid in the gaps is water at $30°C$? What is the uncertainty in the torque value?

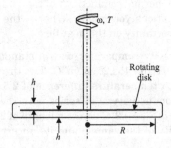

Figure IV.1: *Rotating disk for Exercise IV.10*

Ex IV.10: A thin rotating disk of radius R is driven by an electric motor at a constant angular speed ω when it is surrounded by a fluid of constant thickness h as indicated in Figure IV.1. The torque experienced by the rotor is measured independently and is given by T. Making reasonable assumptions obtain an expression for the fluid viscosity.

In a particular arrangement the disk radius is 30 mm, angular speed is 225 RPM and the fluid layer thickness is 1 mm. If the torque may be measured when it is in the range $0.05 - 0.25$ N m, determine the range of fluid viscosity that may be measured. Discuss the influence of uncertainties in ω, R, h and T on the uncertainty in the estimated value of viscosity. If the uncertainty in viscosity is not to be more than 1% suggest allowable uncertainties for the influencing parameters.

IV.2 Radiation properties of surfaces

Ex IV.11: Measurement of radiation properties of surfaces is considered in greater detail by D.K. Edwards in Chapter 10 of the book - "Eckert, E.R.G. and Goldstein, R.J., (Editors), Measurements in Heat Transfer, Hemisphere, Washington, 2nd Edition, 1976". Read the chapter and at least one reference quoted therein to make a report on the topic.

IV.3 Gas concentration

Ex IV.12: California Analytical Instruments supply Model 600 HCLD - Heated NO/NOx Analyzer. Study* the details about this instrument and prepare a note giving the salient features of the instrument along with its capabilities.
   *http://www.gasanalyzers.com

Ex IV.13: California Analytical Instruments supply Model 600 NDIR/O2. Information on this instrument is available at the web resource mentioned in Exercise IV.12. Study the details about this instrument and prepare a note giving the salient features of the instrument along with its capabilities.

IV.4 Force, acceleration, Torque and Power

Ex IV.14: *E*lectronic balances with milligram resolution are available in the market. Study the working principle of an electronic balance and describe how milligram resolution is obtained. Does an electronic balance indicate the mass or weight? Explain.

Ex IV.15: Wind tunnel balance is used to measure forces and moments experienced by a model over which flow takes place. Using library resources discuss how they are made.

Ex IV.16: In acoustics it is common to measure sound intensity based on a reference sound intensity $I_0 = 10^{-16}$ W/m^2 or reference sound pressure level $p_0 = 2 \times 10^{-5}$ Pa, which correspond to the faintest audible sound at a frequency of 1 kHz.

Correspondingly the sound intensity level and the pressure level are represented in dB and given by

$$dB_I = 10 \log_{10} \left[\frac{I}{I_0} \right] \text{ or } dB_p = 20 \log_{10} \left[\frac{p}{p_0} \right]$$

Using library resources explain how the sound intensity and sound pressure levels may be measured. Find out how the distance from the source of sound affects these measurements.

Ex IV.17: Consider plane sound wave in air at $30°C$ and a mean pressure of $1\ bar$. The sound intensity has been measured by a meter and is given to be $80\ dB_I$. What is the sound intensity in W/m^2? Referring to a book on acoustics determine the corresponding rms pressure fluctuation in Pa.

Ex IV.18: Frequency of vibration may be measured by the use of a cantilever beam as a frequency measurement device. For a rod of uniform cross section and length L, the natural frequency is given by the formula

$$\omega_n = 0.55\sqrt{\frac{EI}{\mu L^4}}$$

where ω_n = Natural frequency in $/; Hz$, E = Young's modulus of the rod material, I = Moment of inertia with respect to the centroidal axis of the rod and μ = Mass per unit length. All quantities are in SI units.

When such a rod is in contact with a vibrating surface, the rod will go in to resonance when its natural frequency coincides with the frequency of vibration of the vibrating surface. The length of the rod is varied by a suitable arrangement so that the rod goes in to resonance for a particular L. Thus the length L is the measured quantity and relates to the frequency as an estimated quantity.

A $1.5\ mm$ diameter rod of special spring steel is used as the frequency measuring device. The density of the material is $7500\ kg/m^3$ and the Young's modulus is $195\ GPa$. The length of the rod when resonance was achieved in a particular application was $65 \pm 0.2\ mm$. Estimate the frequency and its uncertainty.

Ex IV.19: Describe how the frequency of a sound wave may be measured using a microphone and an oscilloscope. A spectrum analyzer is required if the wave has a complex non-sinusoidal shape. Describe how a spectrum analyzer works.

Ex IV.20: The response of a second order system consisting of a spring - mass - damper has been presented in Chapter 13. Describe how you would set up an experiment to determine the damping coefficient of such a system. Specify all the instruments that would be needed for conducting the proposed experiment. Hint: If such a system is set to oscillations the amplitude reduction with time is related to the damping constant.

Ex IV.21: Refer to the rotating disk arrangement presented in Exercise IV.10. It may be modified by using a wire to suspend the disk inside a casing containing a viscous fluid. The wire may be given an initial twist and allowed to perform free oscillations. The damping coefficient may be obtained by studying the variation of the amplitude of angular motion with respect to time.

Suggest a means of measuring the twist angle amplitude and hence suggest a method for determining the damping coefficient.

Ex IV.22: A U tube manometer uses water at $25°C$ as the manometer liquid and
has a total liquid length of $1\ m$. The manometer that was indicating a head
of $0.5\ m$ initially is set to oscillation by reducing the impressed pressure to
the atmospheric pressure, thus reducing the difference in head across the
manometer to zero. The maximum height of the liquid in one of the columns
is noted as a function of time and given below in tabular form.

Time s	Amplitude m
0	0.25
1.05	0.036
2.10	0.006

Estimate the damping parameter for the manometer system.

Ex IV.23: Consider the data of Exercise IV.22 again. Make a plot of head indicated
by the manometer as a function of time if it is initially indicating zero head
and a pressure difference of $0.50\ m$ water column is applied suddenly across
the two limbs of the U tube.

Data Manipulation and Examples
from laboratory practice

This module consists of two chapters (15 and 16) that deal with Data Manipulation and Examples from laboratory practice. Data manipulation consists of mechanically, optically or electronically modifying the data by amplification, conversion to a voltage or a current, suppression of noise by filtering etc. In recent times measurements are made using digital devices such as a digital voltmeter or a digital computer. Basics of such devices are described here. Use of a data logger is shown in Chapter 16. Calibration issues are also discussed there.

Chapter *15*

Data Manipulation

Physical quantity of interest to us is generally converted by a transducer to a form suitable for some type of processing before being displayed. The transducer may give an output that is mechanical such as movement of a pointer, movement of a spot of light or a current or voltage that may be displayed by a suitable meter. In each of these cases data may be modified suitably so that an accurate measurement and display of the measured quantity becomes possible. The present chapter describes these in sufficient detail.

15.1   Introduction

The present chapter deals with some of the issues that were raised in the beginning of this book in Module I. These pertain to the following:

- Signal coditioning/manipulation
- Data acquisition
- Data presentation/display

In order to appreciate these issues we look at the schematic of a general measurement process shown in Figure 1.2 in Chapter 1. We see from Figure 1.2 that the measurement process involves several steps between the measured quantity and the data output that is proportional to it. In mechanical analog instruments manipulations are done using mechanical elements like gears, levers, optical magnification and use of vernier scale and such things. Certainly they involve moving parts and these may some times malfunction. They may also involve such things as backlash, parallax error and may be susceptible to external influences due to changes in temperature, pressure, humidity etc. These lead to errors in that the data output has uncertainties that are due to these effects. In the case of transducers that operate with electrical quantities these are done by suitable electrical and electronic circuits. Moving parts are, in general, reduced as compared to the analog mechanical methods of measurement. However electrical interference, pick up of spurious signals in the form of electrical noise will be serious issues. Electronic components may overheat, solder joints may fail, optical sensors are prone to stray radiation and thus have their own problems.

In general the trend in recent times has been to move towards electrical (or electronic) data acquisition and manipulation in view of many advantages. Some of these are:

- Reliable functioning
- Compactness
- Availability of digital processing technology
- Availability of suitable hardware and software
- Ease of transmission of data over long distances with little distortion

15.2   Mechanical signal conditioning

Even though electronic methods of signal conditioning are popular, we still use mechanically manipulated signal conditioning in many laboratory applications. Many of these are as dependable and reliable as electronic methods. One major advantage of the mechanical signal conditioners is that they may not require electrical power excepting for illumination purposes. The instrument will not fail even if the laboratory power fails.

We shall consider two examples representing typical applications. These are:

Measurement of a small pressure difference: Betz manometer
Measurement of twist in a wire: Optical twist angle measurement

15.2.1 Betz manometer

In applications such as the measurement of fluid velocity using a Pitot static tube one has to measure a very small pressure difference. It may be a few millimeters of water column (1 mm water column is approximately 10 Pa). Since the precision of the velocity measurement is intimately connected with the precision of the water column height measurement signal conditioning is required. We have already seen one such signal conditioning method viz. the use of the inclined tube manometer which amplifies the signal by the small inclination of the manometer tube, essentially using the principle of the inclined plane. The inclined plane provides a mechanical advantage. For precision measurement it is normal to employ an optical method of measurement by the use of a Betz manometer.[1] The schematic of a Betz manometer is shown in Figure 15.1.

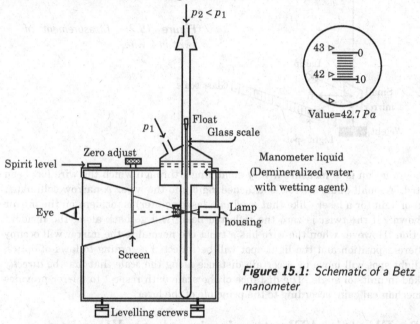

Figure 15.1: *Schematic of a Betz manometer*

The instrument consists of a reservoir that holds the manometric liquid - distilled water with a small amount of wetting agent added to it - and communicates with a vertical manometer tube as shown in the figure. The reservoir has a tubular extension below and a transparent glass scale attached to a float follows the position of the meniscus in the tube. The pressure difference to be measured is communicated between the top of the vertical tube and the top of the reservoir as shown. The float has scale divisions engraved on it. A light source and a lens arrangement completes the instrument. The light pointer is seen by the observer viewing through the lens, along with a scale engraved on the ground glass screen. The position of the pointer against the scale divisions engraved on the glass scale are viewed with

---

[1]Visit for details http://test.acin.nl/EN/products/calibration/instruments/Betz.htm

enough optical magnification to afford a measurement of the manometric height with 0.01 *mm* resolution. The observer, of course, makes an entry of the reading in a note book, for further processing! Note that the Betz manometer is provided with leveling screws for orienting it vertically and is provided with spirit level to verify it.

15.2.2   Optical measurement of twist angle in a wire

This example brings out how a mechanical amplification is possible using essentially an optical arrangement. A long straight wire is kept under tension by a weight hanging as shown in Figure 15.2.

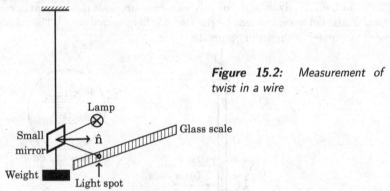

Figure 15.2:   Measurement of twist in a wire

The intention is to measure a very small angle through which the wire has been twisted. A small plane mirror is attached rigidly to the wire. A narrow collimated beam of light (or a laser - like that used in a laser pointer) is incident on the mirror as shown. If the twist is zero the reflected beam will go back along the incident direction. However when the wire has a twist the normal to the mirror will occupy a different position and the light spot will be reflected according to laws of optics. The light spot will move by a certain distance along the scale that may be *directly* marked in units of angle. The distance of the scale with respect to mirror provides the mechanical gain, according to the principle of the lever.

15.3   Electrical/Electronic signal conditioning

By far the more common method employed in modern measurement practice is electrical/electronic signal conditioning. Use of analog meters preceded the recent trend of using digital instruments. Even now it is common to use analog electrical instruments such as voltage, current and power meters in many applications both in laboratory practice and industrial practice. The single disadvantage of analog instruments is that the data needs to be recorded manually. Of course it is possible to interface them with analog recording devices such as the strip chart recorder. However these recorders consume stationery and need constant attention.

It is thus clear that electrical/electronic signal conditioning may either be analog or digital. Digital manipulation and recording of data has many advantages and are more common in modern measurement practice.

15.3.1 Signal conditioning

Following are some of the manipulations that are involved in signal conditioning:

- Amplification of signal - Increase voltage level
- Convert signal to current for industrial $4-20\,mA$ current loop
- Analog to Digital Conversion - Convert analog signal to digital signal readable by a computer
- Amplification after Manipulation - Increase voltage level after some algebraic manipulation
- Communicate signal to a computer or data storage device
- Manipulation using software such as LABVIEW, MATLAB, EXCEL etc.
- Presentation of data in the form of a report using appropriate software

15.3.2 Signal Amplification and manipulation

Amplifiers are analog devices that change the signal level from low levels at the transducer end to a value large enough to be measured by an indicating instrument *without* drawing power from the transducer, that is, without loading the transducer. In recent times amplifiers are usually in the form of integrated chips called operational amplifiers (or Op Amp, for short) with very high open loop gain (without a feed back resistor in the basic circuit shown in Figure 15.3) of the order of a million. These amplifiers are very reliable and have become progressively less and less expensive. Many configurations are possible with operational amplifiers. Some of these are described briefly here. The interested reader should consult books on analog electronics and manufacturer "Application Notes" for more information.

Figure 15.3 shows the basic operational amplifier circuit. The operational amplifier has several pins for connections to be made to the external circuit as shown by the pin diagram of a typical operational amplifier identified as IC 741. The pin diagram shown here is for dual in line package (DIP package) of the operational amplifier. The small dot indicates pin No.1. The operational amplifier is powered from an external source which provides positive and negative voltages shown as V_{cc}^+ (pin No.7) and V_{cc}^- (pin No.4) in the figure. Usually the voltages are $\pm 15\,V$ capable of providing adequate current to run, may be, several operational amplifiers. The operational amplifier has two input terminals; inverting input shown with "-" sign (pin No.2) in the figure and a non-inverting input shown with "+" sign (pin No.3) in the figure. The operational amplifier is a device with a very large input impedance i.e. the impedance between the "-" and "+" terminals. In the simplest configuration the input voltage is applied across the inverting and non-inverting input terminals through a resistor R_1 and the output is available as show in the figure at pin No.6. A feed back resistor R_f connects the inverting input terminal and the output terminal. Pins 1 and 5 are used for nulling of any offset. When operational amplifiers appeared in the market they were usually a single Op Amp in a package. More recently dual and quad Op Amp packages are also available.

Analysis of the basic operational amplifier circuit

We assume that no current flows in to the operational amplifier. Hence the potential difference is zero between the "- " and "+" terminals. Thus $V^+ = V^- \approx 0$.

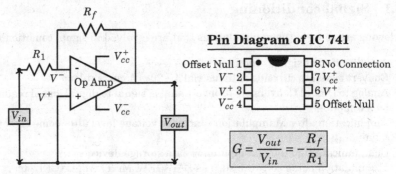

Figure 15.3: Basic operational amplifier circuit showing pin diagram for a DIP package. Also shown is the formula for Gain.

These two terminals act as virtual ground. Consider the node "-" in Figure 15.3. Kirchoff's law is applied to this node to get

$$\frac{V_{in} - 0}{R_1} = \frac{V_{out} - 0}{R_f} \qquad (15.1)$$

This may be rearranged to read

$$G = \frac{V_{out}}{V_{in}} = -\frac{R_f}{R_1} \qquad (15.2)$$

where G is the voltage gain of the amplifier. A single operational amplifier may be used to provide a voltage gain of about 100. If larger gain is required one may use more than one operational amplifier in series. Dual and Quad Op Amp are useful in such cases. The overall voltage gain will be the product of the gain of each amplifier. Operational amplifiers may be used to amplify DC or very low frequency or slow varying signals to AC signals with frequencies of MHz. The frequency response of the operational amplifier is specified by the manufacturer as also the circuit diagram to achieve the desired response. Practical amplifier circuits are built following current practice described in specialized books on electronics and Application Notes provided by the manufacturer of operational amplifiers.

Summing amplifier

The reader should now be able to convince himself that the arrangement in Figure 15.4 will amplify the sum of all the inputs with individual gain factors determined by the respective input resistances.

$$V_{out} = -R_f \left[\frac{V_1}{R_1} + \frac{V_2}{R_2} + \frac{V_3}{R_3} \right] \qquad (15.3)$$

If all the resistances are the same (i.e. $R_1 = R_2 = R_3 = R$) then the summing amplifier will amplify the sum of all the input voltages with a common gain given by $G = -\frac{R_f}{R}$.

It is clear that the summing amplifier output is also the mean of the input voltages multiplied by the common gain factor. At once it is clear, as an example, that a summing amplifier may be used with several thermocouple inputs to measure the mean temperature (see section 4.3.3).

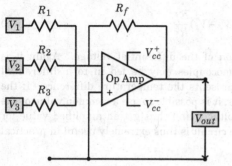

$$V_{out} = -R_f \left[\frac{V_1}{R_1} + \frac{V_2}{R_2} + \frac{V_3}{R_3} \right]$$

Figure 15.4: *Summing amplifier using an operational amplifier*

Differential amplifier

Similarly it is possible to construct a differential amplifier using both the inverting and non-inverting inputs to feed in the input signals. Figure 15.5 shows the arrangement. Nodal analysis may be performed as follows.

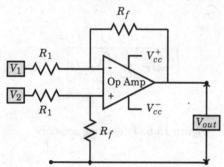

$$V_{out} = \frac{(V_2 - V_1)R_f}{R_1}$$

Figure 15.5: *Differential amplifier*

We note as earlier that $V^+ = V^- \approx 0$. Nodal equation for the "+" node is written as

$$\frac{V_2 - V^+}{R_1} = \frac{V^+ - 0}{R_f} \qquad (15.4)$$

This may be simplified as

$$\frac{V_2}{R_1} = V^+ \left(\frac{1}{R_1} + \frac{1}{R_f} \right) \qquad (15.5)$$

Nodal equation for the "-" node is written as

$$\frac{V_1 - V^-}{R_1} = \frac{V^- - V_{out}}{R_f} \qquad (15.6)$$

This may be simplified as

$$\frac{V_1}{R_1} + \frac{V_{out}}{R_f} = V^- \left(\frac{1}{R_1} + \frac{1}{R_f} \right) \approx V^+ \left(\frac{1}{R_1} + \frac{1}{R_f} \right) \qquad (15.7)$$

The right hand sides of these two expressions (15.5 and 15.7) are one and the same. Hence we have, after minor simplification

$$V_{out} = (V_2 - V_1)\frac{R_f}{R_1}$$
(15.8)

Thus the output is an amplified version of the differential voltage. Again it is apparent that the output of two thermocouples may be fed in to a differential amplifier to amplify the signal that represents the temperature difference. If the two temperatures are close to each other, it is possible to use a common value of the Seebeck coefficient for the two input voltages and thus get an amplified value for the temperature difference directly. The circuit is thus extremely useful in practical applications.

Integrating and Differentiating Amplifier

In some applications one may need the input voltage to be integrated or differentiated. These are possible using the Integrating and Differentiating amplifiers shown respectively in Figures 15.6 and 15.7.

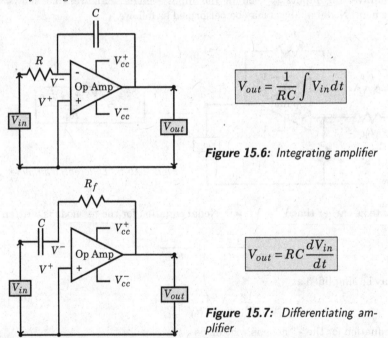

$$V_{out} = \frac{1}{RC}\int V_{in}dt$$

Figure 15.6: *Integrating amplifier*

$$V_{out} = RC\frac{dV_{in}}{dt}$$

Figure 15.7: *Differentiating amplifier*

The input and output are related as follows:

Integraing amplifier: $V_{out} = \dfrac{1}{RC}\int V_{in}dt$
(15.9)

Differentiating amplifier: $V_{out} = RC\dfrac{dV_{in}}{dt}$
(15.10)

Logarithmic amplifier

Some times we require functional manipulations to be made with the signal being measured. One such useful amplifier is the logarithmic amplifier shown schematically in Figure 15.8. In this amplifier the feed back resistor is replaced by a diode (or a transistor). The diode characteristics is such that the forward bias current I_f is related to the reverse bias current I_r according to the relation

$$I_f = I_r \left(e^{\frac{qV_b}{kT}} - 1 \right)$$

where q is the electronic charge, V_b is the reverse bias voltage, k is the Boltzmann constant and T is the temperature in Kelvin. We assume that the input voltage V_{in} is much much larger than $I_r R$. We may replace V_b by V_{out} and perform the familiar nodal analysis to get

$$\frac{V_{in}}{R} = I_f = I_r \left(e^{\frac{qV_{out}}{kT}} - 1 \right) \tag{15.11}$$

This may be rearranged to get

$$e^{\frac{qV_{out}}{kT}} = 1 + \frac{V_{in}}{I_r R} \approx \frac{V_{in}}{I_r R}$$

Taking natural logarithms, we then have

$$V_{out} = \frac{kT}{q} \ln \left(\frac{V_{in}}{I_r R} \right) \tag{15.12}$$

Thus the output of the amplifier is related to the natural logarithm of the input to the amplifier. Equation 15.12 shows that the output is also affected by the temperature. In order to avoid the effect of temperature the diode may be replaced by a transistor.

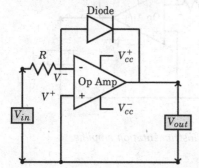

$$V_{out} = \frac{kT}{q} \ln \left(\frac{V_{in}}{I_r R} \right)$$

Figure 15.8: *Logarithmic amplifier*

One possible use for a logarithmic amplifier is in taking the ratio of two signals. This type of manipulation is done very often in radiation measurement, in double beam instruments or in two channel measurements. An example is the two color or ratio pyrometer that was described in section 4.5.4. The schematic of such a circuit is shown in Figure 15.9. The signal from two channels A and B are conditioned by two logarithmic amplifiers. Subsequently the difference of these two is obtained by a differential amplifier. We thus see that the output signal is an amplified version of the logarithm of the ratio of the signals in the two channels. The gain of the differential amplifier may be adjusted such that the output is in dB, defined as $20 \log\left(\dfrac{V_A}{V_B}\right)$.

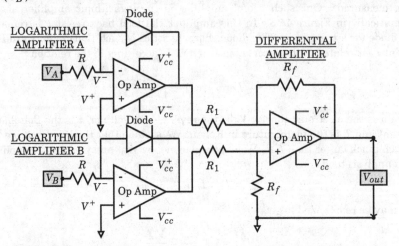

Figure 15.9: *Schematic of a circuit for taking the ratio of two signals*

Instrumentation amplifier

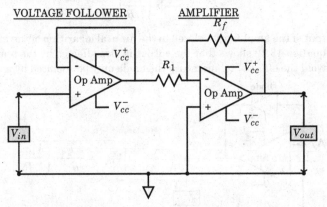

Figure 15.10: *Schematic of an instrumentation amplifier*

In instrumentation practice the amplifier may be used with DC or AC signals. In either case one would not like to load the transducer. This requires the amplifier to have very large (in principle infinite) input impedance. For this purpose we may use special integrated chips called instrumentation amplifiers. One way of achieving high input impedance is to use the arrangement shown in Figure 15.10.

The instrumentation amplifier has two stages, the first one being a unity gain buffer amplifier called the voltage follower. This buffer helps in presenting an input impedance of $\sim 1\,G\Omega$ to the transducer. The second stage is an amplifier with suitable gain. Thus the transducer is not loaded by the amplifier. This means that the current drawn from the transducer is negligibly small. Special instrumentation amplifiers with special input stage like field effect transistor (FET) are available in the form of a single package.

Example 15.1

Strain is measured using a bridge circuit powered by a $10\,V$ stable DC power supply. Resistance of each strain gage is $350\,\Omega$ in the unstrained condition. The strain gages are mounted on a pressure tube as shown in Figure 15.11 and maximum strain is limited to $1500\,\mu-strain$. It is required that the bridge output be conditioned such that it is in the range of $1-5\,V$. Suggest a suitable scheme for this.

Solution :

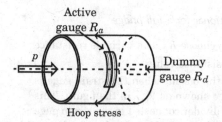

Active
gauge R_a

p

Dummy
gauge R_d

Hoop stress

1. Active gauge responds to hoop stress due to internal pressure
2. Dummy gauge is used for temperature compensation

Figure 15.11: Pressure tube with bonded strain gauge in Example 15.1

Step 1 Refer to Figure 15.11 for the notation. The bridge is constructed with R_a and R_d on one arm while two equal resistances of $R = 350\,\Omega$ each form the other arm of the bridge.

Step 2 We assume that the gage factor is $GF = 2$ and calculate the change in resistance of the active strain gage element with maximum strain $\epsilon_{max} = 1500\,\mu-strain$ as

$$\Delta R_a = R_a GF \epsilon_{max} = 350 \times 2 \times 1500 \times 10^{-6} = 1.050\,\Omega$$

Step 3 Referring to Figure 15.12, we see that the maximum voltage output of the bridge circuit is given by (it is a quarter bridge circuit)

$$(Sig^+ - Sig^-) = V\left(1 - \frac{R_d}{R_d + R_a + \Delta R_a} - \frac{R}{R+R}\right)$$

$$= 10 \times \left(1 - \frac{350}{350 + 350 + 1.05} - \frac{350}{350 + 350}\right) = 0.00749\,V \approx 7.5\,mV$$

Step 4 Schematic of the signal conditioning circuit that may be used is shown in Figure 15.12. The overall Gain of the instrumentation amplifier is given by $F = \frac{R_2}{R_1}\left(\frac{2R_f}{R_g} + 1\right)$. We choose $R_1 = 1\,k\Omega$ and $R_2 = 50\,k\Omega$. Since the maximum signal should yield $5-1 = 4\,V$, we choose the other resistances such that the overall gain is $F = \frac{4}{0.00749} = 534.13$. Hence we should have $\frac{R_f}{R_g} = \left(\frac{FR_1}{2R_2} - \frac{1}{2}\right) = \left(\frac{534.13 \times 1000}{2 \times 50000} - \frac{1}{2}\right) = 4.841$. We conveniently choose $R_f = 5\,k\Omega$

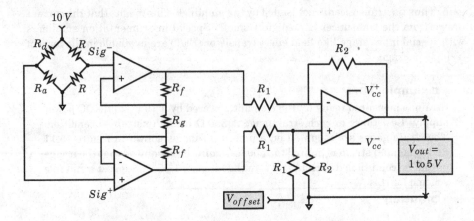

Figure 15.12: *Signal conditioner for a full bridge*

and $R_g = \dfrac{5}{4.841} = 1.033\,k\Omega$. It is usual to choose R_g as a variable resistance with fine adjustment to get the desired resistance value.

Step 5 In order to have $1\,V$ appear at the output when the strain gage is not elongated we inject a reference voltage shown as V_{ref} in the figure. This voltage may be generated using a voltage divider connected to the strain gage power supply. It is easily seen that the reference voltage is given by $V_{ref} = \dfrac{R_1}{R_2} = \dfrac{1}{50} = 20\,mV$.

Other issues

In many applications we operate with ac signals and hence it is necessary to know how the amplifier behaves with signals of different frequencies. An ideal AC amplifier should show gain independent of frequency. This is seldom the case and hence it is necessary to choose the amplifier based on its frequency response. Also the coupling between the signal and the input to the amplifier should be properly made keeping this in mind. It is best to refer to application notes that accompany the devices supplied by the manufacturer.

In addition to amplification one may also choose or reject some frequencies. Electronic circuits are known to be prone to noise being picked up since the laboratory space is full of electrical disturbances. For example the line frequency of AC power from the socket is typically at $50\,Hz$ and this should be discarded by the amplifier circuit. One interposes a suitable "filter" section to avoid picking up such noise. There are many types of filters available. We shall look at the most elementary of these, just to understand how they function. The case being considered is a passive filter consisting of resistance and capacitance.

RC Notch Filter (Twin T)

Figure 15.13 shows a notch filter that helps in removing an unwanted frequency in the input. It is also called a Twin T filter since it has two filters that look like the letter "T". The filter consists of resistors and capacitors arranged in the form of a network consisting of two individual filters - the 'high pass' and the 'low pass'. Before we look at the behavior of the notch filter we look at the low pass and high pass filters. These are first order filters and the analysis is as follows. Consider a

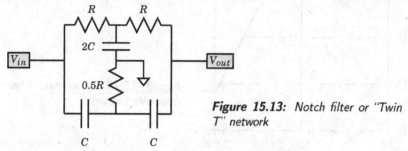

Figure 15.13: Notch filter or "Twin T" network

low pass filter consisting of a resistor and a capacitor as shown in Figure 15.14. The input voltage is a sinusoidal with a frequency f and the output is also a sinusoidal at the same frequency. The impedances are $Z_1 = R$, $Z_2 = X_C = \dfrac{j}{2\pi f C}$ as indicated in the figure. The network is a voltage divider and hence we have

$$\left|\frac{V_{out}}{V_{in}}\right| = \left|\frac{Z_1}{Z_1 + Z_2}\right| = \frac{X_C}{\sqrt{R^2 + X_C^2}} \tag{15.13}$$

The above may be recast in dB units as

$$\text{Gain or attenuation in } dB = 20\log\left|\frac{V_{out}}{V_{in}}\right| = 20\log\left[\frac{X_C}{\sqrt{R^2 + X_C^2}}\right] \tag{15.14}$$

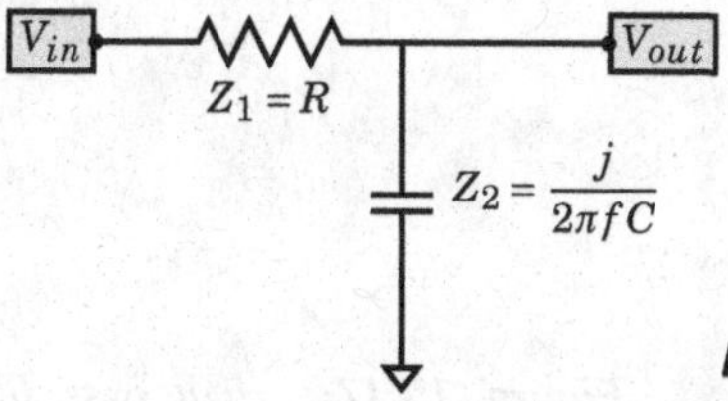

$$V_{in} \quad \overset{Z_1 = R}{\text{———}} \quad V_{out}$$

$$Z_2 = \frac{j}{2\pi f C}$$

Figure 15.14: Low pass filter

A plot of this expression is shown in Figure 15.15 for a network consisting of $R = 1000\ \Omega$ and $C = 1\ \mu F$. Consider now the high pass filter shown in Figure 15.16. It is obtained by simply interchanging the resistor and the capacitor in the low pass filter circuit. We can easily show that Equation 15.13 is replaced by

$$\left|\frac{V_{out}}{V_{in}}\right| = \left|\frac{Z_1}{Z_1 + Z_2}\right| = \frac{R}{\sqrt{X_C^2 + R^2}} \tag{15.15}$$

Equation 15.14 will be replaced by

$$\text{Gain or attenuation in } dB = 20\log\left|\frac{V_{out}}{V_{in}}\right| = 20\log\left[\frac{R}{\sqrt{R^2 + X_C^2}}\right] \tag{15.16}$$

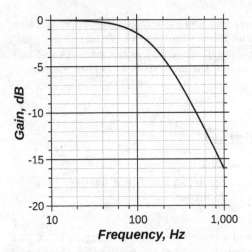

Figure 15.15: Low pass filter frequency response

A plot of this expression is shown in Figure 15.17 for a network consisting of $R = 1000\ \Omega$ and $C = 1\ \mu F$.

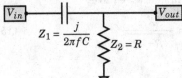

Figure 15.16: High pass filter

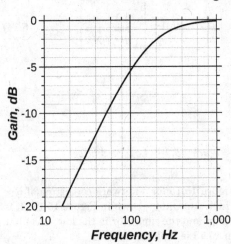

Figure 15.17: High pass filter frequency response

We thus observe that a low pass filter passes on low frequency signal and the high pass filter passes on the high frequency components. The output in either case shows the presence of "knee" at a characteristic frequency for which the resistive impedance and the capacitive impedance are the same. This corresponds to a critical frequency given by

$$f = \frac{1}{2\pi RC} \tag{15.17}$$

In the case illustrated here this frequency is around 160 Hz. At this frequency the gain is 0.5 or $-3dB$.

A notch filter is one that combines the response of low and high pass filters. The twin T network is an example (see Figure 15.13). In practice the circuit is arranged as shown in Figure 15.18.

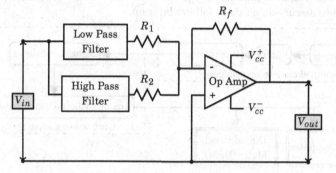

Figure 15.18: *Notch filter followed by a summing amplifier to reject an unwanted frequency component*

Frequency response of a typical notch filter is shown in Figure 15.19. The filter is efficient in blocking input component having frequency close to $50\,Hz$. It allows all other frequencies to pass through with negligible attenuation. The depth of the valley near the frequency that needs to be blocked determines how good the filter is. In the case shown in the figure the blocking is such that it is attenuated by a factor of close to 200.

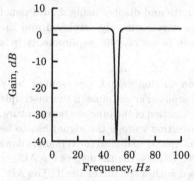

Figure 15.19: *Frequency response of a notch filter designed to filter $50\,Hz$ component*

Schematic of a measuring instrument

We may now give a schematic of any measuring instrument (Figure 15.20). Typically it consists of a transducer followed by a pre-amplifier with possibly a suitable filter. The preamplifier is normally placed very close to the transducer so that noise pick up is avoided. All cables are either coaxial shielded cables or twisted pairs. These also help in reducing noise pick up. In modern laboratory practice it is normal to use a digital data logger or a digital computer to collect and store the data.

For this purpose the signal conditioner is followed by an analog to digital converter or ADC. The signal is then communicated to a data logger or a digital computer; either a PC or a laptop computer. This requires a communication between the ADC and the computer. In case the ADC is done in a board in one of the slots of the PC the communication is through a data cable via a suitable port. Nowadays miniature data loggers are available that connect to the USB (Universal Serial Bus) port of a computer. More discussion on these follows later on.

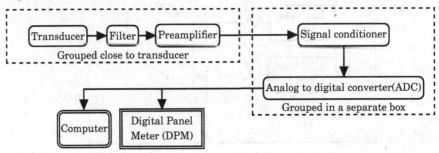

Figure 15.20: Schematic of a general instrument

It is thus clear that the analog circuits are used at the transducer end and connected to a data logger or PC using digital form of data. Digital circuits are basically logic circuits using gates, comparators, ADC and so on. A full discussion of these are beyond the scope of the present book.

15.3.3   Digital panel meter or Digital voltmeter

Data in digital form is easy to store, manipulate and display using digital panel meter (DPM) or a digital voltmeter. These are very commonly employed and the choice of the proper digital meter is to be made to satisfy the requirements in a particular case.

Normally the output of a transducer is an analog signal represented by a continuous variation of output as a function of input. For example, a thermocouple produces a Seebeck voltage that is a continuous function of the junction temperature. If we want to use a digital voltmeter to measure this voltage the signal has to be suitably amplified and converted to digital form. Analog to digital conversion is done in an ADC and transmitted by a cable to the PC. In most digital meters the ADC is an integral part of the instrument and is housed within the meter itself. The ADC converts the input voltage to a count (or number) represented in digital form using a certain number of bits. Consider an example.

Let the maximum and minimum values of the measured voltages be respectively $1\,V$ and $0\,V$. Consider a four bit ADC as a simple example. Count is represented in the form shown in Table 15.1. There are thus 16 levels including 0 and the maximum i.e. 15. Note that the least bit represents 2^0 and highest bit represents 2^3. The total number of levels is given by $2^4 = 16$. Each of these numbers represents a voltage between 0 and $1\,V$. The resolution of the 4 bit digital voltmeter is the Range divided

Table 15.1: *Representation of count by a 4 bit ADC*

Digit	4 bit representation			
	2^3	2^2	2^1	2^0
0	0	0	0	0
1	0	0	0	1
2	0	0	1	0
...	...	...	...	...
...	...	...	...	...
14	1	1	1	0
15	1	1	1	1

by the total number of intervals between the maximum and minimum count. Thus the resolution of a 4 bit digital voltmeter is

$$\text{Resolution of a 4 bit ADC} = \frac{\text{Range}}{\text{counts}} = \frac{1-0}{15-0} \approx 0.067V/count \qquad (15.18)$$

We may generalize the above to the case of n bit DPM. The number of levels is given by 2^n and the resolution by

$$\text{Resolution of a } n \text{ bit ADC} = \frac{\text{Range}}{\text{counts}} = \frac{\text{Range}}{2^n - 1} \qquad (15.19)$$

If n is large the denominator in Equation 15.19 may be approximated by 2^n itself. For example, a 10 bit ADC gives a maximum count of $2^{10} - 1 = 1023$. The resolution is calculated, using Equation 15.19 for a range of $1\,V$, as

$$\text{Resolution of a 10 bit ADC} = \frac{\text{Range}}{\text{counts}} = \frac{\text{Range}}{2^{10} - 1} = \frac{1}{1023} = 9.775 \times 10^{-4}$$

Approximate calculation would yield

$$\text{Resolution of a 10 bit ADC} = \frac{\text{Range}}{\text{counts}} \approx \frac{\text{Range}}{2^{10}} = \frac{1}{1024} = 9.77 \times 10^{-4}$$

For full scale value of $1\,V$ the resolution is very close to a mV. In case we make the full scale exactly equal to $1.023\,V$ the resolution will in deed be exactly equal to $1\,mV$. The instrument is referred to as a 3 digit meter. In general we define the number of digits as the logarithm to base 10 of the number of counts, and of course, round the answer off suitably, to the nearest digit or half digit.

Example 15.2

A T type thermocouple is to be used in an experimental set up. The maximum temperature that will be measured is about $400°C$. The experimenter wants to choose a proper digital voltmeter with a resolution of $0.25°C$. Suggest to him a suitable digital meter.

Solution :

Table of response of T type thermocouple shows that, the output of the thermocouple is $20.872\,mV$, when the measuring junction is at a temperature of $400°C$ and the reference junction is at the ice point. We shall assume, for the present that the experimenter has an ice point reference available in his set up.

We know that the Seebeck coefficient of T type thermocouple is about $40\,\mu V/^\circ C$. The experimenter desires a resolution of $0.25^\circ C$. This corresponds to a Seebeck voltage resolution of $10\,\mu V$. Thus the output of the thermocouple is to be measured with $10\,\mu V$ resolution and is of the form $xx.xx\,mV$. We shall assume that a good amplifier with a gain of 143.7 is available to amplify the thermocouple output. The maximum output after amplification will be $20.87 \times 143.7 = 3.000\,V$. Normally voltmeters have range of 1, 3, 10 V etc. We may choose a digital voltmeter with a range of $3.000\,V$.

The resolution of the voltmeter is thus $0.001\,V$ or $1\,mV$. The counts of the digital voltmeter may now be worked out using Equation 15.18 as

$$\text{Counts} = \frac{\text{Range}}{\text{Resolution}} = \frac{3.000}{0.001} = 3000$$

The number of digits of the voltmeter is simply the logarithm to base 10 of the number count. Thus

$$\text{Number of digits} = \log_{10}(\text{Counts}) = \log_{10} 3000 = 3.4771 \ \text{ round to } 3.5$$

Thus a $3\frac{1}{2}$ digit voltmeter with a range of 3 V is suggested to the experimenter.

Let us look at the number of bits required. The number count for $n = 11$ is 2048 while it is 4096 for $n = 12$. The desired number count is 3000 which can be represented by a binary digit with 12 bits. The maximum count is less than the maximum possible value of 4096.

15.3.4   Current loop

In most industrial applications transducers are far away from the control room where the display of the measured quantities is made. It is difficult to transmit voltages over long distances (without attenuation due to line losses) and one way of circumventing this problem is to transmit the output in the form of a current. It is common practice to represent the zero of the transducer by a current of $4\,mA$ and the full scale as $20\,mA$. The transmitted signal is thus in the analog current form.

An example of a pressure sensor with $4 - 20\,mA$ transmitter is shown in Figure 15.21. The power supply usually is $24\,V$ DC. The power supply is designed so that it will be able to maintain a maximum current of $20\,mA$ in the loop. The signal from the pressure transducer is amplified and converted to a current in the transmitter. The current output is converted to a voltage at the display end by terminating it in a drop resistor, $250\,\Omega$ in the case illustrated in the figure. A pressure of $10\,mm$ water will correspond to a current of $20\,mA$ and a potential difference across the load resistor of $5\,V$. Zero pressure will correspond to current of $4\,mA$ and a potential difference across the load resistor of $1\,V$. Advantage of zero pressure being equal to $1\,V$ output rather than zero output is that it will differentiate it from an open circuit!

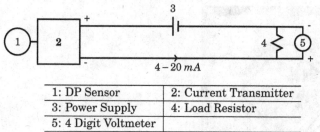

1: DP Sensor	2: Current Transmitter
3: Power Supply	4: Load Resistor
5: 4 Digit Voltmeter	

Figure 15.21: *Differential Pressure (DP) sensor with 4 − 20 mA current loop*

The current transmitter requires at least $8\,V$ to operate properly. If the current loop is very long there will be the line resistance which will drop voltage. Finally the voltage drops across the drop resistor. The power supply voltage should be more than all these put together.

An alternate arrangement for a $4-20\,mA$ current transmitter driven by the signal is ahown in Figure 15.22. Voltage signal is fed to the noninveting input of an operational amplifier as shown in the figure. This $1-5\,V$ signal, may come from a strain gage bridge circuit, discussed earlier in Example 15.1. The voltage drop across the resistor R_1 is equal to the signal and hence produces a current in the $4-20\,mA$ range corresponding to the $1-5\,V$ signal input. This current is fed to the external circuit as indicated in the circuit diagram. The voltahe available as V_{cc}^+ and V_{cc}^- are usually $\pm15\,V$ and hence the load is easily driven by the operational amplifier.

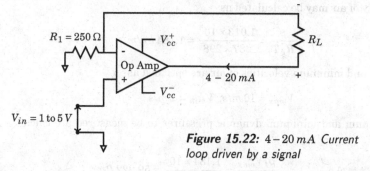

Figure 15.22: *4 − 20 mA Current loop driven by a signal*

Figure 15.23 shows the photograph of a 4 - 20 mA differential pressure transmitter along with a digital display. The display may be dialled to indicate the pressure directly in Pa or the output current in mA. In the case shown the differential pressure is $1000Pa$ or $\approx 100mm$ water column.

Example 15.3

In a wind tunnel the maximum velocity of air at $25°C$ and $1\,atm$ pressure is expected to be $10\,m/s$. The minimum velocity is expected to be $2\,m/s$. A Pitot static tube is to be used for the measurement of velocity. Prescribe a suitable pressure measuring instrument for this purpose.

Solution :

Figure 15.23: Differential pressure transmitter operating on a 4 - 20 mA Current Loop. Display shows pressure in Pa.

For the purpose of design we assume that the coefficient of discharge of the Pitot static tube is 1. The pressure and temperature of flowing air are:

$$p = 1.013 \times 10^5 \, Pa; \; T = 25°C = 298 \, K$$

The gas constant for air is taken as

$$R_g = 287 \, J/kg \, K$$

The density of air may be calculated as

$$\rho_f = \frac{p}{R_g T} = \frac{1.013 \times 10^5}{287 \times 298} = 1.184 \, kg/m^3$$

Maximum and minimum velocities of air are specified as

$$V_{max} = 10 \, m/s; \; V_{min} = 2 \, m/s$$

The maximum and minimum dynamic pressures to be measured are calculated as

$$\Delta p_{max} \quad = \quad \frac{\rho_f V_{max}^2}{2} = \frac{1.184 \times 10^2}{2} = 59.222 \, Pa$$

$$\Delta p_{min} \quad = \quad \frac{\rho_f V_{min}^2}{2} = \frac{1.184 \times 2^2}{2} = 2.368 \, Pa$$

With distilled water as the manometer liquid, the corresponding manometer column heights are given by

$$h_{max} \quad = \quad \frac{\Delta p_{max}}{\rho_m g} \frac{59.222}{1000 \times 9.8} = 6.043 \, mm \text{water}$$

$$h_{min} \quad = \quad \frac{\Delta p_{min}}{\rho_m g} \frac{2.368}{1000 \times 9.8} = 0.242 \, mm \text{water}$$

Thus we require a very sensitive manometer for this purpose. Two possibilities are explored.

Possibility 1: One may use a Betz manometer for this purpose. Betz manometer can resolve pressure changes as small as $0.01\,mm$ of water column. In that case the h will be rounded to 2 significant digits after the decimal point.

Possibility 2: A differential pressure transducer based pressure transmitter may be used for this purpose. A suitable range for the transducer would be a differential pressure range of $0-10\,mm$ water.

With respect to the latter the following will apply:

Differential pressure of $0-10\,mm$ of water column will correspond respectively to $4-20\,mA$. We convert the maximum and minimum values of pressure difference to current as

$$I_{max} = 4 + h_{max}\left(\frac{20-4}{10}\right) = 4 + 6.043 \times 1.6 = 13.669\,mA$$

$$I_{min} = 4 + h_{min}\left(\frac{20-4}{10}\right) = 4 + 0.242 \times 1.6 = 4.387\,mA$$

Assume that a digital panel meter would be used to measure the corresponding voltages by dropping the current across a resistor of 500Ω. The corresponding voltage readings indicated by the DPM are

$$\text{Output}_{max} = I_{max}R = 13.669 \times 10^{-3} \times 500 = 6.835\,V$$

$$\text{Output}_{min} = I_{min}R = 4.387 \times 10^{-3} \times 500 = 2.194\,V$$

We require a digital panel meter that is capable of reading up to $10\,V$ with mV resolution. Appropriate instrument would be a 4 digit voltmeter. This option is shown in Figure 15.21.

Concluding remarks

Any sensor used for measuring a physical quantity produces an output signal of certain type and magnitude. Making this signal have a suitable form and magnitude requires conditioning of the signal - mechanical, optical or electrical. This chapter has considered techniques that may be employed for signal conditioning.

Chapter 16

Examples from laboratory practice

Issues involved in signal conditioning have been considered in Chapter 15. In practice it is necessary to make sure that the measurement using a particular instrument will give readings that are authentic. Authentication of the reading is made by calibration of the measuring instrument i.e. by validating it against measurement by a standard instrument that is traceable to an accepted standard. It is generally insisted that the standard instrument should have an accuracy at least four times that of the instrument that is being calibrated. For example, if the measuring instrument is accurate to within 1% the standard instrument should have an accuracy of at least 0.25%.

16.1   Introduction

In Chapter 14 we have dealt with topics on signal conditioning and manipulation. We have also considered the use of digital voltmeter and the current loop. In most measurement applications the experimental set up would use several instruments to monitor the various process variables, possibly in real time. In most cases the experimenter would like to have a real time (or near real time) tracking of the variables as they change during the experiment. It is quite difficult or impossible to take readings individually, note down in a record book and later use these in obtaining the information that is to be communicated as the outcome of the experiment.

Before the advent of the digital technology the only way one could do these was to use individual recorders for each measured quantity. However, with the digital revolution it is possible to make several measurements *simultaneously* or nearly so using multi-channel digital data loggers. Without going in to too much detail, a multi-channel data logger scans the input in various channels very rapidly using sample and hold amplifiers for each channel. The scanned parameters are converted rapidly to digital form and stored in the computer memory or displayed in real time on the monitor. Thus a multi-channel data logger comprises of suitable interface with the individual instrument or transducer, electronic scanning of each channel (switch from channel to channel in a periodic fashion), sample the input and convert to digital form and finally communicate the data to a computer through a port.

Data loggers are available from various manufacturers like Hewlett Packard (HP), National Instruments (NI) and others. A data logger operates as stand alone or as attachment to a computer. In the case of the former the data logger has an in built processor and memory and also a front panel display. These are supplied with appropriate software with user friendly features. For example, National Instruments supplies a general purpose software called LABVIEW that may be used for creating virtual instruments based on add on cards that reside in the slots available in a PC or by using a miniature logger that connects to a PC through the USB port.

For example, the NI-USB 9211 module is capable of the following:

- **USB Based DAQ from National Instruments**
  4 Channels of 24 bit thermocouple input
  Internal cold junction sensor
  Compatible with J, K, R, S, T, N, E and B thermocouple types·
  Plug and play connectivity with USB
  Bus powered
  Windows 2000/XP, MAC OS X, Linux
  Software supplied by National Instruments (NI)

The interested reader may visit http://www.ni.com/dataacquisition for more details. Most important advantage of the module is the USB connectivity that is quick and does not require opening the PC cover! It has a 24 bit ADC giving a sensitivity of $\sim 0.1\mu V/V$.

16.2   Thermocouple calibration using a data logger

In the laboratory it was necessary to measure several temperatures in an experimental set up and log the data continuously during the experiment runs. It was decided to calibrate each of the thermocouples used over a narrow range of temperatures over which the experiments were run. Typical arrangement for the calibration is shown in Figure 16.1.

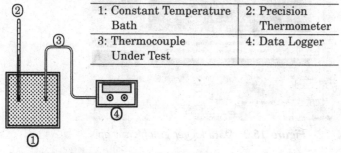

1: Constant Temperature Bath	2: Precision Thermometer
3: Thermocouple Under Test	4: Data Logger

Figure 16.1: *Thermocouple calibration arrangements using a data logger*

The thermocouple to be calibrated and a high precision reference thermometer were immersed in a constant temperature bath. The reference temperature is the room temperature since an ice point cell is not made use of. Simultaneous readings are taken of both the thermocouple output as measured by the data logger and the reading of the precision thermometer. It was found that the thermocouple response within the range of interest was represented very well by a linear relation and this was written directly in to the data logger. Figure 16.2 shows the mean calibration line for a set of 11 thermocouples that were calibrated in the experiment. The data logger channel that records the thermocouple output is set with a gain

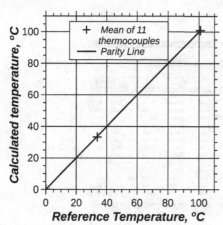

Figure 16.2: *Parity plot showing the relationship between the calculated and reference temperatures*

of 25253 and an offset of 40.718. Thermocouple type is chosen as K type. The thermocouple temperature is then recorded directly and is displayed in °C. The

graphic set up of the data logger (supplied by Agilent) is shown by the entries in
the GUI (Graphic User Interface) as shown in Figure 16.3 for various thermocouple
channels identified by channel numbers. Individual channels in the data logger are

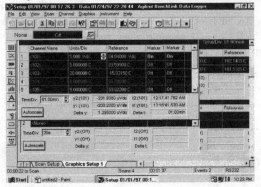

Figure 16.3: Data logger graphic set up

set up for scanning as shown in Figure 16.4. Note that the output of the channels are
scanned electronically and stored in respective memory locations. Since the scanning
is very fast, for the experiments that are being performed here the measurements are
basically simultaneous! The desired resolution may be entered for each channel. The
default value is $5\frac{1}{2}$ digits or 5.5 as entered. This corresponds to a count of 300,000
and hence a resolution of $3\ \mu V/V$.

We also note that the formula for conversion of voltage to temperature is written
as indicated in Figure 16.4. For the linear relation the slope goes in as the Gain and
the intercept in to its own cell. In the case shown the slope as well as the intercepts
are different for the individual thermocouples and are individually entered in to each
of the channels. The figure also shows how to select the thermocouple type using the
scroll bar under "Function".

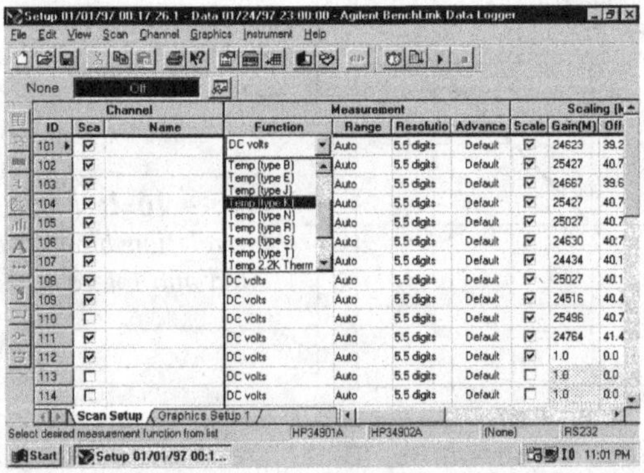

Figure 16.4: Data logger scan set up

16.3 Calibration of a digital differential pressure gauge

The Betz micro-manometer is a primary instrument used for the calibration of pressure measurement instruments. A differential pressure transducer with digital display was calibrated in a wind tunnel using a Betz manometer as the reference. The differential pressure signal was developed by a pitot static tube facing air flow in the wind tunnel. The signal was varied by changing the wind speed systematically. Data obtained in the experiment is shown in the table.

We analyze the data so that a formula for correcting the digital pressure gauge reading to make it agree with the reading obtained using the Betz manometer. Parity plot shown in Figure 16.5 is useful in looking for the proper expression.

h_{Betz} mmH_2O	Δp_{Betz} Pa	Δp_{DP} Pa	h_{Betz} mmH_2O	Δp_{Betz} Pa	Δp_{DP} Pa
0.45	4.4	4.4	4.90	48.1	47
0.90	8.8	8.7	5.20	51.0	50
1.45	14.2	14.2	5.40	53.0	52
1.80	17.7	17.7	6.25	61.3	60
1.90	18.6	18.6	6.40	62.8	62
2.20	21.6	21.6	7.10	69.7	68
2.60	25.5	25.5	7.50	73.6	72
2.75	27.0	27.0	8.10	79.5	78
3.30	32.4	32.4	8.55	83.9	83
3.55	34.8	34.8	9.00	88.3	87
4.30	42.2	42.2	9.90	97.1	94

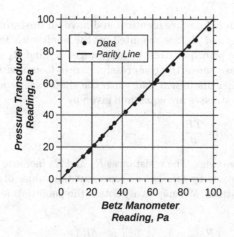

Figure 16.5: Parity plot comparing the DP gauge reading with Betz manometer reading

It is observed that the deviations are reasonably small and it is acceptable to look for a linear relation between the two in the form $\Delta p_{Betz} = a + b\Delta p_{DP}$. Linear regression yields the following expression:

$$\Delta p_{Betz} = -0.301 + 1.024\Delta p_{DP}$$

This relation has an adjusted R^2 value of 0.9997. The left side of above equation is interpreted as giving the correct pressure difference when the digital pressure gauge reading is inserted on the right hand side. If data is processed using a computer the above expression may be keyed in to obtain the correct pressure difference on line.

16.4 Signal conditioning for torque measurement using strain gauges

In a certain experiment it is desired to measure the torque by the use of four strain gauges in a full bridge arrangement. The shaft on which the strain gauge bridge is attached is made of steel and is of 6 mm diameter. The torque is expected to be any value between 1 and 8 Nm. We shall discuss this case from the signal conditioning angle. Also we shall suggest suitable components in the measurement scheme.

A full bridge arrangement has been discussed briefly earlier in section 14.4.3. The full bridge arrangement is shown schematically in Figure 16.6.

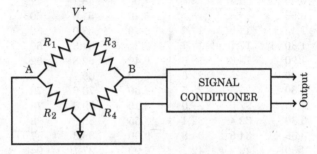

Figure 16.6: Strain gauges in full bridge configuration

We shall assume that all the strain gauges are identical and have an unstrained resistance of 120 Ω. The gauges R_2 and R_3 are mounted on the shaft such that these sense the tensile stress (tensile principal stress) while gauges R_1 and R_4 are mounted on the shaft such that these sense the compressive stress (compressive principal stress). We know from earlier discussion that when the shaft experiences pure torsion the magnitudes of these stresses are equal and given by

$$|\sigma| = \frac{TR}{J} \tag{16.1}$$

where the symbols have the usual meanings. The resistances R_2 and R_3 increase by δR while resistances R_1 and R_4 decrease by δR. Let us represent the values of all strain gauges as R when there is no strain. We may easily obtain the potentials at A and B by using Ohm's law:

$$V_A = V_b \left(1 - \frac{R - \delta R}{(R - \delta R) + (R + \delta R)}\right) = V_b \left(1 - \frac{R - \delta R}{2R}\right) \tag{16.2}$$

$$V_B = V_b \left(1 - \frac{R + \delta R}{(R + \delta R) + (R - \delta R)}\right) = V_b \left(1 - \frac{R + \delta R}{2R}\right) \tag{16.3}$$

The output of the bridge is thus given by

$$V_O = V_A - V_B = V_b \left(\frac{R+\delta R}{2R} - \frac{R-\delta R}{2R} \right) = V_b \frac{\delta R}{R} \qquad (16.4)$$

Note that V_b is the battery supply voltage impressed across the bridge. We shall calculate the output when the impressed torque on the shaft has the upper limit of $T = 8\ Nm$. We are given that $R = 0.003\ m$. Hence the polar moment of inertia is

$$J = \frac{\pi R^4}{2} = \frac{\pi \times 0.003^4}{2} = 1.27 \times 10^{-10}\ m^4$$

The corresponding principal stress magnitude is given by

$$|\sigma| = \frac{TR}{J} = \frac{8 \times 0.003}{1.27 \times 10^{-10}} = 1.886 \times 10^8\ Pa$$

With the Young's modulus of steel taken as $E = 200\ GPa$, the corresponding longitudinal strain may be calculated as

$$\varepsilon = \frac{\sigma}{E} = \frac{1.886 \times 10^8}{200 \times 10^9} = 9.431 \times 10^{-4}$$

Taking the Poisson ratio of steel as $v = 0.3$, the gauge resistance change may be calculated using Equation 7.20 as

$$\frac{\delta R}{R} = \varepsilon(1+2v) = 9.431 \times 10^{-4}(1 + 2 \times 0.3) = 0.001509$$

If we assume that the supply voltage is $V_b = 3\ V$, the output of the bridge will be

$$V_0 = V_b \frac{\delta R}{R} = 3 \times 0.001509 = 4.527\ mV$$

The calculations may be repeated for the lower limit of torque of $1\ Nm$ to get the lowest gauge output as $0.566\ mV$.

Thus the output signal is very weak and needs to be amplified before being recorded. If the shaft is rotating, the signal needs to be taken out either through slip rings with brush contacts or using a high frequency transmitter mounted on the shaft. In either case the signal needs to be boosted by a pre-amplifier. In the latter case the output should also be modulated with a carrier. The signal is to be received by a receiver, demodulated and amplified again before being measured by a voltage measuring instrument. If the voltage is measured by a digital meter the signal must be converted to the digital form using an ADC. The schematic of the transmitter and receiver electronics will be as shown in Figure 16.7.

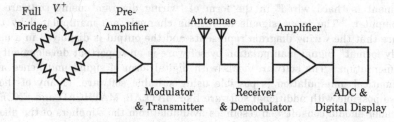

Figure 16.7: *Schematic of circuit for torque measurement*

Finally the signal may be measured by a digital voltmeter. We may assume an overall gain of 2000 such that the signal is in the range $1.132 - 9.054\,V$. This will require a digital voltmeter having a range of $0 - 10\,V$ and a resolution of $0.001\,V$. A 4 digit voltmeter is suitable for this purpose.

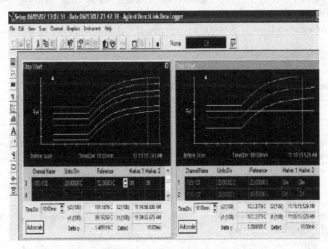

Figure 16.8: *Typical "real time" output*

16.5   Software

Most modern measurements involve the coupling of the experimental set up to a computer via suitable sensors and connecting lines. The connection between the sensor and the computer must be two way so that control information can go from the computer to the sensor and the data from the sensor to the computer. The computer uses suitable software to scan the information available on various channels (example was given earlier with respect to a data logger) and store or save the data in the computer in a file. Software also helps in creating a virtual instrument on the monitor of the computer. A typical example is shown as Figure 16.8. The data is presented in real time in graphical form. Output of each channel is shown graphically with different user selected colors. This data is also available in the form of a file for manipulation using any of the programs like MATLAB, EXCEL etc.

In the case of LABVIEW software provided by National Instruments the virtual instrument is "hard wired" in the form of "wiring diagram" using software in the computer. The input signals in various channels are manipulated by the software that the wiring diagram represents and the output is displayed in a user friendly format. Signal manipulation by software is an important development in instrumentation. The software deals with digital form of signal and hence any mathematical manipulation is possible using suitable software. Many of these are also possible with additional software like MATLAB, MATHCAD and EXCEL. The reader should consult web resources available from the suppliers of the above software for more details.

Concluding remarks

> *This chapter brings us to the end of the book. It has considered a few examples from practice dealing with signal conditioning, calibration and a discussion on useful software. In a way we are back at the beginning in that we can look for many new techniques and methods of measurement in the future. The reader should refer to current literature to go forward from here.*

Exercise V

Ex V.1: Ready to use instrumentation amplifiers are available in the form of IC chips from many manufacturers such as Analog Devices.* Study the specifications of some of them and comment on how they may be used in laboratory practice. Discuss the methods of removing or reducing noise in high gain amplifier circuits.

*(accessible at http://www.analog.com)

Ex V.2: An inexpensive amplifier is the 741* series IC available from various manufacturers. This IC is available with one, two or four amplifiers in a single package. Refer to manufacturer Application Notes and discuss how they may be used for simple laboratory applications.

* A useful tutorial on 741 is available at
http://www.uoguelph.ca/~antoon/gadgets/741/741.html

Ex V.3: Charge amplifiers are used in many instrumentation applications. Prepare a note on these, using library and web resources.

Ex V.4: We have discussed in brief about filters in Chapter 15. more details can be found in appropriate references that deal with signal conditioning. Prepare a note on filters and how they may be used for signal conditioning in instrumentation.

Ex V.5: A stroboscope is useful in the measurement of angular velocity. It consists of a lamp that may be "flashed" at the desired frequency. Use library resources to learn about stroboscope and how it may be used. What are all the uses to which a stroboscope may be put to? Discuss.

Ex V.6: An experimental rig is to be designed to measure the variation of heat transfer coefficient with air flow normal to the axis of a cylinder. Give a schematic of a suitable rig for this purpose. What are the instruments you will need to use for conducting the experiment? Indicate all the precautions you need to take in performing an experiment using the rig. Indicate how you would tabulate the data and also how you would analyze it to arrive at a useful design relationship.

Ex V.7: Data Acquisitions Systems and Digital Data Loggers are used very commonly in the laboratory. These are available from many manufacturers. Prepare a detailed report on these and compare Data Acquisition Systems from at least two different manufacturers. Discuss the factors that go in to the choice of a particular system.

Ex V.8: The mean effective pressure in an engine cylinder is to be estimated using a suitable pressure transducer. The maximum engine speed is $3000 rpm$ while the idling speed is around $800 rpm$. How would you choose a suitable pressure transducer for this purpose? Also how would you analyze the data?

The following reference may be useful:

W. Doggett, Measuring Internal Combustion Engine In-Cylinder Pressure with LABVIEW, accessible at: http://www.ni.com/pdf/csma/us/361574a1.pdf

Appendix *A*

Bibliographic Notes and References

A.1 Bibliographic Notes

References for Module I

The material presented in Module I deals with very general ideas about measurements, the statistical analysis of data, regression analysis and design of experiments. These topics have been dealt with, in this book, to the extent possible at an elementary introductory level. Many important results have been given without proof and described in brief. The interested reader should consult books that deal in more detail with these topics.

Books [1, 2, 3] consider these topics at a comparable level of detail as the present book. More practical aspects may be obtained from books such as [4]. Specialized books such as [5] and [6] consider the statistical analysis of data in more detail. The topic of design of experiments is considered in much greater detail in [7].

References for Module II

We have mentioned earlier that temperature measurement is a topic on which very extensive literature is available. Even though it is a mature field much research is directed towards improvement of accuracy and repeatability of measurements. New applications add new methods and instruments in this field. A very useful web resource for learning the details of ITS90 is [8]. A specialized book on temperature measurement that is a definitive work is that by McGee [9]. Reference [10] deals at length with the models for estimating temperature errors under various measurement conditions. Temperature error estimation relies on a sound knowledge of the fundamentals of heat transfer. The interested reader can refer to a book such as [11]. Reference [2] also gives details about errors in temperature measurement.

Thermocouple reference tables are best accessed at the web resource [12]. A leading manufacturer and supplier of thermocouples and accessories is Omega Engineering, INC.,[13]. Their web site contains extensive documentation on measurements of temperature as well as other quantities. Thermocouple tables are also accessible in their web site. Typical suppliers of temperature measuring hardware in India are [14] and [15].

Heat flux measuring hardware are specialized equipment and are available, sometimes at great expense, from a few manufacturers. Gardon gauges are supplied by Vatell [16]. Through their web site one can access a large amount of appropriate literature also. Indian Space Research Organization (ISRO) is a major user of heat flux sensors and have developed different types of heat flux sensors indigenously. Both IIT Madras and IIT Kharaghpur have developed heat flux sensors for their own use. Axial heat conduction probes have been developed at IIT Madras for BHEL, Trichy. Heat transfer coefficient probes have been developed for in house use at IIT Madras. Descriptions of these are available in the heat transfer literature. Thin film heat flux gauges are made in the laboratory using platinum paint and MACOR substrate. They seldom last long, may be one or two experimental runs, at the most. At present these materials have to be imported.

References for Module III

Manometers for laboratory use are normally made locally using in house glass blowing facilities. Special manometers such as the Betz manometer are procured from the manufacturer [17] at great cost. Many Indian manufacturers supply pressure transmitters covering a very wide range of pressures [18]. Diaphragm gauges are available from [19]. Bourdon gauges are available from many suppliers within the country. They cover a wide range of pressures. They may be easily procured from any hardware shop. Vacuum equipment and gauges (such as Pirani and Penning gauges) are readily available from [20]. Pressure calibration equipment such as the Dead weight tester is available from many manufacturers, such as [21] and [22]. Strain gauges are available from manufacturers such as [23]. They also provide technical support and service. Repair and calibration of strain gauges is a specialized service provided by [24]. Strain gauge amplifiers and electronics are available from many suppliers such as [25].

Flow and velocity measuring instruments are discussed in detail in [2] and [26]. A classic on flow measurement is the book by Ower [27]. Pitot and Pitot static tubes are commercially available from manufacturers such as [18]. Suitable traverses for accurate positioning of these are available from the same source. Obstruction type flow meters are available from many manufacturers, as for example [18,28]. Orifice plates may also be easily machined using workshop facilities in your college at very little cost. Rotameters are manufactured to cover various ranges, fluids and are available from various manufacturers such as [29].

LDV and hot wire systems are expensive and made by only a small number of manufacturers. They are normally imported, for example from [30], since there are no manufacturers in India. Less expensive, hand held hot wire probes for measuring steady velocities of air up to a few m/s are available from many suppliers such as [31]. These are useful in laboratory practice as well in troubleshooting of air handling systems.

References for Module IV

Measurement of thermo-physical properties require special equipments. These are of two types; (a) Apparatus used for performing laboratory experiments by students (b) Apparatus used for research quality measurements. The former are available from many manufacturers while the latter are custom made or designed by the researcher himself. The leader in this category is [32]. Some of the suppliers in India of laboratory equipment are [33] and [34].

Optical instruments are very specialized and are very expensive. Integrating spheres and spectrometers to cover different parts of the electromagnetic spectrum are available from several manufacturers. The cost is directly dependent on the features and specifications. Typical suppliers are [35] and [36]. There are many manufacturers of gas analyzers including NDIR, Gas chromatographs etc. Typical of such manufacturers is [37].

Vibration, force and acceleration measuring instruments are normally very expensive since they require sophisticated electronic processing. The signal levels are low and the output is obtained after electronic processing of the output. World leaders in these equipment are [38] and [39].

References for Module V

Digital meters, electronic equipment and data loggers are manufactured by a large number of manufacturers world wide. There was a time when the practice was to buy discrete components and wire the required circuit in the laboratory. The procedure was cumbersome and reliability was rather poor. In the past three decades the tendency is to purchase ready to use electronics and do only the minimal interconnections required in setting up an experiment. Many electronic circuits are available in modular form so that they may be quickly assembled to cater to different laboratory needs. With the PC taking a central position in all experimental rigs electronic circuits were designed to reside as boards in the PC itself. However, in recent times the boards are replaced by USB connectable modules. The main suppliers of these are [40] and [41].

A.2 References

1. Beckwith, T.G., Buck, N.L. and Marangoni, R.D., 1982, *Mechanical Measurements,* Narosa, New Delhi, *Indian Student Edition.*
2. Holman, J.P., 2001, *Experimental Methods for Engineers,* McGraw Hill, NY, Seventh Edition.
3. Doebelin, E.O., 1990, *Measurement Systems - Application and Design,* McGraw Hill, NY, Fourth Edition.
4. Smith, E., 1984, *Principles of Industrial Measurement for Control Applications,* ISA.
5. Paradine, C.G. and and Rivett, B.H.P.,1953, *Statistics for Technologists,* English Universities Press, London.
6. Freund, J.E., 1973, *Modern Elementary Statistics,* Prentice Hall, INC., NJ, Fourth Edition .

7. Montgomery, D.C.,2004, *Design and Analysis of Experiments*, Somerset,NJ, Sixth Edition.

8. www.its-90.com, The Internet resource for the International Temperature Scale of 1990.

9. McGee, T.D., 1988, *Principles and Methods of Temperature Measurement*, John Wiley, NY.

10. Eckert,E.R.G. and Goldstein, R.J., (Editors) 1976, *Measurements in Heat Transfer*, Hemisphere, Washington, 2nd Edition.

11. Venkateshan, S.P., 2004, *A First Course in Heat Transfer*, Ane Books, New Delhi.

12. www.temperatures.com.

13. www.omega.com, or OMEGA Engineering, INC. One Omega Drive, Stamford, Connecticut 06907-0047

14. www.yolk.in/pda/health-and-medicine/thermocouples/tricouple-devices-20449.html or Hospital Rd, , Tamil Nadu, 600015 Chennai, India

15. www.kflexcables.com/F8311/thermocouple_cables.html, or No.234, 3rd Floor Garudachar Complex, Chickpet, Bangalore 560053.

16. www.vatell.com, or Vatell Corporation, 240 Jamelle Road, Christianburg, VA 24073.

17. www.westenberg-engineering.de, West mountain engineering, Hammerschmidt STR. 114, 50999 Cologne, Germany

18. www.switzerinstrument.com, Switzer Instrument Limited, CHENNAI (Madras) 600017, INDIA.

19. www.vishay.com, diaphragm gauge fundamentals are available here!

20. www.hindhivac.com or Hind High Vacuum Co. Pvt. Ltd., Site No. 17, Phase 1, Peenya Industrial Area, Bangalore 560058.

21. www.wika.com.au, or Melbourne Office: Unit 3, 24 Lakeside Drive, Burwood East VIC 3151

22. www.ravika.com, Ravika Engineers 186, DSIDC Complex, Okhla Industrial Area, Phase - 1, New Delhi - 110 020.

23. www.emersonprocess.com or Unit No. 2A, Second Floor, Old No. 117 (New No. 54) Dr. Radhakrishnan Salai, Mylapore, Chennai ÂŬ 600 004.

24. www.senstechindia.com or Sens Tech (P) Ltd, NO. 91 , 1st Cross, MEI Layout Road, Nagasandra Post, Bangalore - 560073.

25. www.encore-elec.com or Encore Electronics Inc., 4400 Route 50, Saratoga Springs, NY 12866

26. Benedict, R.P., 1969, *Fundamentals of Temperature Pressure and Flow Measurements,*John Wiley, NY.

27. Ower, E., 1949, *The measurement of air flow.* Chapman & Hall, Ltd., England.

28. www.flowsystemsinc.com, or Flow Systems, 220 Bunyan Ave., Berthoud, CO 60553 OH, USA.

29. www.flowtechinstruments.com or 2/A, Pushkraj Society, B/h. T. B. Hospital, Gotri Road, Vadodara-390 021. GUJARAT (INDIA).

30. www.dantecdynamics.com or Dantec Dynamics A/S, Tonsbakken 16 - 18, DK-2740 Skovlunde, Denmark

31. www.airflowinstruments.co.uk or Airflow Instruments, TSI Instruments Ltd., Lancaster Road, Cressex Business Park, High Wycombe, Buckinghamshire, HP12 3QP, England.

32. www.p-a-hilton.co.uk or P. A. Hilton Ltd., Horsebridge Mill, KingÂŠs Somborne, Stockbridge, Hampshire, SO20 6PX.

33. www.dynamicequipments.com or Dynamic Engineering Equipments, Plot No. 38, New Block, MIDC, MIRAJ - 416 410.

34. RCG Instruments, A-1/25, Rambagh Colony, Navi Peth Pune - 411 030, India.

35. www.avantes.com or Soerense Zand Noord 26, NL-6961, RB Eerbeek, The Netherlands, Europe.

36. www.thermo.com or Thermo Fisher Scientific, Inc., 81 Wyman Street, Waltham, MA 02454.

37. www.ametekpi.com or AMETEK Singapore Pte. Ltd., 10 Ang Mo Kio Street 65, #05-12 Techpoint 569059 Singapore.

38. www.bkhome.com or Brüel & Kjaer represented by Josts Engineering Company Limited, C - 39, II Avenue Annanagar, Chennai 600 040.

39. www.endevco.com or Endevco Corporation, 30700 Rancho Viejo Road, San Juan Capistrano, CA 92675.

40. www.home.agilent.com or Agilent Technologies India Pvt. Ltd., Chandiwala Estate, Maa Anandmai Marg, Kalkaji, New Delhi-110019.

41. sine.ni.com or National Instruments Corporation, 11500 N Mopac Expwy, Austin, TX 78759-3504.

Appendix B

Useful tables

Table B.1: Primary units

Primary quantities		
Quantity	*Dimension*	*Name,symbol*
Length	L	meter, m
Mass	M	kilogram, kg
Time	T	second, s
Temperature	θ^a	Degree Celsius, $^{\circ}C$ or K
Electric charge	Q	coulomb, $coul$

[a]Temperature is indicated by θ to avoid confusion with time

Table B.2: Units of physical quantities

Derived quantities				
Quantity	Dimension	Name	Symbol	Unit
Acceleration	LT^{-2}		a	m/s^2
Angular velocity	T^{-1}		ω	rad/s
Area	L^2		A or S	m^2
Capacitance	$M^{-1}L-2T^2Q^2$	Farad	C	$coul/V$
Density	ML^{-3}		ρ	kg/m^3
Dynamic viscosity	$ML^{-1}T^{-1}$		μ	$kg/m \cdot s$
Electrical current	QT^{-1}	Ampere	I	$coul/s$ or A
Electrical potential	$ML^2T^{-2}Q^{-1}$	Volt	V	$N \cdot m/coul$
Energy	ML^2T^{-2}	Joule	E	$kg \cdot m^2/s^2$ or J
Enthalpy	ML^2T^{-2}		h	$kg \cdot m^2/s^2$ or J
Force	MLT^{-2}	Newton	F	$kg \cdot m/s^2$ or N
Frequency	T^{-1}	Hertz	f	$1/s$ or Hz
Gas constant	$L^2T^{-2}\theta^{-1}$		R_g	$J/kg \cdot K$ or $J/kg \cdot {}^\circ C$
Impedance, electrical	$ML^2T^{-1}Q^{-2}$	Ohm	Z	Ω or V/A
Heat Flux	MT^{-3}		q	$J/s \cdot m^2$ or W/m^2
Heat transfer coefficient	$MT^{-3}\theta^{-1}$		h	$W/m^2{}^\circ C$
Kinematic viscosity	L^2T^{-1}		ν	m^2/s
Momentum	MLT^{-1}		p	$kg \cdot m/s$ or $N \cdot s$
Power	ML^2T^{-3}	Watt	P	W

...Continued on next page

	Table B.2 continued from previous page			
Quantity	Dimension	Name	Symbol	Unit
Pressure or stress	$ML^{-1}T^{-2}$	Pascal	p or P	Pa or N/m^2
Ratio of specific heats	non-dimensional		γ	No unit
Resistance, electrical	$ML^2T^{-1}Q^{-2}$	Ohm	R	Ω or V/A
Resistance, fluid due to viscosity	$L^{-1}T^{-1}$		R	$1/m \cdot s$
Resistance, thermal	$M^{-1}T^3\theta$		R	$^\circ C \cdot m^2/W$
Specific heat	$L^2T^{-2}\theta^{-1}$		c	$J/kg \cdot K$ or $J/kg \cdot^\circ C$
Spring constant	MT^{-2}		K	N/m
Strain	Non-dimensional		ϵ	No unit
Thermal conductivity	$MLT^{-3}\theta^{-1}$		k	$W/m^\circ C$
Thermal diffusivity	L^2T^{-1}		α	m^2/s
Velocity	LT^{-1}		V	m/s
Velocity, sonic	LT^{-1}		a	m/s
Wavelength	L		λ	m or μm
Wavenumber	L^{-1}			cm^{-1}
Work	ML^2T^{-2}	Joule	W	$kg \cdot m^2/s^2$ or J

Table B.3: *Physical constants*

Atmospheric pressure	101.325	kPa
Avogadro number, N	6.022×10^{23}	$molecules/mole$
Velocity of light in vacuum, c_0	3×10^8	m/s
Planck constant, h	6.62×10^{-23}	$J \cdot s$
Stefan-Boltzmann constant, σ	5.67×10^{-8}	W/m^2K^4
Universal gas constant, $\Re$		$J/mole \cdot K$
Boltzmann constant, k	1.39×10^{-23}	J/K
First radiation constant, C_1	3.74×10^{-16}	$W \cdot m^2$
Second radiation constant, C_2	14387.7	$\mu m \cdot K$
Standard acceleration due to gravity, g	9.807	m/s^2

Table B.4: *Confidence intervals and the cumulative probability for the normal distribution with zero mean and unit standard deviation*

CI	CP	CI	CP	CI	CP	CI	CP
±0.1	0.0797	±1.1	0.7287	±2.1	0.9643	±3.1	0.9981
±0.2	0.1585	±1.2	0.7699	±2.2	0.9722	±3.2	0.9986
±0.3	0.2358	±1.3	0.8064	±2.3	0.9786	±3.3	0.9990
±0.4	0.3108	±1.4	0.8385	±2.4	0.9836	±3.4	0.9993
±0.5	0.3829	±1.5	0.8664	±2.5	0.9876	±3.5	0.9995
±0.6	0.4515	±1.6	0.8904	±2.6	0.9907	±3.6	0.9997
±0.7	0.5161	±1.7	0.9109	±2.7	0.9931	±3.7	0.9998
±0.8	0.5763	±1.8	0.9281	±2.8	0.9949	±3.8	0.9999
±0.9	0.6319	±1.9	0.9426	±2.9	0.9963	±3.9	0.9999
±1	0.6827	±2	0.9545	±3	0.9973	±4	0.9999

CI = Confidence interval, CP = Cumulative probability

Table B.5: Confidence intervals according to t distribution

	Confidence interval				Confidence interval		
d	90%	95%	99%	d	90%	95%	99%
1	6.314	12.71	63.66	18	1.734	2.101	2.878
2	2.92	4.303	9.925	19	1.729	2.093	2.861
3	2.353	3.182	5.841	20	1.725	2.086	2.845
4	2.132	2.776	4.604	21	1.721	2.08	2.831
5	2.015	2.571	4.032	22	1.717	2.074	2.819
6	1.943	2.447	3.707	23	1.714	2.069	2.807
7	1.895	2.365	3.499	24	1.711	2.064	2.797
8	1.86	2.306	3.355	25	1.708	2.06	2.787
9	1.833	2.262	3.25	26	1.706	2.056	2.779
10	1.812	2.228	3.169	27	1.703	2.052	2.771
11	1.796	2.201	3.106	28	1.701	2.048	2.763
12	1.782	2.179	3.055	29	1.699	2.045	2.756
13	1.771	2.16	3.012	30	1.697	2.042	2.75
14	1.761	2.145	2.977	40	1.684	2.021	2.704
15	1.753	2.131	2.947	50	1.676	2.009	2.678
16	1.746	2.12	2.921	100	1.66	1.984	2.364
17	1.74	2.11	2.898	200	1.653	1.972	2.345

d = Degrees of freedom

Table B.6: Table of χ_α^2 values

d	$\chi_{0.005}^2$	$\chi_{0.01}^2$	$\chi_{0.025}^2$	$\chi_{0.05}^2$	$\chi_{0.1}^2$	$\chi_{0.95}^2$	$\chi_{0.975}^2$	$\chi_{0.99}^2$	$\chi_{0.995}^2$
1	7.8794	6.6349	5.0239	3.8415	2.7055	0.0039	0.000982069	0.000157088	3.92704E-05
2	10.5966	9.2103	7.3778	5.9915	4.6052	0.1026	0.0506	0.0201	0.0100
3	12.8382	11.3449	9.3484	7.8147	6.2514	0.3518	0.2158	0.1148	0.0717
4	14.8603	13.2767	11.1433	9.4877	7.7794	0.7107	0.4844	0.2971	0.2070
5	16.7496	15.0863	12.8325	11.0705	9.2364	1.1455	0.8312	0.5543	0.4117
6	18.5476	16.8119	14.4494	12.5916	10.6446	1.6354	1.2373	0.8721	0.6757
7	20.2777	18.4753	16.0128	14.0671	12.0170	2.1673	1.6899	1.2390	0.9893
8	21.9550	20.0902	17.5345	15.5073	13.3616	2.7326	2.1797	1.6465	1.3444
9	23.5894	21.6660	19.0228	16.9190	14.6837	3.3251	2.7004	2.0879	1.7349
10	25.1882	23.2093	20.4832	18.3070	15.9872	3.9403	3.2470	2.5582	2.1559
11	26.7568	24.7250	21.9200	19.6751	17.2750	4.5748	3.8157	3.0535	2.6032
12	28.2995	26.2170	23.3367	21.0261	18.5493	5.2260	4.4038	3.5706	3.0738
13	29.8195	27.6882	24.7356	22.3620	19.8119	5.8919	5.0088	4.1069	3.5650
14	31.3193	29.1412	26.1189	23.6848	21.0641	6.5706	5.6287	4.6604	4.0747
15	32.8013	30.5779	27.4884	24.9958	22.3071	7.2609	6.2621	5.2293	4.6009
16	34.2672	31.9999	28.8454	26.2962	23.5418	7.9616	6.9077	5.8122	5.1422
17	35.7185	33.4087	30.1910	27.5871	24.7690	8.6718	7.5642	6.4078	5.6972
18	37.1565	34.8053	31.5264	28.8693	25.9894	9.3905	8.2307	7.0149	6.2648
19	38.5823	36.1909	32.8523	30.1435	27.2036	10.1170	8.9065	7.6327	6.8440
20	39.9968	37.5662	34.1696	31.4104	28.4120	10.8508	9.5908	8.2604	7.4338

d = Degrees of freedom

Table B.6 continued from previous page

d	$\chi^2_{0.005}$	$\chi^2_{0.01}$	$\chi^2_{0.025}$	$\chi^2_{0.05}$	$\chi^2_{0.1}$	$\chi^2_{0.95}$	$\chi^2_{0.975}$	$\chi^2_{0.99}$	$\chi^2_{0.995}$
21	41.4011	38.9322	35.4789	32.6706	29.6151	11.5913	10.2829	8.8972	8.0337
22	42.7957	40.2894	36.7807	33.9244	30.8133	12.3380	10.9823	9.5425	8.6427
23	44.1813	41.6384	38.0756	35.1725	32.0069	13.0905	11.6886	10.1957	9.2604
24	45.5585	42.9798	39.3641	36.4150	33.1962	13.8484	12.4012	10.8564	9.8862
25	46.9279	44.3141	40.6465	37.6525	34.3816	14.6114	13.1197	11.5240	10.5197
26	48.2899	45.6417	41.9232	38.8851	35.5632	15.3792	13.8439	12.1981	11.1602
27	49.6449	46.9629	43.1945	40.1133	36.7412	16.1514	14.5734	12.8785	11.8076
28	50.9934	48.2782	44.4608	41.3371	37.9159	16.9279	15.3079	13.5647	12.4613
29	52.3356	49.5879	45.7223	42.5570	39.0875	17.7084	16.0471	14.2565	13.1211
30	53.6720	50.8922	46.9792	43.7730	40.2560	18.4927	16.7908	14.9535	13.7867

d = Degrees of freedom

Table B.7: Properties of dry air at atmospheric pressure

T $^\circ C$	ρ kg/m^3	C_p $kJ/kg \cdot K$	$\mu \times 10^6$ $kg/m \cdot s$	$v \times 10^6$ m^2/s	$k \times 10^3$ $W/m^\circ C$	$\alpha \times 10^6$ m^2/s	Pr
0	1.2811	1.004	17.09	13.34	24.2	18.8	0.709
20	1.1934	1.004	18.09	15.16	25.8	21.5	0.704
40	1.1169	1.005	19.07	17.07	27.4	24.4	0.699
60	1.0496	1.007	20.02	19.08	29	27.4	0.696
80	0.9899	1.008	20.95	21.16	30.5	30.5	0.693
100	0.9367	1.011	21.85	23.33	32	33.8	0.691
120	0.8889	1.013	22.74	25.58	33.4	37.1	0.689
140	0.8457	1.016	23.59	27.9	34.9	40.6	0.688
160	0.8065	1.019	24.43	30.29	36.3	44.1	0.687
180	0.7708	1.023	25.25	32.76	37.6	47.7	0.686
200	0.7381	1.026	26.04	35.29	38.9	51.4	0.686
240	0.6803	1.034	27.58	40.54	41.5	59	0.687
280	0.631	1.043	29.05	46.04	43.9	66.8	0.689
320	0.5883	1.051	30.45	51.76	46.3	74.8	0.692
360	0.551	1.06	31.8	57.71	48.5	83	0.695
400	0.5181	1.069	33.09	63.87	50.6	91	0.699
440	0.489	1.078	34.34	70.23	52.7	100	0.703
480	0.4629	1.087	35.55	76.8	54.7	109	0.707
520	0.4395	1.096	36.73	83.58	56.7	118	0.71
560	0.4183	1.105	37.88	90.56	58.6	127	0.714
600	0.3991	1.114	39.01	97.75	60.6	136	0.717

Table B.8: Thermal properties of saturated water

t $^\circ C$	ρ kg/m^3	C_p $kJ/kg^\circ C$	$\mu \times 10^6$ $kg/m \cdot s$	$\nu \times 10^7$ m^2/s	k $W/m^\circ C$	$\alpha \times 10^6$ m^2/s	Pr
0	999.8	4.217	1752.5	17.53	0.569	1.35	12.988
10	999.7	4.193	1299.2	13	0.586	1.398	9.296
20	998.3	4.182	1001.5	10.03	0.602	1.442	6.957
30	995.7	4.179	797	8.004	0.617	1.483	5.398
40	992.3	4.179	651.3	6.564	0.63	1.519	4.32
50	988	4.181	544	5.506	0.643	1.557	3.537
60	983.2	4.185	463	4.709	0.653	1.587	2.967
70	977.7	4.19	400.5	4.096	0.662	1.616	2.535
80	971.6	4.197	351	3.613	0.669	1.641	2.202
90	965.2	4.205	311.3	3.225	0.675	1.663	1.939
100	958.1	4.216	279	2.912	0.68	1.683	1.73
110	950.7	4.229	252.2	2.653	0.683	1.699	1.562
120	942.9	4.245	230	2.439	0.685	1.711	1.425
130	934.6	4.263	211	2.258	0.687	1.724	1.309
140	925.8	4.285	195	2.106	0.687	1.732	1.216
150	916.8	4.31	181	1.974	0.686	1.736	1.137
160	907.3	4.339	169	1.863	0.684	1.737	1.072
170	897.3	4.371	158.5	1.766	0.681	1.736	1.017
180	886.9	4.408	149.3	1.683	0.676	1.729	0.974
200	864.7	4.497	133.8	1.547	0.664	1.708	0.906
220	840.3	4.614	121.5	1.446	0.648	1.671	0.865
240	813.6	4.77	111.4	1.369	0.629	1.621	0.845
260	783.9	4.985	103	1.314	0.604	1.546	0.85
280	750.5	5.3	96.1	1.28	0.573	1.441	0.889
300	712.2	5.77	90.1	1.265	0.54	1.314	0.963
320	666.9	6.59	83	1.245	0.503	1.145	1.087
340	610.1	8.27	74.8	1.226	0.46	0.912	1.345
360	528.3	14.99	64.4	1.219	0.401	0.506	2.407

Table B.9: Thermal properties of metallic solids*

Metal	ρ kg/m^3	C $J/kg°C$	k $W/m°C$	$\alpha \times 10^5$ m^2/s
Aluminum	2702	903	237	9.71
Alumel	8574	464	29.2	0.734
Brass	8470	377	116	3.63
Bronze	8830	377	52	1.36
Cast Iron (4% C)	7272	420	51	1.67
Chromel	8730	448	17.3	0.493
Copper	8900	385	287	8.38
Constantan	8900	390	22.7	0.654
Duralumin	8933	385	401	11.7
Gold	19300	129	317	12.7
Iron	7870	447	80.3	2.28
Nickel	8900	444	90.7	2.3
Platinum	21450	133	71.6	2.51
Silver	10500	235	429	17.4
Stainless Steel 304	7900	477	14.9	0.395
Tin	7310	227	66.6	4.03
Titanium	4500	522	21.9	0.932
Tungsten	19300	132	174	6.83

*Tabulated values are at room temperature and representative

Table B.10: Emissivities of surfaces at room temperature

Surface	Emissivity
Aluminium (anodized)	0.77
Aluminium (oxidized)	0.11
Aluminium (polished)	0.05
Aluminium (roughened with emery)	0.17
Aluminium foil	0.03
Aluminized Mylar	0.03
Anodized black coating	0.88
Asbestos board	0.94
Brass (dull)	0.22
Brass (polished)	0.03
Brick (dark)	0.9
Concrete	0.85
Copper (oxidized)	0.87
Copper (polished)	0.04
Fire-clay	0.75
Glass	0.92
Gold	0.02
Iron oxide	0.56
Paper	0.93
Plaster	0.98
Porcelain (glazed)	0.92
Silver - pure (polished)	0.020-0.032
Steel galvanized (new)	0.23
Steel galvanized (old)	0.88
Steel Oxidized	0.79
Steel Polished	0.07
Tile	0.97
Tungsten (polished)	0.03
Water	0.95

Index

Printed in the United States
By Bookmasters